IMPROVING AGRICULTURAL EXTENSION

A REFERENCE MANUAL

Edited by

Burton E. Swanson
Robert P. Bentz
Andrew J. Sofranko

Prepared under the guidance of the

Extension, Education and Communication Service
Research, Extension and Training Division
Sustainable Development Department

2005

Published by Arrangement with the

Food and Agriculture Organization
of the United Nations

by

DAYA PUBLISHING HOUSE

Delhi - 110 035

for Sale and Distribution only in South-East Asia

"This book was originally published by the Food and Agriculture Organization of the United Nations (FAO).

"The designations employed and the presentation of material in this publication do not imply the expression of any opinion whatsoever on the part of the Food and Agriculture Organization of the United Nations concerning the legal status of any country, territory, city or area of its authorities, or concerning the delimitation of its frontiers or boundaries."

"The designations "developed" and "developing" economies are intended for statistical convenience and do not necessarily express a judgement about the stage reached by a particular country, territory or area in the development process."

First Indian Edition 2005

ISBN 81-7035-408-0

Published by : **Daya Publishing House**
1123/74, Deva Ram Park
Tri Nagar, Delhi - 110 035
Phone: 27383999
Fax: (011) 23244987
e-mail : dayabooks@vsnl.com
website : www.dayabooks.com

Showroom : 4762-63/23, Ansari Road, Darya Ganj,
New Delhi - 110 002
Phone: 23245578, 23244987

Printed at : **Chawla Offset Printers**
Delhi - 110 052

PRINTED IN INDIA

Contents

Preface

In the context of sustainable agricultural development, agricultural extension has a very crucial role to play. The tasks and responsibilities of extension service will need to be broad-based and holistic in contents and scope, thus beyond agricultural technology transfer. Its normal task of transferring and disseminating to farmers appropriate agricultural technologies and good farm practices would not be sufficient. Extension agencies, services, and workers will need to exercise a more proactive and participatory role and serve as knowledge/information "brokers", initiating and facilitating mutually meaningful and equitable knowledge-based transactions among agricultural researchers, trainers, and primary producers. To improve its cost-effectiveness, proper strategies to advocate favourable and explicit agricultural extension policies are needed. Modern strategic planning and quality management tools and approaches should be utilized in developing or restructuring extension organizations or institutions. More innovative methods must be developed to identify systematically farmers problems and felt needs, and to help formulate and set agricultural research agenda based on such needs and problems. In short, there is a need to develop and improve the conceptual, technical, and operational methods and tools in order to strategically plan, efficiently manage, and scientifically evaluate a problem-solving, demand-driven and needs-based agricultural extension programmes.

It is thus imperative that the body of knowledge on agricultural extension be continuously improved, updated, and disseminated. It is also important that the review and inventory of the "State of the Art" in agricultural extension theory and best practices should be shared widely among agricultural extension practitioners, educators and scholars. FAO's Extension, Education and Communication Service (SDRE) has now completed the Third Edition of a classic reference manual on agricultural extension. The First Edition of such a Reference Manual was published in 1973, and the Second Edition in 1984. This Third Edition entitled: *Improving Agricultural Extension: a Reference Manual* is completely new, and reflects many of the issues raised and discussed during a FAO-sponsored Global Consultation on Agricultural Extension held in Rome in December 1989.

It is not easy to undertake a normative task such as in trying to compile generic agricultural extension "know-how" that can be applicable globally. However, FAO/SDRE has tried to minimize such a problem by involving extension practitioners, educator, and scholars from many countries with different educational backgrounds and working environments. A total of 38 authors from 15 countries contributed to the 23 chapters of this publication. We hope that this book which is aimed at agricultural extension planners, managers, trainers, educators, and field practitioners could be useful in improving the quality of agricultural extension and in generating new ideas and methods for increasing further the cost-effectiveness of agricultural extension programmes.

The preparation of this publication was coordinated by Prof. Dr. Burton E. Swanson, in collaboration with, and under the guidance from, the FAO's Extension Education and Communication (SDRE) staff. Prof. B.E. Swanson was assisted by two of his colleagues from the University of Illinois at Urbana-Champaign, USA, Dr. Robert P. Bentz and Dr. Andrew J. Sofranko, and together they edited the contributions of the 38 international authors who wrote the 23 chapters. We hereby would like to express our gratitude to these authors and editors, especially to Prof. B.E. Swanson for his untiring efforts in coordinating with authors in 15 different countries.

This third edition of this Reference Manual would not have been completed without the strong support and useful contributions from FAO's Extension, Education, and Communication Service (SDRE), especially its Chief, Dr. Tito E. Contado, and its Extension Education and Training Methodology Specialist, Dr. Ronny Adhikarya, who has taken the initiative, and provided the leadership, in conceptualizing and producing this publication. Both of them have also written chapters for this Reference Manual.

Rome, August 1996
FAO of the United Nations

Stein W. Bie
Director
Research, Extension and Training (SDR)
Division Sustainable Development Department

Introduction

This new book on *Improving Agricultural Extension* is the third in a series that was started in 1972. In the first edition of *Agricultural Extension: A Reference Manual*, Addison H. Maunder authored and edited a wide range of material that first defined "extension" as a field of study and practice, then went on to explain about extension programme development and different extension teaching processes, and ended with a section on extension management. At the time this first volume was written, many developing countries were just in the process of establishing their agricultural extension organizations. Therefore, this first reference manual became both an essential handbook to extension managers and programme leaders, as well as a basic textbook for students preparing to enter the extension profession.

The second edition of *Agricultural Extension: A Reference Manual* was an edited volume published in 1984. It built on many of the same themes found in the first manual, but gave more attention to the role of extension in agricultural development, especially extension's role in technology transfer. This trend reflected, in part, the concern of many countries and donors with the problem of expanding food production to feed the world's growing population and extension's role in getting improved technology to farmers. By that time, investment in national and international research was expanding, and there was a growing concern about how to get improved technology to nearly 1 billion farmers, more than 90 per cent of whom were small and marginal farmers who live and work in developing countries.

This third volume reflects the continued development of extension within a rapidly changing world. Temporary food surpluses, especially in the industrially developed countries, have shifted the attention of many governments and most donors away from further investments in agricultural sector. Governments are being urged to cut public sector expenditures in agricultural research and extension and to shift more of these costs to farmers themselves. In the case of better-off commercial farmers, the eventual goal is to privatize the technology system while shifting more responsibility for small, marginal, and landless farmers to nongovernmental organizations (NGOs).

For agricultural extension directors to guide their organizations through this demanding and difficult period, they will need to direct their attention to three main issues or concerns. First, they should concentrate the work of extension on those activities where it has a comparative advantage. For example, extension should focus its efforts on those *knowledge-based* technologies that are central to farmers' concerns *and* that will maintain the natural resource base. In general, these are subject-matter areas that will not be taken up by the private sector. Examples include dissemination of production management technologies that are specific to different crop and livestock systems; natural resource management technologies, such as soil and water management, integrated pest management, agroforestry, and other technologies associated with sustainable development; and farming systems technologies, including farm management skills, that will enable farmers to improve their efficiency, increase their cropping intensities, and to diversify into more high value commodities.

Second, extension directors must improve management procedures within their organizations. In short, they must (1) do a better job of justifying their annual budgets by detailing programme thrusts and expected impacts, (2) decentralize management decisions to improve the efficiency and effectiveness of extension operations, and (3) monitor, measure, and report on extension's accomplishments, including impacts on agricultural productivity, farm incomes, and natural resource management.

Third, escalating pressures for increased accountability are being fueled by a growing frustration with *top-down*, bureaucratic behaviour that makes extension appear unresponsive to the needs of its clientele. Therefore, extension leaders, technical specialists, and the field staff must increase farmers' participation in assessing needs, setting priorities, and implementing programmes. An important reason why most farmers, especially small and marginal farmers, are not more fully involved in research and extension programmes is that they are not organized. Therefore, extension should play a more active role in helping farmers get organized into functional organizations, including commodity groups, credit societies, cooperatives, and other types of farmer associations (FAs). Such FAs can increase extension's efficiency in disseminating messages, and these groups become the logical feedback mechanism for input into specific programmes and activities. In addition, representatives of these different FAs should serve on local, district, provincial, and national extension advisory panels to provide formal feedback at different levels in the extension system. In short, getting farmers organized and directly involved in shaping extension policy and setting extension priorities is an essential element in creating a demand-driven extension system that will gain the public's confidence and support.

These current trends and perspectives were used in preparing this new volume on *Improving Agricultural Extension: A Reference Manual*. In the view of the editors, agricultural research and extension must become more *demand driven* if these institutions are to maintain or regain the public trust and to compete effectively for public funds. Therefore, the book's first section, Overview of Extension in Agricultural and Rural Development, which is comprised of four chapters, lays out the evolution of agricultural extension over time, outlines different extension approaches that have emerged over the past thirty to forty years, revisits the role of extension in the development process, and summarizes the economic contribution of extension to agricultural and rural development.

The second section, Improving Extension Programmes and Processes, is composed of seven chapters and concentrates on improving extension programme development and delivery. The first chapter in this section concentrates on the needs assessment process, and it is followed by a chapter that outlines different methodologies of rapidly assessing clientele needs using participatory approaches. Chapter 7 deals with the process of developing extension programmes around these identified needs, followed by a chapter on selecting extension content and methods appropriate in meeting these identified needs. Rural women have been routinely ignored by many extension systems; therefore, chapter 9 gives specific attention to organizing extension programmes to reach this clientele group. The last two chapters in this section highlight the need for extension managers to develop a strategic extension plan, including the use of mass media, in organizing extension programmes, and the need to evaluate the impact of extension on improving programme development and delivery.

The third section, Improving Extension Management, is composed of eight chapters and deals directly with improving the organization and management of extension systems. Chapter 12 outlines the importance of extension policy in specifying the mission, goals, and clientele for extension, including how extension will be financed. The next chapter outlines key principles necessary to improve the organization and management of extension. The next two chapters deal with managing extension personnel and the process of professional staff development, and they are followed with a chapter on the process of acquiring and managing financial resources. The need to strive for improved performance and accountability within extension is highlighted in the next two chapters, which focus on the importance of monitoring extension programmes and establishing a management information system to monitor extension resources, activities, and outputs. The final chapter in this third section deals with one of the most persistent management problems confronting most research and extension systems: the need to improve research-extension-farmer linkages.

The final section, Current Trends and Developments, deals with current themes in the field of agricultural extension as we look toward the future and its implications for extension. First, the expanding world's population, including higher income levels among many former developing countries, is increasing the global demand for more and better quality food. This demand, in turn, is putting increased pressure on the planet's natural resources and leading to environmental degradation. Therefore, chapter 20 addresses extension's role in disseminating sustainable, *knowledge-based* technologies, such as integrated pest management (IPM). The next chapter recognizes the need for extension (and NGOs) to help organize and empower farmers to solve more of their own problems. In addition, getting farmers organized into commodity and related farmer associations is essential if different groups of farmers are going to become effective in identifying and articulating their needs to the research-extension system.

Chapter 22 looks at the process of privatizing extension as one means of shifting the cost of (research and) extension to farmers as countries progress along the development continuum. In shaping the future of extension, it is important that policy makers differentiate between *public and private goods* to ensure sustainable development, both in political and environmental terms. The final chapter of this volume deals with the emerging and important role of nongovernmental organizations in the extension process, including their success in organizing small, marginal, and landless farmers, and in addressing problems of poverty alleviation and sustainable development in a world where fewer resources are being invested in agricultural and rural development.

Many people played an instrumental role in preparing this new FAO extension manual. First, we would like to acknowledge the Food and Agriculture Organization (FAO) of the United Nations for providing financial support for this project, and especially the guidance and assistance of Dr. Tito Contado and Dr. Ronny Adhikarya, Extension, Education and Communication Service, in helping move this project from concept to final publication, as well as in writing their respective chapters. Next, we would like to recognize the contributions of all 38 authors, from 15 countries around the world, who wrote the 23 chapters for this new publication. In addition to the editors, several INTERPAKS staff played an instrumental role in editing and producing the final, camera-ready manuscript. Ms. Sheila Ryan served as technical editor for this book and was instrumental in ensuring quality standards and consistency across the different manuscripts. Ms. Chen Hui served as project manager to keep chapters moving through the writing, review, editing, and formatting process. Finally, Ms. Lori Snipes was responsible for key-stroking and formatting the book for final publication.

To each person involved in the development of this new book on *Improving Agricultural Extension*, we wish to extend our sincerest thanks. We hope that this third volume builds on the strengths of the past and that it will become a valuable resource to extension professionals in the future as we work together to continue the process of improving agricultural extension to meet the needs of farm families in a rapidly changing world.

Burton E. Swanson, Robert P. Bentz, and
Andrew J. Sofranko
Editors, *Improving Agricultural Extension: A Reference Manual*
July 30, 1996

Burton E. Swanson is Professor and Director of the International Programme for Agricultural Knowledge Systems (INTERPAKS); **Robert P. Bentz** was formerly the Associate Director of the Illinois Cooperative Extension Service and is currently Senior INTERPAKS Adviser; and **Andrew J. Sofranko** is Professor of Rural Sociology in the Departmentof Human and Community Development at the University of Illinois at Urbana-Champaign.

Part I

Overview of Extension in Agricultural and Rural Development

Chapter 1
The History, Development, and Future of Agricultural Extension

Gwyn E. Jones Chris Garforth

Agricultural extension work has a venerable, albeit largely unrecorded, history. It is a significant social innovation, an important force in agricultural change, which has been created and recreated, adapted and developed over the centuries. Its evolution extends over nearly four thousand years, although its modern forms are largely a product of the past two centuries. Today, the organizations and personnel engaged in agricultural extension encompass a diverse range of socially sanctioned and legitimate activities which seek to enlarge and improve the abilities of farm people to adopt more appropriate and often new practices and to adjust to changing conditions and societal needs.

The Term "Extension"

The use of the word "extension" derives from an educational development in England during the second half of the nineteenth century. Around 1850, discussions began in the two ancient universities of Oxford and Cambridge about how they could serve the educational needs, near to their homes, of the rapidly growing populations in the industrial, urban area. It was not until 1867 that a first practical attempt was made in what was designated "university extension," but the activity developed quickly to become a well-established movement before the end of the century. Initially, most of the lectures given were on literary and social topics, but by the 1890s agricultural subjects were being covered by peripatetic lecturers in rural areas (Jones, 1994). The growth and success of this work in Britain influenced the initiation of similar activity elsewhere, especially in the United States. There, in many states, comparable out-of-college lectures were becoming established by the 1890s (True, 1900, 1928). During the first two decades of this century, the extramural work of the land-grant colleges, concerned with serving the needs of farm families, was to expand dramatically and become formally organized; but the use of the term "extension" continued and has persisted as the designation for the work.

The overt use of the notion of "extending" relevant and useful information to the adult population at large, however, predates the university extension movement. Earlier in the nineteenth century, a British politician, Lord Henry Brougham, an influential advocate of formal eduction for the poor and of mass adult education, founded the Society for the Diffusion of Useful Knowledge in 1826. Its objective was "imparting useful information to all classes of the community, particularly to such as are unable to avail themselves of experienced teachers, or may prefer learning by themselves." The society sought to do this largely through producing low-priced publications and establishing local committees throughout the country "for extending the object of the Society" (Society for the Diffusion of Useful Knowledge 1827). During its twenty years' existence, agricultural topics were well covered in the society's publications. Similar, albeit short-lived, societies were also established before 1840 in several other European countries, India, China, Malaysia, and the United States (in Virginia) (Grobel, 1933; Smith, 1972).

The Distant Origins

The dissemination of relevant information and advice to farmers, however, has a long if chequered history prior to the emergence of modern forms of agricultural extension in the nineteenth century.

The first known example was in Mesopotamia (roughly, present-day Iraq) around 1800 B.C. Archaeologists have unearthed clay tablets of the time on which were inscribed advice on watering crops and getting rid of rats—important for mitigating any potential loss of taxation revenue from farmers (Ahmed, 1982, as quoted in Bne Saad, 1990). Some hieroglyphs on Egyptian columns also gave advice on avoiding crop damage and loss of life from the Nile's floods. An important advance was the beginning of agricultural writings. Though few have survived, the earliest were written during the ancient Greek and Phoenician civilizations, but some of them were adapted by Roman writers. From the second century B.C. to the fourth century A.D., several important Latin texts were written, frequently drawing on practical farming experience, which aimed to help Roman landowners to maintain and improve their estates and their revenues (White, 1970, 1977).

At around the same period in imperial China, early forms of advancing and disseminating agricultural information also began. That landowners and their tenants should improve their production was a matter of concern to the state since, from the sixth century B.C.

Gwyn E. Jones is Senior Lecturer, Agricultural Extension and Rural Development Department, and **Chris Garforth** is Senior Lecturer and Head of the Agricultural Extension and Rural Development Department, both at The University of Reading, United Kingdom.

onwards, it relied heavily on land taxes for its revenues. The support of relevant agricultural research and the dissemination of information and advice had certainly begun by the late Han Dynasty (25-220 A.D.). The oldest fully surviving Chinese agricultural treatise, *Essential Techniques for the Peasantry,* dating from 535 A.D., aimed to show landowners how to improve their estate management through the advice they gave to their tenants. The Sung and Yuan Dynasties (960-1368) with their firm local government administrations were notable in organizing and promoting agricultural research, extension work, and the teaching of agriculture and sericulture, much facilitated by the invention of woodblock printing, which allowed agricultural treatises and practical handbooks to be widely distributed. Similar activities continued during the succeeding Ming (1368-1644) and Chi'ing (1644-1912) Dynasties, driven not only by the growing population and periodic threats of famine, but also by the state's recognition of the importance of well-coordinated extension work on agricultural recommendations if the most benefit was to be achieved (Perkins, 1969; Elvin, 1973; Bray, 1984; Delman, 1991).

Necessary Conditions for Agricultural Extension to Evolve

Apart from the importance of farmers and agriculture in the society and economy concerned, several conditions appear to be necessary for the initiation and organized development of agricultural extension work.

The prime condition is that information has been assembled, systematized, and made available on good or progressive or new agricultural practices suited to a particular environment, and is based on either (or both) the accumulation of experience or findings from research (however rudimentary). Second, this information is used, among other things, to educate professional agriculturists who may further enlarge or refine this body of knowledge or become active promoters and disseminators of it. Third, an appropriate administrative or organizational structure exists by and within which the dissemination activities may be established and conducted. Fourth, there is a legislative or some other official mandate or influential proponent which prescribes or enables that agricultural extension work is desirable and must occur. Fifth, there are invariably a variety of antecedents which have attempted protoforms of agricultural information and advice dissemination. In addition, the incidence of critical situations, such as famine, crop failure, soil exhaustion, or altered economic conditions or relationships, may create an immediate cause for initiating the organization of extension work. All or several of these conditions have been present in the evolution of modern forms of agricultural extension.

Towards the Modern Era

The direct antecedents of organized agricultural research and dissemination of its results which occurred in nineteenth century Europe and North America can be traced back to the "renaissance" which began in the fourteenth century. Between 1300 and 1700, European society became transformed from its medieval feudal forms into recognizably modern social systems. It was a period of complex, multistranded development. Along with the growth of national states and European exploration and "discovery" of the rest of the world was the "new learning." This involved not only a fresh appreciation of rediscovered classical writings and art forms, but also many novel ideas and activities, a spirit of humanism, and rational enquiry. All of this was considerably facilitated by the invention of printing using movable type, usually attributed to Gutenberg around 1450, and the rapid diffusion over Europe of the printing press, for whose output there existed a ready market.

The earliest known renaissance agricultural text was written in Latin by Pietro de Crescenzi in 1304 and was translated into Italian and French. This became the first book on agriculture to be printed in the mid-fifteenth century. Others soon followed, often based on the old Latin texts or on the collected wisdom of farmers and their families. A well-known example, a compendium of helpful advice in simple verse and a best-seller in Tudor England, was Thomas Tusser's *A hundredth goode pointes of husbandrie*, published in 1557 and expanded in 1573 to five hundred good points with as many on "goode housewiferie" (Tusser, 1580). Less popular, but of greater significance, were Francis Bacon's writings early in the next century based on his observations and scientific experiments on his estate north of London—the beginnings of the application of science and scientific method to agriculture (Russell, 1966).

By the mid-eighteenth century, throughout much of Europe, progressive landowners (frequently aristocrats) and their agents and a few similarly minded farmers were being known as "improvers." These, along with some "men of science," were the main proponents of agricultural clubs or societies. At their regular meetings and demonstrations, locally and regionally, landowners and leading farmers exchanged ideas and information and discussed farming improvements. Two main forces underlay the movement. First, many landowners were eager to learn of ways to improve their estates and the production capabilities of their tenants so as to increase the value of their estates and their rental incomes. Secondly, progress was being made towards modern science and its application to agriculture, especially in agricultural chemistry and plant physiology (Russell, 1966). These societies sought to alter radically the traditional modes of farming by initiating experiments, arranging demonstrations, disseminating information, and advocating the adoption of innovations. It was considered almost a duty by their elite membership to make their initiatives and activities known to "the generality" of farmers through publishing their proceedings and reporting their meetings in newspapers (Hudson, 1972). Although such agricultural societies initially spread slowly—the first had been formed at Rezzato near Milan in 1548 (Coletti, 1900)—they had become common throughout much of Europe by 1800, and a small number had been established by that year in the young United States and eastern Canada.

It is not possible, here, to enter into detail on the interactions between a growing scientific knowledge of agriculture and its application in practice, the many examples of increasingly widespread agricultural improvement, and the numerous personalities involved in Europe and North America during the century or so after 1750. Reference must, however, be made to one figure whose ideas and activities were of pivotal significance to the developments of the time, and later. This was Philipp Emanuel von Fellenberg (1771-1844), who in 1799 purchased the estate of Wylhof, which he renamed Hofwyl, near Bern in Switzerland (Gray, 1952; Guggisberg, 1953). Over the next decade or so, he established agricultural schools at Hofwyl for the children of peasants and of the poor and for the aristocracy and their agents. Although not the first agricultural schools in Europe, those of von Fellenberg became a model for many more which were established before 1850, especially in Denmark, Germany, France, and the United Kingdom, thus assuring a cadre of trained agriculturists.

At Hofwyl, von Fellenberg also established an experimental-cum-model farm to test and develop suitable husbandry practices and technology. He publicised the work at this veritable "educational colony" through a journal and agricultural festivals (shows) at Hofwyl and by welcoming a large number of visitors from all over Europe and maintaining a voluminous correspondence with these and others. Many of his visitors became active proselytes of his methods, recognizing their practical value in disseminating useful information on agriculture—and other topics. One such notable visitor was Lord Henry Brougham, referred to earlier, who became the main publicist of von Fellenberg's work in Britain and whose Society for the Diffusion of Useful Knowledge was an early form of organized "extension."

By the 1820s, most of the elements for creating modern forms of agricultural extension were in being, although each was to develop considerably during the nineteenth century. A crucial missing element, however, was an effective means by which the "generality" of farmers could be directly given information, advice, and encouragement. This required itinerant agriculturists who could meet farmers in their home localities, give instructional talks and demonstrations, advocate superior or new practices, and have discussions with the farmers. The notion of "itinerancy" was not new: since late medieval times, tradesmen and proto-professional men had travelled through rural areas to serve their clients. The first examples of itinerant agricultural lecturers-cum-instructors were in parts of New England and New York in the 1820s (True, 1928) and in France, where a first migratory agricultural teacher was appointed in the Gironde in 1837, followed by nine more in various areas of the country in succeeding years (Boulet n.d.). During the 1840s, further sporadic developments also occurred in the United States, particularly in New York, Ohio, and Maryland (True, 1928), while in Württemberg, in southwest Germany, a pasture specialist (*Wiesenbaumeister*) together with a staff of eighteen technicians was employed by the state agricultural society to advise farmers, landowners, and town administrations on land drainage, irrigation, and improved pasture management (CLVS, 1845).

In Europe, agricultural science was evolving rapidly by the 1840s, with notable strides being made in Germany by Justus von Liebig at Giessen, and with the establishment of agricultural experiments at Rothamsted in England in 1843 by John Bennet Lawes and Henry Gilbert. Agricultural societies and their shows were flourishing. Numerous publications and periodicals were aimed at farmers. Agricultural schools, if not commonplace, had been established in most European countries. Thus a small minority of younger landowners and farmers had received a formal education in their calling, while purposely trained agriculturists were available to be engaged as estate agents or teachers. Many of the more progressive landowners employed agents to travel around their estates to urge improved methods on their tenants. The main element necessary to create modern agricultural extension services was for legitimate authorities to establish the necessary organizations—and the germ of this had already been present in France, Germany, and the United States.

The Birth of Modern Agricultural Extension Services

The first agricultural extension service of a modern kind came into existence as the result of a crisis and the initiative of the occupant of a high office of authority. The crisis was the outbreak of potato blight in Europe in 1845. In Ireland its effects were particularly severe because the predominantly peasant population relied on potatoes in their diet, and "the potato famine" persisted until 1851. The new British viceroy appointed to Ireland in 1847, the Earl of Clarendon, soon after his arrival in Dublin wrote a letter (Jones, 1982) to the president of the Royal Agricultural Improvement Society of Ireland (founded in 1841), which acted as the central society for numerous local agricultural societies. This letter, no less than an official directive, urged the society to appoint itinerant lecturers to travel around the most distressed districts to inform and show small farmers, in simple terms, how to improve their cultivation and how to grow nutritious root crops other than potatoes. "Lord Clarendon's practical instructors in husbandry," as they became known, were centrally appointed, deployed, and paid and reported weekly to the society in Dublin, with some local control of their activities being exercised by the major landowners in their areas. Over the four years of its existence, the scheme was funded to about half its total cost by landowners and charitable donations, with the remainder coming from government-controlled funds (Jones, 1979, 1981).

The potato famine also led to consideration being given in Württemberg to employing itinerant farm advisers, but the proposal failed to gain approval (R. Bühler, personal communication). However, from the mid-1850s, first in Württemberg, Hesse, and western Prussia, itinerant agricultural teachers (*Wanderlehrer*) began to be appointed under the auspices of central agricultural societies. After some ten years, the system grew rapidly, influenced in part by the crisis among vine growers resulting from the devastation caused by *phylloxera* aphid infestations, and became formalized (Jones, 1981). Normally, the *Wanderlehrer* spent the summer half of the year travelling around their districts giving talks, demonstrations, and advice to farmers; during the remainder of the year they taught farmers' sons at winter agricultural schools. Although officially they were part of the activities of the agricultural associations, their work was in all cases supported heavily by state funds, and their advice was free to farmers. When the system was adopted in the kingdom of Bavaria in 1896, it was as an integral part of the state civil service; the extension workers were grandly titled Royal Agricultural Teachers (*Königliche Landwirtschaftslehrer*) (Maier-Bode, 1910).

By the close of the nineteenth century, agricultural extension systems modelled to a considerable extent on the German *Wanderlehrer* had spread: to Denmark from 1870 onwards; to the Netherlands, where a few extension workers (*wandelleraren*) had been appointed by agricultural societies in the late 1840s and 1850s, but had then disappeared before being revived as a government system in the 1890s; to Italy, where the first itinerant agricultural teacher (*cattedra ambulante di agricoltura*) was appointed in 1886 at Rovigo, near the estuary of the River Po, with many others following in the next decade and funded largely by public donations, the church, and the banks; to Switzerland; to much of the Austro-Hungarian Empire; and to Russia.

Meanwhile, in France the first national, wholly state-funded agricultural extension service was established in 1879. The few itinerant agriculturalists appointed before 1848 (referred to earlier) had continued, but they served in only a very small minority of the country's *départements*. In 1874, the minister of public instruction in the reforming Third Republic issued a circular letter strongly commending the system and advocating its extension (*J. d'Agric. Prat.*, 1874, p. 257-258). This resulted in an additional thirty-three itinerant agricultural teachers being appointed by *départements* over the next five years, and a law passed in 1879 officially instituted the office of a department-level itinerant agricultural teacher (*professeur départmental d'agriculture*). This law was given practical effect by a decree in 1880 and an explanatory ministerial circular early in 1881 (Min. de l'Agric., 1882, p. 8-9). From then on, each *professeur* was a state-appointed civil servant. His duties included giving agricultural instruction to trainee primary school teachers. Mainly, however, under the responsibility of the Ministry of Agriculture, he was to be "nomadic" within his *département*, "to keep farmers informed regarding modern discoveries and new inventions which could be applied economically and with advantage," "to be a populariser (*vulgarisateur*) of progress," "to carry enlightenment into the heart of the countryside." The number of these extension workers grew rapidly, and by the end of the 1880s the whole of France was being served (Jones, 1981).

The growth of agricultural education and extension work in continental Europe was to have a strong impact on the emergence of comparable activity in the United Kingdom. An official commission on technical education in the early 1880s included a detailed review of the European developments (Jenkins, 1884). At the end of the decade, a cluster of enactments, which established county-based local government, created a board of agriculture, promoted technical (including agricultural) education, and allocated funds for the purpose, enabled agricultural extension work to be initiated. It was to be part of the services provided by the local government authorities. They either employed their own agricultural officer or more commonly sponsored lectures and travelling schools on agriculture (especially dairying) as part of the university extension system. This meant drawing on the staffs of the agricultural departments which were being created in new institutions of higher education. Government funds were available to support these activities, but funding also had to be provided by the local county authorities (Jones, 1994). By the turn of the century, such work existed throughout Great Britain.

This system and its underlying legislation, however, did not apply to Ireland (then entirely a part of the United Kingdom). There, agricultural extension work became established in 1900 as a result of the initiative of Horace Plunkett, well known for his advocacy of agricultural cooperation. An official committee in 1896, chaired by Plunkett, reviewed the developments in Europe and North America (Report, Recess Committee, 1896) and set out to adapt the various systems to suit Irish conditions. In 1900, a Department of Agriculture and Technical Instruction was established in Dublin, governed by a board of representative Irishmen. This initiated itinerant agricultural instruction, organized within each county as in Britain and similarly resourced partly from local and partly from central funds. A vague recollection existed of Lord Clarendon's "practical instructors" half a century earlier, and the title "itinerant instructors" was applied to the new extension workers, who were expected to provide information and advice, each to be "the guide, philosopher and friend of the existing farmers" (Plunkett, 1901-02, p. 26).

Many visitors and several official delegations from North America to Europe, particularly from the mid-nineteenth century onwards, reported back on the progress in agricultural research and education, including the itinerant teachers. In the United States and eastern Canada, agricultural societies had become common during the first half of the century and, usually supported by their state or provincial legislatures, some had at times sponsored itinerant lecturers in agriculture. However, two other developments after 1850 were of more significance to the evolution of agricultural extension in the United States. First was the Morrill Act of 1862, signed by President Lincoln during the Civil War, which was seminal in the creation of state colleges "of agriculture and the mechanic arts" in the northern United States; its land-grant provisions enabled the states to establish and fund their colleges. Second was the beginning at about the same time of the farmers' institute movement. These institutes organized one-or two-day (and later longer) meetings, which became popular after 1860, arranged by and for farmers.

Both developments had been widely discussed during the previous decade, and their growth over the next half century was closely interwoven. The visiting speakers at the institutes were largely professors at the state colleges of agriculture, and both depended on the formal support of their state legislatures and of farmers, especially through their agricultural societies (True, 1895, 1928; Kile, 1921). Over the next forty years, these activities were influenced also by the university extension movement in Britain and the growing interest in adult self-improvement (inspired, for example, by the Chautauqua adult education institution in New York State). By 1890, when the second Morrill Act granted federal funds for the establishment of agricultural colleges in the remainder of the United States, the farmers' institutes had spread throughout and become a national institution with federal support and supervision, further stimulated by the formal establishment of experimental work at the state colleges of agriculture under the 1887 Hatch Act. A comparable development of farmers' institutes began in Ontario, Canada, in 1885. These were financially supported by the provincial legislature and spread rapidly with lecturers mainly from the Ontario Agricultural College at Guelph (founded in 1874). A somewhat similar system began in Prince Edward Island (Province of Ontario, 1900; Blackburn & Vist, 1984).

Thus, by the end of the last century, a system of agricultural extension work had become well established in a large part of North America. In the United States, the colleges and their leading professors, including several notable proponents of more practical extension work, progressively took over the initiation and organization of the activity. This culminated in 1914 with the passage of the Smith-Lever Act, establishing the Cooperative Extension Service—a tripartite cooperation of federal, state, and local county governments, with the state college as the extension agency—"in order to aid in diffusing among the people of the United States useful and practical information on subjects relating to agriculture and home economics, and to encourage the application of the same."

In the Southern Hemisphere, extension work also became established along the wide coastal belts of southern and eastern Australia. Several agricultural ("show") societies were formed in the second half of the nineteenth century, although their effect was slight, but as the state administrations became more organized, departments of agriculture were established in the 1870s and 1880s with the aim of developing the potential of their territories. They recognized the importance of agricultural education, influenced by British, Irish, and some American examples whose activities were widely reported in the Australian press. Before the end of the century, under specific state legislation, the departments of agriculture had established agricultural colleges and experimental work in Victoria, South Australia, New South Wales, and Queensland (Black, 1976).

Associated with this development was the official appointment in these states of the first itinerant agricultural instructors in the late 1880s. At the same time, because of the potential importance of milk products, travelling dairy schools were begun, while state exhibitions, especially the Centennial Exhibition in Melbourne in 1888, showed what was possible and gave considerable impetus to farming improvements. The few "government experts," some from the United Kingdom or the United States, grew in number during the 1890s and the first decade of this century, developing the range of the extension work. Its impact and that of the agricultural colleges in their early years was probably slight, but the basis had been laid for further development (Logan, 1984).

Agricultural extension work had also started before 1900 in Japan. Following the Meiji Restoration in 1868, new administrative structures and various modernizing policies were adopted. Two agricultural colleges were established in the mid-1870s, staffed by Western (mainly European) teachers. At these colleges and government farms, experimental work was conducted and new practices were tested and developed. At the same time, agricultural fairs and exhibitions were begun, and progressive Japanese farmers gave talks and demonstrations at them. These led to the development of many agricultural societies from 1881 onwards, a "movement" formalized by legislation in 1899. In 1885, the government also initiated, at national and prefectural levels, a system of appointing experienced farmers as itinerant agricultural lecturers (because the Western "experts" knew little about rice husbandry). Supported by the work at government experiment stations, established from 1893, these farmers formed the basis of agricultural extension work. This activity, including the establishment of demonstration farms, was allocated in 1903 to the numerous agricultural societies which, with state funds, appointed agricultural technicians. In 1910, the 1899 law was strengthened; thereafter, farmers were required to belong to a village agricultural society which was linked to a national network and hierarchy of societies, and farmers were compelled to adopt the technical guidance and recommendations of the societies' extension workers—what became known as "forced extension" (Tajima, 1991; Ministry of Agriculture, 1993).

The development and organization of agricultural extension work was not entirely confined to temperate countries. In a variety of ways, it had also begun in tropical areas, especially in colonial territories. The European colonial powers looked to their overseas territories as a source of tropical agricultural products.

Despite a long connection with some of the colonial areas, the Europeans remained largely ignorant of many tropical agricultural plants. The solution was to establish experimental and demonstration "botanical gardens." The earliest was opened in 1821 at Peradeniya, Sri Lanka (Ceylon), and two others were established in the country later in the nineteenth century. Smaller ones were also created in several Caribbean islands and some West African territories. During the early years of this century, some of these developed considerably, although others were short-lived. Those which succeeded provided important sources of agricultural knowledge and innovation and formed the basis for an interest in agricultural societies and agricultural instruction. Some attempts were also undertaken to improve "native agriculture." This was often associated with the creation, as part of the administration, of departments of agriculture and the appointment of professional agriculturists as directors of agriculture.

A central department of agriculture was established in India after the 1866 Orissa famine, and the government of India soon after resolved to establish departments in each province. However, it was 1905 before a central government directive ordered every province to appoint a fulltime director of agriculture who should organize agricultural research and demonstration farms with staff who could advise farmers (Mook, 1982). The first British colony to appoint a director of agriculture was Zanzibar in 1896. Of more significance, however, was the creation in 1898 of the Imperial Department of Agriculture for the West Indies, with headquarters in Barbados. Before 1914, such departments of agriculture had been created in several African and Southeast Asian territories, as well as in several Caribbean islands (Masefield, 1950). In Sri Lanka, a few agricultural instructors had been appointed about 1880 to work alongside government agents. When in 1904 the Ceylon Agricultural Society was formed to promote experimental work, it also began an agricultural extension service with the objective of reaching native cultivators (Arasasingham, 1981). Along with school gardens (Willis, 1922), the extension workers were considered an effective way of demonstrating improved cultivation practices to villagers. Similar developments also occurred in the Caribbean.

In most tropical African territories, the European interaction with native agriculture was minimal before 1914. The "scramble for Africa" had been mainly in the late nineteenth century, and the young departments of agriculture, where they existed, were largely involved in administrative duties. Before 1914, however, agricultural instruction was given in most government-assisted schools and at four agricultural stations in Ghana (the Gold Coast) (Lucas, 1913). In addition, missionaries often undertook agricultural education, with demonstration and improvement activities, alongside their religious work. The church farms (*fermes-chapelles*) begun in 1895 by Jesuits in the then Belgian Congo (de Failly, 1970) were copied by missionaries of other persuasions in many other areas.

Modern Agricultural Extension

In the early years of this century, extension services were in their formative stage; they were relatively small in scale and limited in the scope of their work and contact with farmers, and their organization was often somewhat haphazard even though based on legislation. They were organized predominantly either by central or local governments, or by agricultural colleges, usually in close association with experiment stations, or by farmers' organizations (agricultural societies, cooperatives, farmers' unions, or chambers of agriculture), or combinations of these parent bodies. As the century has progressed, the organizations have matured. Changes have often occurred to their parent affiliations, government funding has become relatively more important, their objectives have become broader, especially in "the North," and the extension workers have become better trained and more professional. In addition, several other kinds of organizations have developed comparable work: agriculture-related commercial companies; agricultural commodity marketing boards, concerned to assure the supply and quality of their specific product; agricultural development projects, many of considerable territorial scale; and a variety of nongovernmental organizations (especially religious and charitable) involved in agricultural and rural development.

As agricultural extension organizations have grown and changed, they have invariably become more bureaucratic with distinct hierarchical structures. The work of dispersed extension workers had to be administered and controlled so that one or more levels of intermediary structure (for example, district, region) have been created between the field-level agents and their headquarters. Thus the management of extension activities has become a major preoccupation, and many organizations have been open to the criticism of being top heavy and top-down in their approach. However, with funding derived largely from national revenues (or international donors), senior managers have necessarily had to account for and justify their organization's activities. This has been equally pronounced in the North as in the South where, after colonial territories gained their independence, extension work has commonly been reinvented and staffed by nationals under the aegis of their new administrations (usually ministries of agriculture).

During the past quarter century, the work of extension services has often become more diversified. In the less developed countries, the main focus remains on agricultural (mainly food) production, but there has been a growing recognition of the need to reach, influence, and benefit the multitudes of small, resource-poor farmers. Strong efforts have been made in this direction, notably through the training and visit system. Among the commercial farmers of the North, a major problem has become surplus production, with farmers facing economic and policy pressures to restrict it. Associated with intensive production methods, many issues and problems regarding environmental deterioration and livestock welfare have also arisen. Thus these have become important aspects of extension work, particularly socioeconomic guidance which focusses both on means by which farmers might maintain their income levels from their resources (for example, introduction of novel crops or livestock and involvement in various rural enterprises) and on the ways of assuring the longer term welfare of farmers and their families. Agricultural extension services are thus adding a strong social dimension to their activities.

Agricultural extension has now become recognised as an essential mechanism for delivering information and advice as an "input" into modern farming. Since commercial farmers can derive direct financial benefits from these inputs, there is a trend towards the privatization of the extension organizations, often as parastatal or quasigovernmental agencies, with farmers being required to pay for services which they had previously received free of charge. This trend is strong in the North, and there are examples of it beginning in the South.

The pace of change in the organization, functions, strategies, and approaches of agricultural extension is clearly accelerating.

The Future

The need for agricultural and rural information and advisory services is likely to intensify in the foreseeable future. In much of the world, agriculture faces the challenge of keeping pace with rapidly increasing population with few reserves of potentially cultivable land. Farmers will have to become more efficient and specialized. From government perspectives, whatever priority is given to production, extension will remain a key policy tool for promoting ecologically and socially sustainable farming practices.

Some of the most promising recent developments in extension methodology have occurred where the key agenda is environmental or is concerned with equity, for example in the need for the joint management of forests by professionals and local forest users and in integrated pest management. A consistent theme running through the innovative approaches being used, such as participatory rural appraisal (Chambers, 1993), is a fundamental change in what are the respective roles of extension agent and clients. The agent is no longer seen as the expert who has all the useful information and technical solutions; the clients' own knowledge and ingenuity, individually and collectively, are recognized as a major resource; solutions to local problems are to be developed in partnership between agent and clients. Since the scale at which extension support is required is thus often larger than the individual farm, extension workers need new skills of negotiation, conflict resolution, and the nurturing of emerging community organizations (Garforth, 1993; Smith, 1994).

The future is also likely to witness a reversal of recent trends towards bureaucratization within hierarchical extension services and a reduction in their levels of public funding. Moreover, a rapid increase can be expected in the use of information technology in support of extension. The forces for change in these areas (see Rivera & Gustafson, 1991) will come from four main directions.

Economic and Policy Climate

With the collapse during the past decade of socialist forms of economic organization, the (dominant) role of the public sector in national economies has become questionable, with a strengthening trend to reduce levels of public spending. Thus government extension services and those which are largely publicly financed are, and will continue to be, under pressure to become more efficient, to reduce their expenditure and staff, and to pass on (some of) the costs of provision to their clients who directly benefit financially. This is particularly the case in countries where the farm population forms a small minority and agricultural production is in surplus. The case is weaker, but not absent, in less developed countries where farming households form a high proportion of the total population and where increasing food production is still important. Thus charging clients for services is likely to become more widespread, while governments will find it attractive to contract out the operation of services to the private or the voluntary sector.

Social Context in Rural Areas

In the future, rural populations will undoubtedly be progressively better educated, while their exposure to the mass media will continue to reduce their isolation and detachment from information, ideas, and an awareness of their situation within a national and international context. However, this exposure will not reduce the need for extension. Rather, given the changing demands on agricultural producers from population growth, increasing urbanization, legislative changes, and market requirements, the more knowledgeable farming population will require different kinds of extension services. Social and economic trends within rural areas will therefore necessitate more highly trained, specialized, and technically competent workers, who also know where to obtain relevant information and problem solutions and various provision and organizational forms (Moris, 1991; Hayward, 1990) to replace monolithic government extension agencies. These agencies will need to recognize and serve different types of clients defined not in terms of "adopter categories" but of access to markets, degree of commercialization, and relative dependence on agriculture for family income and welfare.

Systems Knowledge

A recognition of the locale-specific nature of farming systems and the agricultural information systems which support them is an important source of the pressure towards the debureaucratization and devolution of extension services. This recognition also implies that extension workers and farmers be jointly involved in the verification and adaptation of new technology, and thus that the extension workers respect farmers as experimenters, developers, and adapters of technology and devote more energy on communication within their local areas. The devolution of extension services to become local organizations is a reasonable corollary of this. Developments in mass media technology, already apparent over a decade ago (Garforth, 1986), will continue to support this localization of extension effort.

Information Technology

The continuing rapid development of telecommunications and computer-based information technology (IT) is probably the biggest factor for change in extension, one which will facilitate and reinforce other changes. There are many possibilities for the potential applications of the technology in agricultural extension (FAO, 1993; Zijp, 1994). IT will bring new information services to rural areas over which farmers, as users, will have much greater control than over current information channels. Even if every farmer does not have a computer terminal, these could become readily available at local information resource centres, with computers carrying expert systems to help farmers to make decisions. However, it will not make extension workers redundant. Rather, they will be able to concentrate on tasks and services where human interaction is essential—in helping farmers individually and in small groups to diagnose problems, to interpret data, and to apply their meaning (Leeuwis, 1993).

The future will call for more able, more independent, more client-oriented extension workers. The emphasis will be on the quality of interaction between agent and client rather than on the movement of "messages" through a hierarchical system.

Flexibility and adaptability will be seen as virtues rather than aberrations. Paradoxically, these trends will bring us full circle to the early manifestations of modern extension in Europe. The itinerant agricultural teachers, unencumbered by large bureaucracies and tall hierarchies, will find their modern counterparts in the computer-carrying extension workers who are at ease helping farmers to identify the information they need in order to realize the potential of their farming operations. Looking back, we can regard the period from 1970 to 1995 as a necessary but expensive stage in the evolution of extension systems, after which extension agents were able to settle down to their main task—bringing together the expertise of farmers and the best available scientific knowledge to develop farms and local agricultural economies.

References

Ahmed, A. (1982). *The role of the information system in development.* Studies Series, No. 314. Baghdad, Iraq: Ministry of Culture and Information.

Arasingham, S. R. (1981). Agricultural extension in Sri Lanka. *Tropical Agriculturist,* 137, 41-54.

Black, A. W. (1976). *Organizational genesis and development: A study of Australian agricultural colleges.* St. Lucia: University of Queensland Press.

Blackburn, D. J., & Vist, D. L. (1984). Historical roots and philosophy of extension. In D. J. Blackburn (Ed.), *Extension handbook* (1st ed.). Guelph, Ontario: University of Guelph.

Bne Saad, M. H. A. Al-h. (1990). *An analysis of the needs and problems of Iraqi farm women: Implications for agricultural extension services.* Unpublished doctoral thesis, University College, Dublin.

Boulet, M. (n.d.). Un type original d'enseignant: Le professeur départmental d'agriculture. Unpublished paper.

Bray, F. (1984). *Joseph Needham's science and civilization in China: Vol. 6. Biology and biological technology. Pt. II: Agriculture.* Cambridge: Cambridge University Press.

Chambers, R. (1993). *Challenging the professions: Frontiers for rural development.* London: IT Publications.

CLVS (1845). Wiesenbewässerung. *Wochenblatt für Land- und Hauswirthschaft, Gewerbe und Handel* (Centralstelle des Landwirthschaftlichen Vereins zu Stuttgart), *24*, 133.

Coletti, F. ([1900] 1985). Le associazioni agrarie in Italia della metà del secolo XVIII al XIX. Reprinted in *Il Dottore in Scienze Agrarie e Forestali, 35*(1985, October), 17-30.

de Failly, D. (1970). Histoire de l'enseignement agricole au Congo (Pt. 1). *Cahiers Congolais, 13*(1), 100-133.

Delman, J. (1991). *Agricultural extension in Renshou County, China: A case-study of bureaucratic intervention for agricultural innovation and change.* Aarhus, Denmark: Institute of East Asian Studies.

Elvin, M. (1973). *The pattern of the Chinese past.* London: Methuen.

FAO (1993). *The potentials of microcomputers in support of agricultural extension, education and training.* Rome: FAO.

Garforth, C. (1986). Mass media and communications technology. In G. E. Jones (Ed.), *Investing in rural extension: Strategies and goals* (p. 185-192). London and New York: Elsevier Applied Science Publishers.

Garforth, C. (1993). Seeing the people for the trees: Training for social forestry in Karnataka, India. *Rural Extension Bulletin*, 2, 33-39.

Gray, E. M. (1952). *The educational work of Philipp Emanuel von Fellenberg (1771-1844).* Unpublished master's thesis, The Queen's University, Belfast.

Grobel, M. (1933). *The Society for the Diffusion of Useful Knowledge, 1826-1846, and its relation to adult education in the first half of the nineteenth century.* Unpublished master's thesis, University of London.

Guggisberg, K. (1953). *Philipp Emanuel von Fellenberg und sein Erziehungsstaat* (2 vols.). Bern, Switzerland: Verlag Herbert Lang.

Hayward, J. (1990). Agricultural extension: The World Bank's experience and approaches. In FAO *Report of the Global Consultation on Agricultural Extension* (p. 115-134). Rome: FAO.

Hudson, K. (1972). *Patriotism with profit: British agricultural societies in the eighteenth and nineteenth centuries.* London: Hugh Evelyn.

Jenkins, H. M. (1884). *Report on agricultural education in north Germany, France, Denmark, Belgium, Holland, and the United Kingdom* (C.3981-I). London: Eyre and Spottiswoode.

Jones, G. E. (1979). The original agricultural advisory service: Lord Clarendon's practical instructors in mid-nineteenth century Ireland. In *Proceedings of the 4th international seminar on extension education.* Dublin: University College, Department of Agricultural Extension.

Jones, G. E. (1981). The origins of agricultural advisory services in the nineteenth century. *Social Biology and Human Affairs, 46*(2), 89-106.

Jones, G. E. (1982). The Clarendon letter. In G. E. Jones and M. J. Rolls (Eds.), *Progress in rural extension and community development: Vol. 1. Extension and relative advantage in rural development* (p. 11-19). Chichester: John Wiley & Sons.

Jones, G. E. (1994). Agricultural advisory work in England and Wales: The beginnings. *Agricultural Progress, 69*, 55-69.*Journal d'Agriculture Pratique* (1874). 38th year, vol. 2.

Kile, O. M. (1921). *The farm bureau movement.* New York: Macmillan.

Leeuwis, C. (1993). *Of computers, myths and modelling: The social construction of diversity, knowledge, information, and communication technologies in Dutch horticulture and agricultural extension.* Wageningen Studies in Sociology, 36. Wageningen: Agricultural University.

Logan, G. N. (1984). Man and the land in Queensland education 1874-1905: Agricultural education as a solution to man-physiography dissonance in colonist Queensland. Unpublished M.Ed. thesis, University of Queensland.

Lucas, C. P. (1913). *A historical geography of the British colonies: Vol. III. West Africa* (3rd ed.). Oxford: Clarendon Press.

Maier-Bode, F. (1910). *Die Organization und die Erfolge des landwirtschaftlichen Wanderunterrichts im Königreich Bayern.* Landsberg am Lech: Georg Verza.

Masefield, G. B. (1950). *A short history of agriculture in the British colonies.* Oxford: Clarendon Press.

Ministère de l'Agriculture (1882). *Bulletin*, 1st year, vol. 1.

Ministry of Agriculture (1993). *Outline of cooperative agricultural extension service in Japan.* Tokyo: Ministry of Agriculture, Forestry and Fisheries (Extension and Education Division).

Mook, B. T. (1982). *The world of the Indian field administrator.* New Delhi: Vikas Publishing House.

Moris, J. (1991). *Extension alternatives in tropical Africa.* London: Overseas Development Institute.

Perkins, D. H. (1969). *Agricultural development in China, 1368-1968.* Chicago: Aldine Publishing; Edinburgh: Edinburgh University Press.

Plunkett, H. (1901-02). Agricultural education for Ireland. *Department of Agriculture and Technical Instruction for Ireland, Journal, 2*(1), 18-41

Province of Ontario (1900). *Report of the superintendent of farmers' institutes of the Province of Ontario, 1899-1900.* Toronto: L. K. Cameron.

Report of the Recess Committee on the establishment of a Department of Agriculture and Industries for Ireland (1896). Dublin: Browne and Nolan.

Rivera, W. M., & Gustafson, D. J. (Eds.). (1991). *Agricultural extension: Worldwide institutional evolution and forces for change.* Amsterdam and New York: Elsevier.

Russell, E. J. (1966). *A history of agricultural science in Great Britain, 1620-1954.* London: George Allen and Unwin.

Smith, H. (1972). *The Society for the Diffusion of Useful Knowledge, 1826-1846.* London: Vine Press.

Smith, V. (1994, Summer). A learner centred approach to social skills for technical foresters. In *From the field.* Rural Development Forestry Network Paper 17e (p. 12-20). London: Overseas Development Institute.

Society for the Diffusion of Useful Knowledge (1827). *Rules of the Society for the Diffusion of Useful Knowledge*. London: William Clowes.

Tajima, S. (1991). Processes of development of agricultural extension in Japan. In *Agricultural extension in Asia and the Pacific* (p. 19-36). Tokyo: Agricultural Productivity Organization.

True, A. C. (1895). Education and research in agriculture in the United States. In *Yearbook of the U.S. Department of Agriculture, 1894* (p. 81-116). Washington, DC: U.S. Government Printing Office.

True, A. C. (1900). Agricultural education in the United States. In *Yearbook of the U.S. Department of Agriculture, 1899* (p. 157-190). Washington, DC: U.S. Government Printing Office.

True, A. C. (1928). *A history of agricultural extension work in the United States, 1785-1923*. USDA, Miscellaneous Publication No. 15. Washington, DC: U.S. Government Printing Office.

Tusser, T. ([1580] 1984). *Five hundred points of good husbandry*. Oxford: Oxford University Press.

White, K. D. (1970). *Roman farming*. London: Thames and Hudson.

White, K. D. (1977). *Country life in classical times*. London: Elek Books.

Willis, J. C. (1922). *Agriculture in the tropics: An elementary treatise* (3rd ed.). Cambridge: Cambridge University Press.

Zijp, W. (1994). *Improving the transfer and use of agricultural information: A guide to information technology*. World Bank Discussion Paper 247. Washington, DC: The World Bank.

Chapter 2
Alternative Approaches to Organizing Extension

Uwe Jens Nagel

Extension, as the organized exchange of information and the purposive transfer of skills, is a rather recent phenomenon. Obviously, transfer of information and skills has existed since the emergence of permanent agriculture. Today's practice is different in that the process is dominated by organizations, and its scope has extended from disconnected local events to a complicated, large-scale, and even worldwide activity.

In this chapter, extension approaches are presented in terms of their most important organizational forms and their respective goals. The goal system reflects the power positions of various groups of actors. Therefore, without an understanding of the historical development and of the interest groups involved, present achievements and shortcomings of extension approaches cannot be evaluated. It is assumed that different forms of organizing extension are per se neither "good" nor "bad." Rather, extension services must be judged against their proper goals. The one universal yardstick, however, is their service function to the rural communities. Extension which is not in touch with and does not significantly contribute to improving the life situation of its clientele has lost its legitimization.

Extension Goals

Goals lead the actions of individuals, groups, and organizations. While pointing towards a future state, they are influenced if not determined by past experiences. They reflect the interests of their stakeholders and differ, therefore, according to specific life situations, power positions, and development philosophies. The prominent features of a system, such as its organizational structure, the choice of clientele, its operational design, and the methods used, are directly influenced by its set of goals and must be evaluated in terms of their contribution to goal achievement.

Main actors within the extension system are the members of rural communities, extension and other development personnel, researchers, and staff of commercial or public service and support organizations. Empirical evidence shows a variety of forms in which interaction among these groups is institutionalized. The variety of forms suggests a similar variety of goals, and either could be used to classify extension approaches. In practice, however, one finds an almost inseparable mixture of goals inhibiting a clear-cut classification. It seems more appropriate, then, to use a broader category, namely, selectivity with regard to clientele, and treat the respective goals as a continuum. The two end points of this continuum would be marked as technology transfer and human resource development, suggesting either a rather narrow technical or a broader socioeconomic view of development.

Technology Transfer

Until the end of the eighteenth century, farming techniques developed gradually and steadily over centuries with few qualitative leaps. Colonialism and imperial expansion introduced innovations—the spread of maize, tobacco, potatoes are striking examples—but experimentation and dissemination of knowledge were basically at the local farm level. The rise of agricultural sciences has induced dramatic changes in this respect. Increasingly, new technology has been created outside the actual farming sector by public sector research organizations. More recently, private firms in industrialized economies find agricultural technology research and development a highly profitable business.

For decades the research-extension-farmer linkage, especially in developing countries, was based on a rather simple model. In order to achieve development, "modern" research results had to be transferred to the "traditional" farmer, and extension seemed to be the appropriate means to do so. The general faith in science and the commitment to modernization led to discrediting indigenous knowledge. Although this view is still held by many administrators, researchers, and extension agents, it is now being seriously questioned. Farming systems research and the "rediscovery" of farmers' knowledge (Chambers, Pacey, & Thrupp, 1989) have shown that "improved technology is a package of inputs and practices that usually comes from many sources" (UNDP, 1991, p. 2). The reexamination of the conventional view on agricultural knowledge cannot, however, result in questioning the important role of research as the source of new technology. For developing countries, one observes that the accelerated growth and spread of problems—such as the degradation of marginal land—surpass the problem-solving capacities of the local population. What is called for is a setting of new priorities and the building of knowledge systems based on problem solving rather than on information transfer.

Uwe Jens Nagel is a Professor of Agricultural Extension and Communication Science, Faculty of Agriculture and Horticulture, Humboldt-Universität of Berlin, Germany.

Human Resource Development

The concept of human resource development is much broader than that of technology transfer, though both are closely interrelated. Increasing complexity not only of technology but also of the life situation of farmers even in remote areas demands new skills. With the help of these skills, rural women and men "acquire a better insight into the network of problems and recognize the alternative solutions available" (Albrecht et al., 1989, p. 34). Traditionally, teaching the basic skills of literacy and numeracy has not been an extension activity. The limited success of literacy programmes in poor countries has drawn attention to nonformal education in which extension has an important part to play (Coombs and Ahmed, 1974). Whereas in most cases this would require a coordinated effort of different organizations of which extension is but one, human resource development may also be regarded as a genuine extension content.

Extension may substitute over a certain period activities such as vocational education that are not yet in place, but more important will be the teaching of managerial and organizational skills that will enable farmers to increasingly solve their own problems. Human resource development thus aims at what may be called "critical competence." Extension clients know what to ask for, they can evaluate the appropriateness of technical information, they are responsible decision makers. Persons with this qualification exist in every rural community, and they will be the ones who actively seek further assistance. One important task of any extension system will therefore be to extend human resource development to underprivileged groups with less access to formal or vocational education—women farmers, rural youth, and generally small farmers in remote areas.

Alternative Ways of Organizing Extension

The goals of extension may vary, as was shown, within the overall system as well as between different extension organizations. In addition, specific objectives may sometimes contradict each other. While smaller systems may come close to pursuing a consistent set of objectives or reconciling conflicting interests, large-scale organizations must work on a compromise basis. In this respect, Axinn's principal observation is of particular importance: "The success of an agricultural extension programme tends to be directly related to the extent to which its approach fits the programme goals for which it was established" (Axinn, 1988, p. 135).

The alternatives to organizing extension demand choices on various levels:

- Public versus private
- Government versus nongovernment
- Top-down (bureaucratic) versus bottom-up (participatory)
- Profit versus nonprofit
- Free versus cost-recovery
- General versus sector
- Multipurpose versus single purpose
- Technology driven versus need oriented

In practice, extension organizations everywhere pursue the overall goals of technology transfer and human resource development, though the emphasis will differ. Within each organization there is a mix of objectives, and within countries there is often a mix of organizational patterns. When presenting an overview on the most important patterns, we will be using a well-established terminology (Axinn, 1988; UNDP, 1991), though the grouping is different. We will differentiate between approaches that, at least in principle, target all persons in rural areas engaged in farming and those that purposely select clientele according to specific criteria.

General Clientele Approaches

Ministry-Based General Extension. Shortly before or after independence, organizing agricultural extension work under the wings of the ministry of agriculture seemed to be an ideal solution for many African and Asian governments. All options for reaching large numbers of clients and serving their needs in terms of quality information and assistance appeared to be open. The original colonial model combined research and extension within the same organization. All important aspects of small-holder agriculture—plant production, animal husbandry, home economics—could be attended to as the ministry established respective sections under its jurisdiction. The fact that the ministerial hierarchy followed the country's territorial subdivision allowed the systematic expansion of the system "down" to the village. The generalist nature of field extension staff functions corresponded to the set of problems faced by noncommercial growers. To cater to specific needs—in terms of technology or in terms of target groups—specialists could be employed. Thus clientele included in principle all persons engaged in agriculture. Commercial service and support organizations lacking, village-level extension staff could be expected to supplement information by rendering services necessary to apply it productively. A uniform and nationwide organizational pattern seemed to facilitate information flow—including the infusion of expatriate expertise—and corrective measures whenever weaknesses were identified. Public interest was to guide goal setting, programme formulation, and the implementation of fieldwork.

A review of the last thirty years of extension work in Africa and Asia shows that reality is quite far for failure are complex and manifold and cannot be removed from this vision (Moris, 1991). The reasons for failure are complex and manifold and cannot be reduced simply to incompetence or the ill-will of national governments.

One reason is the contradictory nature of goals. Public interest implies serving farmers *and* the urban population, securing subsistence production *and* promoting cash crops for export, reaching the masses of rural households *and* serving the needs of specific groups, extending assistance to high-potential *and* disadvantaged producers. In short, priorities will have to be set, and these are all too often pro urban in terms of price policy, favouring innovative individuals within the modern sector, neglecting poorer strata, and forgetting about women farmers.

In many ways, the hierarchical and highly bureaucratic way in which the services are organized hampers a full realization of their potential. Priority setting for research is rarely based on extension field evaluations because the system does not foster critical upward communication.

The way in which technical (and other) knowledge is transformed into field messages frequently leads to distorted and outdated information.

In the eyes of the ministry, extension has never been a purely educational activity. This is a legitimate view as long as the different functions to be performed by extension personnel are compatible and basically client oriented (such as helping to organize input supply). Noneducational activities may include anything from statistical data collection to attending to foreign visitors. Incompatible with and clearly detrimental to regular extension work are such activities as supervising credit repayment, policing disease control measures, organizing "voluntary" community work, and electioneering.

Ministry-based extension has been unable to reach a majority of its potential clientele for economic, sociopsychological, and technical reasons. Even dramatic quantitative increases in personnel—more staff closer to the farmer—have not produced manageable client-to-agent ratios. In recent years, the trend has even been negative. Financial constraints have produced a strong pressure to reduce staff, and the field level has been hit hardest. Those remaining have little if any material resources left to maintain mobility.

In addition, many extension workers select the more responsive section of their clientele. They may have to fulfil production plans, they may want to improve job satisfaction or status, or they may simply be prejudiced against certain target groups. Lastly, extension often has little to offer in terms of messages to large sections of the rural population. Adequate and location-specific answers to a farmer's problem are often not available because it has not been a research concern or the solution has simply not reached the field.

Today's situation is aggravated by two additional aspects which refer to the internal structure of the service: management problems and lack of control from below. Ministry extension employs thousands of persons working under a wide variety of circumstances. Decision making and management are highly centralized and formalized. Extension fieldwork, on the other hand, demands location-specific, flexible, and often quick decisions and actions. Managing the "invisible" man or woman (Chambers, 1974) must be highly ineffective as long as he or she is expected to receive and execute orders.

All these problems are well known, and criticism has come both from within and outside the ministry. What has been lacking is organized feedback from clientele. Farmers may show their discontent by refusing to cooperate with extension, but they have virtually no way of influencing institutional reforms.

Training and Visit Extension (T&V). In the strict sense of the word, T&V is not a separate approach but one way to organize ministry-based extension. The controversial debate on the merits of T&V tends to obscure the fact that it was originally meant to solve some very specific problems of conventional extension services.

Benor and Harrison's original paper—one of the most influential extension publications ever—critically evaluates the ministry-based extension system of the 1970s (Benor & Harrison, 1977, p. 6-9). They found:

- An inadequate internal organizational structure
- Inefficiency of extension personnel
- Inappropriateness or irrelevance of extension content
- Dilution of extension impact

Whichever impact is reached serves "only a few favored farmers in favored areas rather than the bulk of the farming community" (p. 9).

When first being introduced, T&V seemed to be strikingly original and promising because it combined a set of rather convincing simple elements in a plausible way. Rather than trying to reach all farmers directly and thus preprogramming constant failure, the system concentrates on *contact farmers* expected to pass information on to fellow farmers with similar problems. To ensure regular field contacts, facilitate supervision and communication, and set clear and attainable objectives, fixed *visits at regular intervals* are prescribed. Similarly, regular sessions for extension workers to receive *training* and discuss administrative matters are held. Thus costly refresher courses are avoided, knowledge may be enhanced step-by-step, and up-to-date information can be fed into the system.

In addition, T&V operates under the assumption that its extension workers are exclusively engaged in educational activities and that a unified extension service exists. Agricultural research must not only be effective but also work in close collaboration with extension. Both external and internal evaluations are to be used to constantly modify and adapt the system to changing conditions.

Simple as the prescriptions seemed, implementation proved to be difficult. First, the contact farmer concept—implying a two-step flow of information from the extension worker to the contact farmer and from there to other farmers—has frequently failed. Extension workers have been blamed for "wrong selection," but the root of the problem lies within the purely technical philosophy of T&V. Other aspects such as communication skills, leadership, and organizational capacities are neglected. In practice, T&V has been a top-down approach leaving little possibility for participation and initiative, both for farmers and village extension workers. Too little emphasis has been put on critical feedback based on self-evaluation. As a result, rigidity rather than flexibility characterizes local fieldwork.

Secondly, Benor's fear that extension services may "rapidly run out of anything to extend" (Benor & Harrison, 1977, p. 8) characterizes many T&V field situations. The standardized messages passed on are often of little relevance to local conditions. Once T&V was extended to less favoured regions, it soon became clear that technology of the green revolution type showing quick and visible results is not available. Still, training sessions were held and visits made according to schedule, leaving behind disinterested farmers and demotivated extension workers.

The limited success of T&V in its present form as a nationwide extension system should not discredit the quality and appropriateness of many of its elements. Applied less rigidly and combined with the tools of human resource development as well as with the concept of participation, these elements may constitute a valuable base for reforming extension organizations, large or small (Nagel et al., 1992).

The Integrated (Project) Approach. Integrated approaches aim at influencing the entire rural development process. Extension is only one though often crucial element in this strategy which targets the entire population in a given area but emphasizes work with disadvantaged groups. Integrated approaches are generally implemented in the form of large-scale and foreign-funded projects aiming at alleviating mass poverty in rural areas on the basis of "a simultaneous improvement in the utilization of natural resources and of human potential" (Rauch, 1993, p. 6). Measures to promote production are coupled with a strong emphasis on self-help. The underlying concept is typically multisectoral.

Evaluations of more than a decade of integrated rural development (IRD) projects have revealed serious shortcomings in reaching the goal of mass poverty alleviation (IBRD, 1987; BMZ, 1990). Sizeable numbers of the poor were not reached by project activities, nor were positive effects consolidated on a sustainable basis. Project deficiencies were in part management related and very often due to a serious underestimation of the great complexity of multisectoral programmes with ambitious goals. The disregard of the target group principle and of due consideration for framework conditions (economic and institutional) played an even more important role, as did the lack of compatible technical solutions.

Recent efforts to improve regional rural development (RRD) projects and enhance chances for a broad and sustainable impact (Rauch, 1993) are relevant for all general extension approaches. The key concept is the availability of locally adapted solutions established on a common basis. This requires not only participatory technology identification, test, and dissemination, but also an active role by the change agency in mediating between different institutions involved and their interests. A particular emphasis is laid on dealing with adverse framework conditions, explicitly taking them into account and attempting to influence them in favour of clients. Finally, in order to achieve these improvements, new efforts must be made to specify and operationalize (extension) objectives and concepts (sustainability, participation, gender-specific target-group approach, and poverty alleviation).

University-Based Extension. While the Cooperative Extension Service (CES) of the United States is still the only system in which the main extension function remains within the university, some developing countries, notably India, have integrated educational institutions into practical extension work. Within the United States of America, state universities have traditionally cooperated with local counties and the U.S. Department of Agriculture in doing extension besides education and research. Within the last 130 years, extension goals of the land-grant colleges have shifted from practical education to technology transfer and, more recently, to a much broader concept of human resource development.

With the emergence of strong private and other public sector research and development organizations and dramatic changes within the agricultural production sector, CES is facing new challenges with regard to coordination and cooperation. Apart from its traditional roles, *networking* will become a primary role (Bennet, 1990, p. 16). In this model, industry as well as intermediate and end users of knowledge become part of the extension system.

While in most countries, the main contribution of educational institutions to extension will be the training of qualified, dedicated, and responsible personnel, some Indian agricultural universities have come close to the U.S. model without taking over the full load of extension work. In the field, they have taken over functions which are only inadequately performed by the ministry, thus supporting general extension work. Remarkable features are direct assessment of clients' needs, user-oriented research, quality training for state personnel, and a strong linkage between academic education and field practice. Models vary from state to state. The Punjab Agricultural University (PAU) has its own multidisciplinary extension team in each district, engaged in adaptive research, training, and consultancy. Backed up by extension specialists on campus, they are transmitters and receivers of experiences from researchers, farmers, and state extension workers. At PAU, a unique system of processing these experiences is practised. Regular workshops are held which unite university and department staff from research and extension together with outstanding farmers. New findings and feedback are presented, evaluated, and published as a "Package of Practices" to be used by all extension staff for the next season (Nagel, 1980).

In the Philippines, which works with ministry-operated extension, university field contacts have been combined with practical development work. The University of the Philippines at Los Baños (UPLB) has its own "social laboratory" in rural areas. Transfer of ideas is not limited to production technology, but includes the testing of communication strategies as well as helping farmers to organize themselves. Experiences are channelled back into UPLB teaching and research (Axinn, 1988, p. 102-103).

Animation Rurale. For a historically rather short period, the concept of Animation Rurale (AR) gained importance in francophone African countries such as Senegal, Ivory Coast, and Madagascar (de Wilde, 1967, p. 391-414; Joerges, 1967). Though the original approach is no longer pursued, some of its elements are now being reintroduced into rural development programmes.

Animation Rurale was an answer to the authoritarian and often repressive nature of intervention before independence. Developed originally by the French Institut de Recherches et d'Application des Méthodes de Développement (IRAM), it shows many parallels to the Brazilian experiments of Paolo Freire.

Integration of rural areas into the national system was to be achieved by initiating a dialogue between rural communities (*collectivités*) and the state. In a dialectical way, increasing competence of villagers to express their own needs was to liberate them from colonial dependence. In order to initiate and perpetuate this process, AR relied on a large number of voluntary collaborators, so-called *animateurs*. Selected by the villagers themselves these *animateurs* had to be experienced and well-respected farmers but not traditional leaders. Training, supervision, and support of *animateurs* were organized by the Ministry of Rural Development. Their task was to initiate discussions within the community on local needs and objectives, thus empowering rural people for a dialogue with the state. At the same time they were to "interpret" government plans to the villagers and acquaint them with services available. The long-term perspective was a replacement of traditional institutions and the creation of "development cells" able to negotiate contracts with the state bureaucracy.

Sülzer and Payr (1990, p. 34) maintain that AR "did not fail as a philosophy of extension . . . [although] . . . it did not achieve a large-scale breakthrough on a national level." Lack of sustainable impact was due to internal as well as external factors. The objectives of AR were extremely difficult to operationalize and, as a result, the role of *animateurs* remained unclear. In addition, lack of rewards and selection mistakes contributed to the fact that many *animateurs* soon lost interest in their work. Farmers, as it turned out, were more interested in receiving qualified technical assistance, and even if *animateurs* had successfully initiated village projects, it was the "technicians" who reaped the benefits. Lastly, it is highly questionable whether the administration was seriously committed to creating a system which would curtail its own power.

What has remained is the philosophy of empowerment and many of the practical experiences. Many NGOs use the ideas of Animation Rurale often without realizing their roots. The present discussion on participatory extension shows its lasting influence.

Extension to Selected Clientele

Commodity Based Extension. Next to the ministry-operated general approach, commodity-based extension run by government, parastatals, or private firms is the most frequent extension organization. Clients may be dispersed over a large area or closely connected, as in the case of large, centrally operated irrigation projects. Commodity-based extension is the predominant feature in many francophone countries of Africa (Schulz, 1973), but is also strong in other countries with commercial or export crops.

The original rationale was the generation of revenue as well as the assured supply of tropical products for the colonial powers. Today, goals are still clearly and intentionally production and profit oriented. All aspects of producing and marketing a particular crop are vertically integrated, spanning the whole range from research, advice, and material support given to farmers, to organizing marketing and even exports. Proponents of the approach argue that, by infusing modern technologies and monetary incentives into traditional farming, a cumulative chain of effects is triggered, thus contributing to overall development.

Advantages in terms of organizing the extension function seem obvious. One generally works with well-tested technologies. Objectives and targets can be clearly defined and the organizational structure kept simple. The focus on only one or two crops facilitates training of extension workers who are agents of the society or board concerned. Control of agents and farmers is easy, because they are judged in terms of defined targets.

A closer look at these advantages reveals that they are largely defined from the perspective of the commodity organization. This poses no problem as long as organizational and clients' goals are identical, as was the case for coffee, tea, or sisal boards in the private plantation sector. For small farmers, the situation may be quite different. The rigidity of the system leaves little room for incorporating farmers' needs. The border between control and coercion is often crossed, for example, when farmers are forced to plant commercial crops at the expense of traditional subsistence crops. Extension workers are regarded as successful once they have brought farmers to producing "what and how" the organization wants. The obvious advantage of guaranteed marketing does not automatically entail security for the agricultural producer. Farmers cannot react quickly to price fluctuations, and in some cases quality standards are arbitrarily set in order to increase personal or organizational profits. Many governments have used the approach to excessively extract revenue by dictating low farm-gate prices.

Strengths as well as limitations of the commodity approach lie in its narrow focus. It is useful in terms of technology transfer but leaves out important public interestissues (such as environmental protection), as well as target groups (such as noncommercial producers). A successful combination of general and commodity-based extension at the national level, as practised in East Africa, demands clear policy goals and highly efficient management.

Extension as a Commercial Service. Commercial extension is a rather recent phenomenon and typical of either industrialized forms of agriculture or the most modern sector of an otherwise traditional agriculture. It may be either part of the sales strategy of input supply firms or a specialized consultancy service demanded by an agricultural producer. In both cases, the goal of the organization or the individual is profit earning, which in turn is tied very closely to customer satisfaction. Most directly this is the case for private consultants who will be retired only if their clients feel that expenses made have been profitable. Large input supply firms or rural banks that use their own extension workers as sales personnel must also have a long-term perspective with regard to the competitiveness of their products and services. Negative effects of incorrect application or use will be attributed to the product itself. The clients of commercial extension will also be profit oriented. Their objective is the optimal utilization of purchased inputs or contracted expertise.

The emergence of commercial extension has influenced the debate on who should bear the costs of extension. With escalating budget deficits, the idea of extension as a free public service is no longer being generally accepted. It is argued that those who can afford it should actually pay for advisory services. In the case of commercial input suppliers, the solution is very simple: the costs of extension are included in the product price, as are the costs for research or advertisement. Private consultancy, on the other hand, is costly and affordable only to either large-scale or highly specialized producers.

As a general trend, one observes that public extension in industrialized countries has been under pressure to introduce cost sharing or altogether commercialize advisory work. An approach which combines commercial and public elements is at present being introduced in some of the eastern states of Germany. For example, the Ministry of Agriculture in Brandenburg subsidizes consultancies once they have actually taken place. Farmers have the option either to organize themselves in "extension rings" and employ their own extension workers or to choose an extension consultant who is officially accredited by the ministry once he or she organizes at least twenty clients in an "extension association." In both cases, up to 80 per cent of extension costs within a certain limit are reimbursed to the farmer.

Privatization and cost sharing are propagated in the name of greater effectiveness and efficiency, but are largely motivated by financial constraints. It is obvious that the private sector will be active only in case of reasonable returns, and they will not be concerned with public interest issues:

> Because of the selective participation of the private sector, the provision of public good types of information will have to remain a public sector responsibility . . . public and non-profit organizations . . . will have to work together to satisfy the needs of those in "orphan" areas. (Umali & Schwartz, 1994)

Client-Based and Client-Controlled Extension. One way of dealing with the shortcomings of large extension systems has been to localize extension and utilize the self-help potential of rural groups. Often organized by outsiders, these decentralized approaches are in a better position to serve the needs of specific target groups, notably those in disadvantaged positions. Close contact with their clients and intimate knowledge of their life situations are essential for the planning of problem-oriented extension activities. Local personalities are identified who take over leader functions once the external (nongovernmental) organization withdraws. The principles of these organizations (awareness, empowerment, participation, self-help) are close to the philosophy of Animation Rurale without the national dimension.

The impact of client-based approaches must be seen on two levels. Directly, they provide benefits to their clients. The diversity and large number of small projects forbid a general statement on their effectiveness in terms of human resource development. It appears, however, that their weakness lies more in the technical field (UNDP, 1991, p. 22). Besides, they can reach only a very limited number of people. Apart from this, they perform an important role as organizational innovators. They have proved that participation can work in practice and that many farmers are highly competent partners in technology development. Government extension services have been forced to rethink their top-down approach, to accept human resource development as an equally important extension goal, and to address the problems of rural women.

A rather unique approach has been practised in Taiwan, where a large share of extension work is done through farmers' associations (Lionberger & Chang, 1970; Axinn & Thorat, 1972). Organized at provincial, county, and township levels, membership totaled 90 per cent of Taiwanese farmers. Extension education is done by agents employed by the farmers' associations at the township level and financed largely by the farmers themselves.

Unlike the small self-help groups discussed above, there are strong and institutionalized linkages with research and other services. The overall extension policy is defined by the government. On the other hand, the clientele is quite different: farms are highly modernized and extension advice is demand driven.

Present and Future Role of Extension Staff

Person-to-person communication has traditionally been the most important form of information transfer. Print media as well as radio and television were of a supplementary nature because they frequently lacked a target group or location specificity and information was not up-to-date. Revolutionary changes in communication technology have dramatically increased the speed and quality of information transfer and changed the role of extension workers in industrialized countries. Electronic communications systems may in part replace personal visits, and one of the major tasks of any agent will be to link her or his clients with other suppliers of information.

Notwithstanding the fact that their use is not restricted to industrialized countries, the fascination with modern communication means tends to obscure the fact that most extension personnel—[illegible] per cent of which are located in developing countries (FAO, 1990)—are working under extremely difficult and disadvantageous conditions. In fact, little has changed during the last two decades to remedy the basic ills, and the field agent is still the weakest link within the system.

There is a wide discrepancy between organizational goals and the potential of even well-trained and dedicated field staff. Early Indian community development activities covered close to fifty areas (from reclamation of wasteland to improved rural housing on a self-help basis), all to be administered by the local village-level worker (Dube, 1958, p. 19-21). Fieldwork in most developing countries is characterized by conditions that foster low morale: lack of mobility, virtually no equipment, and extremely low salaries. For many extension workers, tapping additional income sources is a question of physical survival. Quality performance is further impeded by low educational qualifications and lack of advancement possibilities (Swanson, Farner, & Bahal, 1990, p. 55-64).

While working conditions of extension personnel have deteriorated, expectations with regard to their role are increasing. They are no longer to be simply transmitters of technical knowledge. They are to practise participatory methods, recognize and respect gender issues, identify indigenous needs and problem solutions, and serve as a link to the world outside the village, to name but a few of the present topics. The emerging role is closer to that of a "socio-economic community worker" (Blackburn & Flaherty, 1994, p. 16) than a technical expert, but their training is insufficient for either.

The situation sketched above is well known and documented. The sheer dimension of the problem surpasses, however, the capacities of poorer countries, notably in Africa. Foreign-funded projects have addressed the issue in a piecemeal fashion and have often drained nonproject areas of personnel. Staff reductions on a national level have not even secured the status quo. Neither approach has solved the basic dilemma: insufficiency of material and human resources to reach universally accepted societal goals. Having to count on their own resources for extension, many countries will not be in a position to implement technology transfer, much less the more demanding strategy of human resource development. Regardless of specific extension approaches, there is no alternative to a strong international commitment to strengthening and revitalizing extension personnel resources.

References

Albrecht, H., Bergmann, H., Diederich, G., Grosser, E., Hoffmann, V., Keller, P., Payr, G., & Sülzer, R. (1989). *Agricultural extension: Vol. 1. Basic concepts and methods*. Eschborn: Deutsche Gesellschaft für Technische Zusammenarbeit (GTZ).

Albrecht, H., Bergmann, H., Diederich, G., Grosser, E., Hoffmann, V., Keller, P., Payr, G., & Sülzer, R. (1990). *Agricultural extension: Vol. 2. Examples and background material*. Eschborn: Deutsche Gesellschaft für Technische Zusammenarbeit (GTZ).

Axinn, G. H. (1988). *Guide on alternative extension approaches*. Rome: FAO.

Axinn, G. H., & Thorat, S. (1972). *Modernizing world agriculture: A comparative study of agricultural extension education systems*. New Delhi: Oxford & IBH Publishing.

Bennet, C. (1990). *Cooperative extension roles and relationships for a new era: Summary*. Washington, DC: Extension Service, U.S. Department of Agriculture.

Benor, D., & Harrison, J. Q. (1977). *Agricultural extension: The training and visit system*. Washington, DC: The World Bank.

Blackburn, D. J., & Flaherty, J. (1994). Transitions and directions in extension. In D. J. Blackburn (Ed.), *Extension handbook: Processes and practices* (p. 8-17). Toronto: Thompson Educational Publishing.

BMZ (1990). *Querschnittsanalyse von Projekten der Ländlichen Regionalentwicklung*. Bonn: Bundesministerium für Wirtschaftliche Zusammenarbeit.

Chambers, R. (1974). *Managing rural development*. Uppsala: Scandinavian Institute of African Studies.

Chambers, R., Pacey, A., & Thrupp, L. A. (Eds.). (1989). *Farmer first: Farmer innovation and agricultural research*. London: Intermediate Technology Publications.

Coombs, P. H., & Ahmed, M. (1974). *Attacking rural poverty: How nonformal education can help.* Baltimore and London: Johns Hopkins University Press.

De Wilde, J. C. (1967). *Experiences with agricultural development in tropical Africa: Vol. 2. The case studies.* Baltimore: Johns Hopkins University Press.

Dube, S. C. (1958). *India's changing villages.* London: Routledge.

FAO (1990). *Report of the global consultation on agricultural extension.* Rome: FAO.

IBRD (1987). *World Bank experience with rural development, 1965-1986.* Washington, DC: The World Bank.

Joerges, B. (1967). Animation Rurale in Africa—Die Methoden der IRAM. *Zeitschrift für ausländische Landwirtschaft, 6,* 293-309 .

Lionberger, H. F., & Chang, H. C. (1970). *Farm information for modernizing agriculture: The Taiwan system.* New York: Praeger.

Moris, J. (1991). *Extension alternatives in tropical Africa.* London: Overseas Development Institute.

Nagel, U. J. (1980). *Institutionalization of knowledge flows: An analysis of the extension role of two agricultural universities in India.* Frankfurt (Main): DLG-Verlag.

Nagel, U. J., Baas, S., Chiyanika, P., Eckert, S., Edsen, J., Geiger, M., Laue, R., Lübke, G., & Marbach, H. (1992). *Developing a participatory extension approach: A design for Siavonga district, Zambia.* Berlin: Centre for Advanced Training in Agricultural Development.

Rauch, T. (1993). *Regional rural development RRD update: Elements of a strategy for implementing the RRD concept in a changed operational context.* Eschborn: Deutsche Gesellschaft für Technische Zusammenarbeit (GTZ).

Schulz, M. (1973). *Landwirtschaftliche Neuerungsverbreitung an der Elfenbeinküste.* Saarbrücken: Verlag der SSIP-Schriften.

Sülzer, R., & Payr, G. (1990). Socio-economic development approach: "Animation Rurale" in francophone Africa.In H. Albrecht et al. (Eds.), *Agricultural extension: Vol. 2. Examples and background material,* p. 31-35. Eschborn: Deutsche Gesellschaft für Technische Zusammenarbeit (GTZ).

Swanson, B. E., Farner, B. J., & Bahal, R. (1990). The current status of agricultural extension worldwide. In FAO, *Report of the global consultation on agricultural extension,* p. 43-76. Rome: FAO.

Umali, D. L., & Schwartz, L. (1994). *Public and private agricultural extension: Beyond traditional frontiers.* Washington, DC: The World Bank.

UNDP (1991). *Agricultural extension programme advisory note.* New York: United Nations Development Programme.

Chapter 3
The Context of Extension in Agricultural and Rural Development

Warren Peterson

Agricultural extension, whether public or private, operates in a context or an environment that influences the organization, form, and content of transfer activities (Moris, 1991). The dominant characteristic of that context is change. Because the changes affect all aspects of extension, the context should be examined and understood so that extension can be better managed.

This chapter first places extension within a system of agricultural technology generation and transfer. Two major sections follow. One describes the macro-context factors: agroecological, infrastructural, policy, political-economic, and sociocultural. The other looks at the institutional context, namely, other institutional actors involved in activities related to agricultural extension such as research and transfer, education and training, input supply, and credit.

A Systems Perspective

The usefulness of a systems approach for understanding and analysing agricultural technology generation and dissemination is widely acknowledged (among others, Nagel, 1979; Swanson & Peterson, 1991; Röling, 1991; Elliott, 1994). Although they are sometimes criticized because they are so abstract, systems analyses offer holistic vantage points for understanding the factors that impede or enhance the two-way flow of technology and information between farmers and the public organizations that constitute the system.

The model used here identifies an agricultural knowledge system consisting of four components set in a larger context (Figure 1). The components are technology generation, technology transfer (knowledge and input transfer), technology utilization, and agricultural policy (Swanson, Sands, & Peterson, 1990). The organizations that constitute the components, as well as others in the system environment, influence each other in complex ways. Such publicly funded systems are established by governments to improve the conditions of life and well-being of rural and urban populations and to increase agricultural productivity.

The functions and linkages related to the flow and feedback of technology and information in the system define the components. In the public sector, agricultural research organizations (technology generation) and extension (technology transfer) are major actors,although commercial companies and NGOs(the private sector) are also involved. The way these are divided among organizations differs from country to country.

Technology generation consists of planning, administration, and implementation of research activities that develop, assess, adapt, and test improved agricultural technology for farmers and other users. In the public sector, these tasks, as well as some dissemination work, are carried out by agricultural research organizations.

Technology transfer further evaluates and adapts research outputs for users and then widely disseminates the knowledge and inputs to different target adopters—farmers of different categories, private companies, and so on. Figure 1 shows two parts of the transfer component, namely, knowledge and inputs. In many countries, government extension does both types of transfer, although the emphasis is on knowledge. Private organizations are also involved in both types of transfer, although farmer coverage is more limited and input transfer may play a bigger role.

The *technology utilization* component encompasses the users of the agricultural technology, mainly farmers. User awareness, adaptation, and adoption of improved technology from various sources affect farm-level productivity and profitability and, ultimately, economic growth at the national level. Interaction and feedback between users and research and transfer organizations improve cooperation and the relevance of technology.

The *policy* component relates to government development goals and strategies, market and price policies, and the levels of resource investments in the system. Various government bodies play a role in setting development policy. Technology development and transfer organizations are affected by the policy in fundamental ways.

The division of generation and transfer tasks organizations and sectors differs from country to country, as does the nature of the policy and utilization components. However, the flow of technology can be mapped as it moves between or is influenced by organizations constituting the agricultural technology system. Flow analysis shows the impediments to generation and transfer and the influences of other institutional actors and highlights contextual issues.

The agricultural technology system context in which government extension organizations operate can be described under two headings: macro-factors and institutional factors. The discussion that follows introduces the more important ones in each category that need to be

---

Warren Peterson is a Senior Fellow with the International Service to National Agricultural Research, The Hague, The Netherlands.

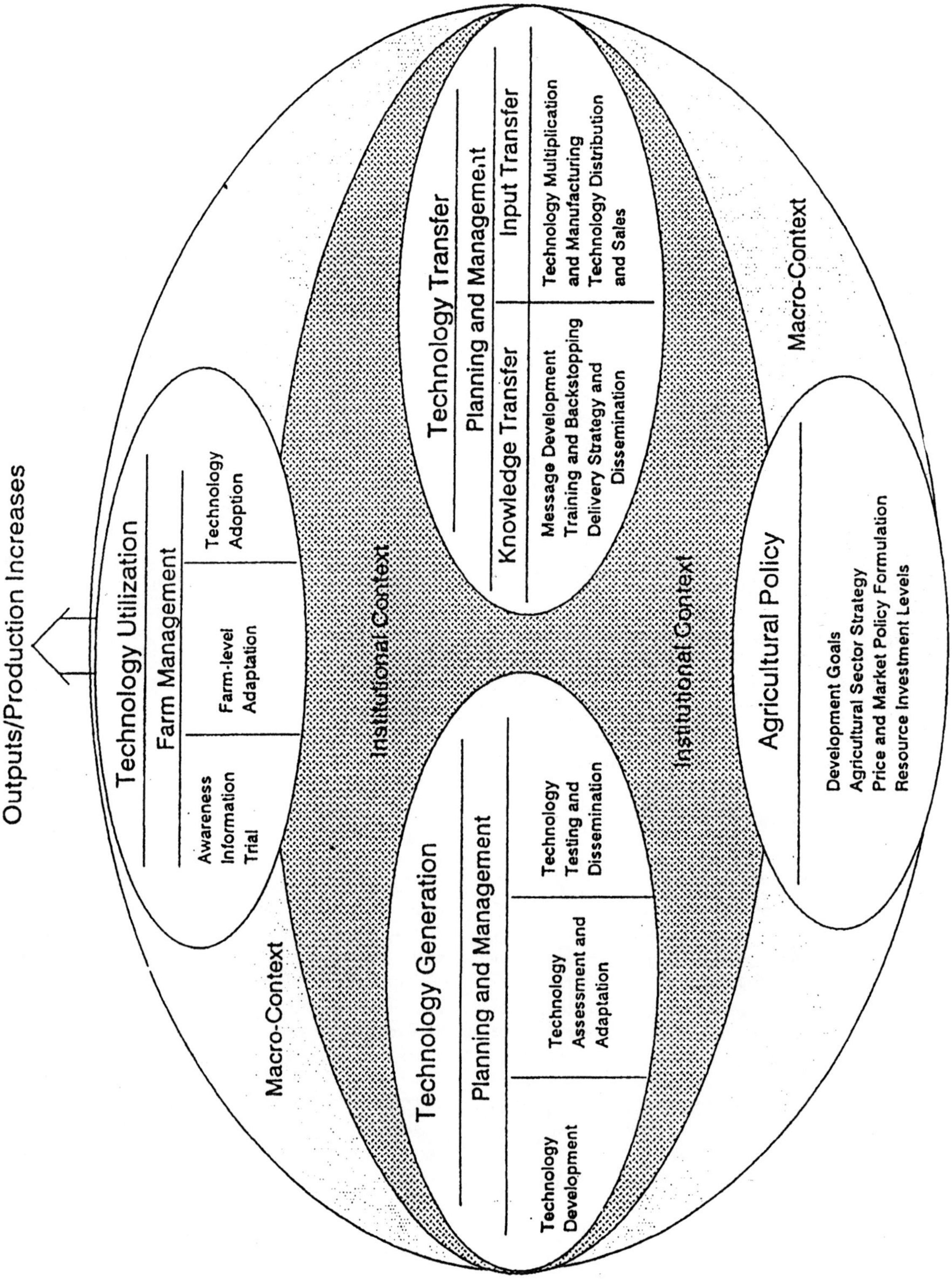
Outputs/Production Increases
Technology Utilization
Farm Management
Awareness
Information
Trial
Farm-level
Adaptation
Technology
Adoption
Macro-Context
Technology Generation
Planning and Management
Technology
Development
Technology
Assessment and
Adaptation
Technology
Testing and
Dissemination
Institutional Context
Institutional Context
Technology Transfer
Planning and Management
Knowledge Transfer
Message Development
Training and Backstopping
Delivery Strategy and
Dissemination
Input Transfer
Technology Multiplication
and Manufacturing
Technology Distribution
and Sales
Agricultural Policy
Development Goals
Agricultural Sector Strategy
Price and Market Policy Formulation
Resource Investment Levels
Macro-Context

understood by extension planners and managers. Any of these factors can impede or influence the flow of technology and information between farmers and extension.

Macro-Factors

Agroecological

Because the natural environment strongly influences extension planning and operations, extension should respond to the technological needs of farmers in different agroecological zones. The variation represented by agroecological zones in a given country can be substantial. Differences in temperatures, rainfall, soil types, evapotranspiration, and so forth are reflected in the diversity of farming conditions and production systems found.

Extension planners face some difficult choices because of the need to respond to the diverse technology and information needs of farmers from many different zones and, at the same time, to satisfy a requirement for extensive countrywide coverage of the rural population. For example, extension resource investments can be determined by farmer population concentrations, potential productivity of selected agroecological zones, or a combination of both.

Political-Economic

The political and economic environment affects extension in many ways. One of the most significant factors is a country's stage of economic development. Another is the level of government investment in public sector extension. This is influenced by the presence or absence of a structural adjustment programme, the degree of economic dependence on agriculture, and the proportion of the population economically active in agriculture as opposed to industry. The percentage of resource-poor smallholder farmers influences the type of technology to be transferred, particularly if government is concerned with the equitable coverage of all categories of farmers.

Politics has another, infrequently identified impact on public extension organizations. Political shifts at the national level often result in changes in extension personnel, management, and programmes. High turnover of top managers undermines management experience and continuity in leadership. In addition, political unrest and war often make extension efforts impossible (Gustafson, 1994).

Sociocultural

In many countries, sociocultural factors are leading constraints to the effectiveness of extension. Language differences and illiteracy can impede the communication of improved technology unless they are taken into account. The division of labour between the sexes can differ along cultural lines and influences the nature of farming systems in different regions. In many countries, the men are employed off-farm, leaving the farm operations to women. In extension organizations, under representation of women on the extension force means that the production responsibilities and needs of women at the farm level may not be adequately addressed.

Cultural differences among farmers, as well as differences in their resource endowments, also need to be taken into account. In particular, these are reflected in land-use strategies. Pastoral herders, for example, will require different types of subject-matter expertise, and extension will need to use different strategies to transfer technology to them than to permanent field agriculturalists. The resource endowments of different categories of farmers also affect technology adoption levels. Subsistence farmers adopt mainly low-cost technologies. For this reason, extension work that focusses on cultural practices and affordable technologies may be more appropriate in countries with large numbers of resource-poor farmers.

Policy

The policy component of an agricultural technology system can enable or limit extension in ways beyond the reach of extension managers. The principal areas of influence are price signals to farmers and decisions by government that affect public agricultural development organizations.

Agricultural Development. Policy-making bodies of government set development goals and objectives such as achieving food security or surplus agricultural production to stimulate economic growth, or providing health care and education for rural development. It is the task of government extension organizations to help meet these goals by formulating subordinate objectives and strategies to achieve them.

The amounts invested by government and the development community in agricultural development influence the pace and scale of effort. In many instances, spending ceilings for public institutions have been imposed by a structural adjustment programme agreed to by government. Extension planning needs to take these limits into account, especially when dealing with the question of staffing, which has major budget implications.

The endowment levels for public agricultural extension are set by government policy and planning bodies and impose limits on expenditures. This can hamper extension efforts in basic ways. The operating as compared to salary budget amounts are often inadequate, with negative effects. For example, a lack of money to buy vehicles and fuel undermines the mobility of extensionists. In addition, if salaries are too low, extension cannot attract or hold qualified staff, and services will suffer. Careful planning, guided by a strategy and programme of action, can improve the use of resources and the effectiveness of extension work within government funding limits.

Market Intervention. Governments set policies on consumer and producer commodity prices, subsidies for inputs, credit availability, import substitution, export earnings, food self-sufficiency, and natural resource management. These send direct and indirect price signals to farmers and influence their production decisions. High fertilizer prices, for example, cause farmers to use less of this input. Extension organizations should be aware of market signals to ensure that they are recommending technologies that meet farmers' current needs. They also require feedback from farmers to ensure the continued relevance of their activities.

Commodity prices influenced by government policy act as incentives or disincentives for farmer production. If there is no profit incentive for production of a specific crop, there is little point in developing or transferring improved technology related to that crop. For example, policies that favour the import of cereal grains at concessionary prices on the international market discourage incountry production of those crops. Technology generation and transfer outputs that focus on such crops are not likely to interest farmers. The examination of the policy context is important and allows organizations to avoid wasted effort and resources.

Infrastructure

Infrastructure, particularly the condition of transport, communication, and market facilities, affects both farmers and extension. The capacity to move people, inputs, and produce and to send and receive information influences extension activities and capacity. Market infrastructure itself can be lacking or inadequate.

Transportation. There may be many areas that cannot be reached by road, or transport vehicles may be in short supply. In either case, farmers under these conditions are difficult to reach with improved technology, and they will have problems transporting inputs and farm produce. For example, input supply outlets can be far apart, and excessive transport distances can make it difficult or impossible for traditional farmers to adopt improved technology.

Communications. Communication infrastructure can impose additional constraints for extension organizations. Farmer access to mass media such as publications, radios, or television may be limited, thus reducing the options available to extension for communicating its messages. At the same time, extension itself may have little or no access to telephone and radio services for long-range communications. This can severely hamper its ability to organize and carry out field operations.

Two aspects of a country's media organizations, both print and audio-visual, affect the flow of extension messages to farmers (Mathur & Sinha, 1991). One is the attitudes and subject interests of media managers responsible for programming for rural audiences. The other is the organizational climate, especially morale. It is crucial that extension consider the capacity of mass media organizations (newspapers, magazines, radio, and television) as part of its strategy and plans for communicating with farmers.

Institutional Factors

Public agricultural extension organizations in most countries have the task of providing a two-way flow of improved technology and information between research and users, primarily farmers. They operate in an institutional environment that includes other public and private organizations active in agriculture. In particular, those other actors involved in generating and transferring agricultural technology must be examined and understood to improve extension's effectiveness and efficiency.

Research

Agricultural research organizations are extension's closest institutional partners in technology generation and transfer. While these functions are also undertaken by private organizations, public sector organizations have the task of generating technology relevant for all categories of producers and agroecological zones. The way research is structured and organized, and the planning and management of research-extension linkages, can limit or enhance extension's effectiveness.

Agricultural research organizations in developing countries confront many problems (Idachaba, 1987; Oram & Bindlish, 1984). These include lack of financial resources, acute shortages of well-trained scientists, lack of farmer feedback to ensure relevance of research results, lack of access to external sources of knowledge, inadequate research facilities and equipment, low staff morale, and inadequate operating budgets, staff incentives, and remuneration. Few of these can be addressed by extension managers, but they can impede the generation of technology, resulting in fewer research outputs for extension to transfer. An understanding of research's problems is an important step in planning extension activities and coordinating them with research.

Poor linkages between research and extension are major constraints in technology flow in many countries (Kaimowitz, Snyder, & Engel, 1990; Eponou, 1993). The linkage problems are of two basic types: those affecting feedback from farmers to research and extension and those relating to coordination and cooperation between research and extension. In both cases, extension managers can improve the situation by developing a linkage strategy and allocating responsibilities and adequate resources for linkage tasks.

Education and Training

Educational organizations that train extensionists are important elements in the institutional context for extension. The work of universities and training institutes in particular has a significant impact on extension organizations. The content of their curricula as well as the numbers and qualifications of their graduates are limiting or enabling factors in any country. In many cases, communication between extension and education organizations is poor. As a result, extension commonly has staffing problems. It is not unusual for extension organizations to have posts that are either vacant or filled by underqualified personnel.

Inadequate numbers and qualifications of staff remain a difficult problem for public sector extension organizations. Salaries and benefits are rarely competitive with those of comparable private and public enterprises, resulting in low morale and high staff turnover. Education levels may be quite low, especially for farmer contact staff. The ability to attract and retain qualified extension staff is limited in most countries by civil service salary scales established by other agencies of government.

The situation can be improved by establishing staffing and training plans. The staffing plan inventories current human resources, identifies staffing gaps, and projects staffing needs over a specified time. The training plan identifies specific types of training (in-service and formal) required to fill skill gaps in human resources and to cover staffing needs for planned operations. The additional step of coordinating training needs with the educational organizations is needed.

Input Supply

Farmers need inputs to increase production, but access to these is often poor in less developed countries. While inadequate transport and marketing infrastructure are often at the root of the problem, there are certain aspects that can be addressed by extension.

Genetic Technology. Among the major outputs of the technology generation subsystem is improved genetic material. New plant and animal varieties with higher yields or resistance to pests or diseases become available. However, farmers need to know when the variety is released, how it performs under farm conditions, and where to obtain seed or breeding material. Extension is responsible for disseminating this information through appropriate mass media and contact methods. But a common constraint on the flow of technology is the availability of genetic material. In some countries, extension is also involved in the multiplication and distribution of seed. Establishing effective linkages with others involved in the process can also help ensure that genetic material is available for farmers.

Agrochemicals and Other Inputs. The performance of new varieties is often improved by, or even dependent on, the availability of agrochemical and other inputs at the farm level. Recommendations for fertilizer types and amounts suited to local soil conditions, for animal feed mixtures and practices, and for the control of plant and animal pests and diseases constitute an indispensable part of extension messages. Lack of access to this information and materials prevents yield maximization, so extension organizations need to ensure that farmers are informed on availability and use. If access is limited because of external factors, extension should plan its campaigns accordingly, with attention given to low-input recommendations.

Private companies and nonprofit organizations also provide advice to farmers on agrochemical and other input use. Extension organizations need to develop communication and coordination linkages with these actors because excessive use of agrochemicals can harm human health and the environment. Conservative recommendations and alternative approaches, such as integrated pest management, are in the interests of the public and can be promoted by extension through its farmer contacts and other linkages.

Credit

Access to credit is one way to improve farmer access to new production technology and increase productivity. Farmers' ability to purchase inputs such as improved seed and fertilizer is particularly important. If appropriate technology is available but not being used by farmers, then the way credit is handled by government may be part of the problem.

Understanding the credit context—government and bank policies, availability of credit, and the institutional relationships involved in its delivery—is important for extension. At a minimum, the existing credit situation should be examined so that factors affecting the adoption and use of technology can be identified. These include inequitable access to credit, insufficient amounts of it, and overlap of transfer activities due to credit institution involvement in extension work. The knowledge is useful for extension managers in targeting farmers and in coordinating extension objectives with credit institutions.

Farmer Organizations and Other NGOs

There is growing involvement of the private sector, both nonprofit and commercial organizations, in agricultural research and extension. While such organizations may have limited research objectives and restricted regional coverage targets, they are partners in technology generation and transfer. Private corporations such as seed and agrochemical companies play a key role in developing some types of technology, providing inputs, and advising farmers on their use. Ideally, extension should reach agreements with these actors so that duplication of effort is minimized and conflicting messages to farmers are avoided. At a minimum, their activities should be noted and an attempt made by extension to catalogue and use their successful methods and outputs.

Farmer organizations, particularly grass-roots organizations, are part of the utilization component. They offer an effective channel for extension contact with large numbers of farmers, as well as opportunities for participatory interaction with extension organizations. Feedback on farmer needs, production problems, and the results of adoption from such groups will be increasingly important considerations.

Concluding Observations

This overview of an agricultural technology system, and the place of extension within it, has stressed the importance of understanding the dynamic context in which the system operates. The agricultural technology systems model given here places extension work in a conceptual framework that underscores the significance of contextual factors for extension planning and strategies. Systematic consideration by managers of the particular outside forces influencing an extension organization allows it to plan its resource use, approaches, methods, and linkages in ways that are responsive to farmer needs and the roles of other organizations.

References

Elliott, H. (1994). Applying the agricultural technology systems approach: Lessons from ISNAR's experience. In R. M. A. Loyns & F. Datepa-Mupondwa (Eds.), *Management for researchers*. Department of Agricultural Economics and Farm Management, University of Manitoba. Winnipeg, Manitoba: Friesen Printers.

Eponou, T. (1993). *Partners in agricultural technology*. ISNAR Research Report No. 1. The Hague: ISNAR.

Gustafson, D. J. (1994). Developing sustainable institutions: Lessons from cross-case analysis of 24 agricultural extension programmes. *Public Administration and Development, 14*,4.

Idachaba, F. (1987). Agricultural research in Nigeria:Organization and policy. In V. W. Ruttan & C. Pray (Eds.), *Policy for agricultural research*. Boulder, CO: Westview Press.

Kaimowitz, D., Snyder, M., & Engel, P. (1990). A conceptual framework for studying the links between agricultural research and technology transfer in developing countries. In D. Kaimowitz (Ed.), *Making the link: Agricultural research and technology transfer in developing countries*. London: Westview Press.

Mathur, P. N., & Sinha, B. P. (1991). Extension and communication strategies for rainfed agriculture—Indian experience. In C. Prasad & P. Das (Eds.), *Extension strategies for rainfed agriculture*. New Delhi: Indian Society of Extension Education.

Moris, J. (1991). *Extension alternatives in tropical agriculture*. London: ODI.

Oram, P. A., & Bindlish, V. (1984). *Investment in agricultural research in developing countries: Progress, problems, and the determination of priorities*. Washington, DC: IFPRI.

Röling, N. (1991). Knowledge systems in their political context. In *Proceedings of the international workshop: Agricultural knowledge systems and the role of extension*. Bad Boll, May 21-24, 1991.

Swanson, B. E., & Peterson, W. E. (1991). Strengthening research-extension linkages to address the needs of resource-poor farmers in rainfed agriculture. In C. Prasad & P. Das (Eds.), *Extension strategies for rainfed agriculture*. New Delhi: Indian Society of Extension Education.

Swanson, B. E., Sands, C. M., & Peterson, W. E. (1990). Analyzing agricultural technology systems: Some methodological tools. In R. Echeverria (Ed.), *Methods for diagnosing research system constraints and assessing the impact of agricultural research: Vol. I. Diagnosing agricultural research system constraints*. The Hague: ISNAR.

World Bank (1994). *Agricultural extension: Lessons from completed projects*. Washington, DC: World Bank.

Chapter 4
The Economic Contributions of Agricultural Extension to Agricultural and Rural Development

Robert Evenson

Agricultural extension programmes are quite diverse from an international perspective. Most are managed as public sector agencies, usually located in the ministry of agriculture, but some are located in other ministries such as education or rural development. Many are managed by nongovernmental organizations (NGOs). Many private firms and private organizations (for example, coffee-growers' associations) conduct extension programmes. Even within the most typical organizational structure, where extension is part of the government's ministry of agriculture, there is great variation in the degree of decentralization of management of extension services. In some countries, extension is decentralized, as in India, where it is a state subject. In most developing countries, however, governmental services are highly centralized, with varying forms of regional and subregional units designed to serve local areas.

Further, there is great variation in the skill level and agricultural competence of field staff. In some systems, field staff have little formal technical training in the agricultural sciences. In some cases, this is dictated by a village worker philosophy, in others by local language demands. But, in most cases, it simply is the result of the decisions to expand agricultural extension programmes rapidly during the 1950s and 1960s, when few highly trained agriculturalists were available (see Bindlish & Evenson, 1993 and Bindlish, Gbetibouo, & Evenson, 1993 for African studies; and Swanson & Claar, 1984 for a general history).

Finally, this diversity of skills, management systems, and objectives has changed over time in many countries. Perhaps the major changes in the management and design of agricultural extension systems over the past four decades is associated with the training and visit (T&V) system introduced in the 1970s by Benor, Harrison, and Baxter (1984) and implemented in many countries with World Bank lending support.

Given this diversity, broad generalizations about the economic contribution of agricultural extension to agricultural development are not feasible. Many situation-specific factors impinge on the effectiveness of extension programmes. The fact that substantial reform and redesign of many extension programmes has taken place indicates that some of them were perceived by their supporters to have been less than fully effective. However, we now have a substantial body of economic studies of extension services in a number of countries; 75 studies of economic impacts of extension systems have been published to date. My task in this chapter is to review the findings of 57 of these studies and to draw out some of the lessons they have to offer.

I begin this review with a brief summary of investment patterns for both agricultural research and extension. This is designed to provide historical perspective and to call attention to some of the economic and institutional diversity in which extension systems must function. In the second section, I review the conceptual foundation for measuring the economic impact. Statistical procedures are reviewed in the third section, and in the fourth I summarize the findings of the studies under review and attempt to relate these to some of the differences in economic and institutional settings. In the final part, I summarize policy lessons.

Investment Indicators: Agricultural Research and Extension

Several relevant indicators of investment in "technological infrastructure" are reported in Table 1. Agricultural extension, agricultural research, and human capital investments are included. The country groups are formed to reflect diversity in levels of "technology infrastructure." These country group categories will be used throughout this paper as a means of recognizing that extension programmes are conducted in different settings, and that their design, management, and effectiveness are conditioned by these settings. It is thus important to describe these country groups in some detail (for more detail, see Evenson & Westphal, 1994).

There are three Type 1 categories and three Type 2 categories for developing countries. The Type 1 categories cover countries which have not yet mastered production of a full range of goods and services using modern technology.

Type 1a includes approximately 20 countries (including Yemen, Laos, Surinam, Zaire) that lack basic infrastructure of all types. Governments have limited influence, and little manufacturing capacity exists in these countries.

Type 1b includes approximately 30 countries (including Nepal, Papua New Guinea, Haiti, Ethiopia, Burkina Faso) with rudimentary technological infrastructure. Some direct foreign investment has taken place, and this provides some access to foreign technology. Agricultural policy in these countries is often dominated by parastatal organizations.

Robert Evenson is a Professor at Yale University, and serves as Director of The International and Development Economics Program.

Table 1. International Technology Investment Indicators: Investment Intensities (R&D/GDP, Extension/GDP, Expenditures/staff, 1990)

	Technological Infrastructure Type							
	Type 1 Developing Countries			Type 2 Developing Countries			Industrialized Countries	
	1a	1b	1c	2a	2b	2c		
Indicators	Traditional Technology	First Emergence	Islands of Modernization	Mastery of Conventional Technology	Transition to NIC-Hood	NIC-Hood	Recently	OECD
R&D/GDP								
Agriculture								
Public-NARs	.002	.004	.005	.006	.007	.010	.010	.015
Public-IARCs	.0005	.0005	.0005	.0005	.0005	0	0	0
Private	0	0	.0002	.002	.003	.005	.005	.015
Industry								
Public	0	.0001	.0001	.0002	.0005	.001	.003	.003
Private	0	0	.0002	.005	.007	.010	.015	.023
Extension/GDP								
Agriculture								
Public	.005	005	.010	.010	.010	.010	.010	.010
Private	0	0	0	0	.001	.002	.005	.010
Expenditures/staff (1980 000 dollars)								
Research	47	47	47	40	45	50	70	95
Extension	2	2	4	4	10	15	35	35

Some higher education is provided. Most graduates are employed by government agencies, including agricultural extension services.

Type 1c includes 25 countries that have achieved partial modernization (including Sri Lanka, Tunisia, Kenya, Ivory Coast, Bangladesh). Modern agricultural practices have been introduced in most of these countries, and most have well-developed agricultural research and extension services. Most workers are literate. Universities have begun to train scientists and engineers. Graduates have begun to work outside the government. No significant private sector R&D capacity has yet been built in these countries.

The 20 Type 2 developing countries, on the other hand, have made sufficient investments in technology infrastructure to realize "followers" or "catch-up" growth. In other words, these countries are catching up to the developed countries.

The Type 2a countries (including India, Colombia, Mexico, Argentina, Turkey) have achieved modern engineering capabilities. Significant private sector R&D is undertaken. Universities are advanced. Most have not achieved the industrial, trade, technology, and macroeconomic policy regimes to realize rapid growth.

The Type 2b countries (including Indonesia, Thailand, Malaysia, Chile, China) have achieved transition to newly industrialized country (NIC) status and the rapid growth associated with it. These countries have advanced technological capabilities and effective policy environments. Some of these countries (for example, Indonesia) have managed to move through the Type 2a stage in a relatively short period of time.[1]

The Type 2c countries (including Hong Kong, South Korea, Singapore, Taiwan) are well-established NICs.

Table 1 reports two sets of indicators of investment intensities (investment/GDP) and one set of "price" indicators for these types of countries.

Agricultural research indicators show that public sector research capacity exists in most countries; perhaps 10 or so of the Type la countries would not have such capacity. In contrast, private sector R&D relevant to agriculture or to industry is not important in Type 1 countries. Such R&D becomes increasingly important as countries become more advanced (that is, for Type 2a to Type 2c).

Research capacity for the industrial sector clearly lags behind research capacity for agriculture in all except the most advanced developing countries. Only the Type 2 countries have significant industrial R&D capacity.

Agricultural extension programmes, by contrast, serve almost all countries. For the Type la and lb countries, these programmes represent the only modernizing investments of significance. Extension spending intensity exceeds research spending intensities until NIC-hood is reached.[2]

The expenditures-staff data show that the real costs of supporting a research scientist are roughly constant across developing country groups and are roughly half the level prevailing in industrialized countries. For extension, however, the real cost of supporting extension field staff is very low in the poorer countries, in fact too low for efficiency in many cases. This ratio explains why many poor countries have very large extension staffs and large staff-farmer ratios. Extension staff are perceived to be low-cost producers of economic growth relative to researchers, and to some degree they are.[3]

The Conceptual Foundation for Extension Impact

Two conceptual themes are relevant to extension impact. The first is the awareness-knowledge-adoption-productivity (AKAP) sequence. The second is the "growth gap" interrelationship between extension, schooling, and research

The AKAP Sequence

It is convenient to visualize extension as achieving its ultimate economic impact by providing information and educational or training services to induce the following sequence:

A: Farmer awareness
K: Farmer knowledge, through testing and experimenting
A: Farmer adoption of technology or practices
P: Changes in farmers' productivity

Changes in farmer behaviour will be reflected in quantities of goods produced, the quantities of inputs used, and in their prices. These, in turn, can be measured as "economic surplus," which is the added value of goods produced from a given set of inputs made possible by the extension activities.

Studies of extension impacts have measured farmer awareness (and sources of awareness), knowledge (and testing of practices), adoption, and productivity. Not all studies have examined all parts of the sequence. Most have shown a statistical relationship between the quantity of extension services made available to farmers and increases in awareness, knowledge, adoption, and productivity.

While the AKAP sequence has a natural ordering, it is clear that real resources in the form of skills and activities by both extension staff and farmers are required to move along the sequence. Awareness is not knowledge. Knowledge requires awareness, experience, observation, and the critical ability to evaluate data and evidence. Knowledge leads to adoption, but adoption is not productivity. Productivity depends not only on the adoption of technically efficient practices, but of allocatively efficient practices as well. Productivity also depends on the infrastructure of the community and on market institutions.

Extension services affect each part of the sequence. They can be seen as both substitutes for and complements to the acquired skills of their clientele farmers. Empirical evidence indicates that they are, on balance, net substitutes for farmers' skills as reflected in farmers' schooling. For example, extension services are typically not the only sources of information (awareness). Skilled farmers can seek information on their own. Farmers with few skills may not do so. Extension information then may have a higher impact on farmers with less schooling. It appears, however, that the awareness-knowledge part of the sequence is where extension services are strong substitutes for farmer schooling. Through organized frequent contact, they "teach" farmers, and this is more than simply informing farmers.4

The teaching versus informing distinction is also relevant to the "newness" of the information (that is, of the recommended practice or other technology) and of the nature of the practice or new technology. When technology is new (as for example with a recently released variety of rice) and is also "simple" to evaluate and adopt (where it is a matter of using new seed without altering other practices), information-awareness is relatively easily converted to knowledge and adoption. Farmers with few skills usually adopt such technology with a time lag. When the technological practice is more complex and requires substantial changes in activities and sometimes capital investment, teaching is required. Repeated messages clearly stated, followed up by field staff and often community organization, are required to proceed through the AKAP sequence in this case.

Productivity Gaps and Extension

The AKAP sequencing is, as noted above, related to the flow of new technical information and to the existing state of unadopted technology. We can see this interrelationship more clearly in the context of productivity "gaps." Figure 1 portrays what is meant by gaps and relates these to the technological infrastructure types used in Table 1.[5]

Figure 1 depicts crop yields, adjusted for fertilizer and other inputs, for five technology infrastructure types. Four yield levels are depicted for each type:

A: Actual yields
BP: Best practice yields
BPBI: Best practice, best infrastructure yields
BPBIRP: Best practice, best infrastructure, research potential yields

These yield levels in turn define three "gaps":

G(P): A practices gap between the best practice (BP) yield and actual (A) farmers' yields
G(I): An infrastructure-institutions gap between the best institutions, best practice (BPBI) yield and best practice (BP) yield
G(R): A research gap between the research potential yield (BPBIRP) and the best practice, best institutions (BPBI) yield

These gaps provide a way to classify the contribution of extension activities and to show how research and extension are linked. A stylized sequence across technology types is depicted. This could also be visualized as a time sequence.

Extension programmes are designed to reduce both the practice gap, G(P), and the institutions gap, G(I). Extension programmes are not the only activities that reduce these gaps. Providing market information to farmers and developing organized farm groups reduce G(I). Information and teaching reduce G(P). Research programmes are generally required to reduce G(R), although extension programmes can facilitate the reduction of G(R) via facilitating the importing and local modification of improved technology developed elsewhere. Research programmes in most developing countries also modify and adopt imported technologies and germplasm.

Two of the gaps are closely linked. When G(R) is closed (that is, when the BPBI yields go up), G(P) is opened.[6] (This may happen with G(I) also, but to a lesser extent.) Further, it should be noted that the size of the gap is an index of the potential impact of research or extension. As extension succeeds in closing G(P), diminishing returns set in. Successful research opens up new potential by increasing G(P). The relative mix of teaching versus informing is also related to these gaps. When the BPBI yield level has been constant for some time, the G(P) gap is closed mostly by teaching. When BPBI is increased, as by "green revolution" rice and wheat varieties, information and testing advice play a larger role.

The pattern of gaps and yield levels across country groups is intended as a stylized pattern. It is roughly based on experience. For Type la countries, both G(I) and G(P) are depicted as large.[7] These are traditional economies with not much new technology being produced that is relevant to them.

As economies move to Type lb, improvements in institutions allow BP (and A) to rise even without new technology; BPBI remains unchanged. Extension can contribute to reducing both G(I) and G(P), and these contributions are qualitatively different from those required in more advanced country groups. There is little new technology (few new practices) in these countries, farmers have little schooling, and infrastructure is poor. The teaching and organizing activities of extension dominate here.

As economies move to the Type 1c category, some new technology has been introduced (BPBI has risen), and the institutions gap has been further reduced. The practices gap, G(P), has been both opened (because BPBI increased) and closed because of continued teaching and because new practices now can be extended.

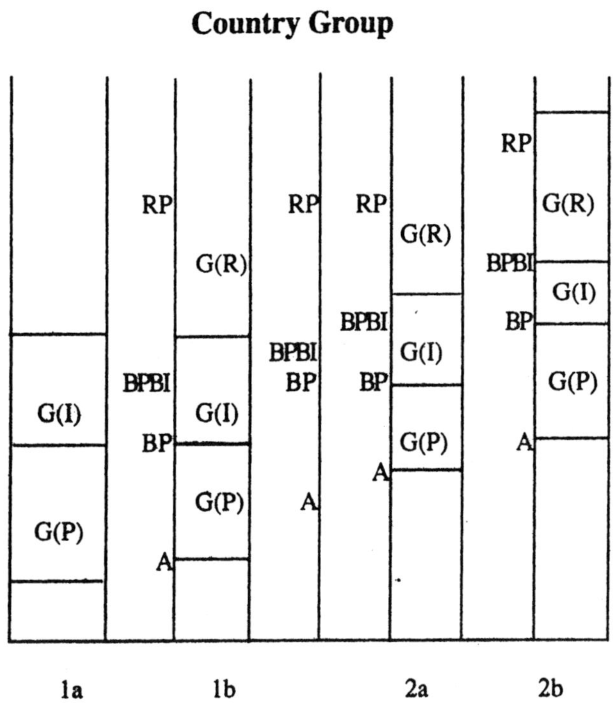

Figure 1. Sequence of crop yields for five technology infrastructure types. Yield levels for each type areas follows: A = actual; BP = best practice; BPBI = best practice, best infrastructure; BPBIRP = best practice, best infrastructure, research potential. These yield levels define the following gaps: G(P) =practice gap between the best practice (BP) yield and actual (A) farmers' yields; G(I) infrastructure-institutions gap between the best institutions, best practice (BPBI) yield and best practice (BP) yield; G(R) = research gap between the research potential yield (BPBIRP) and the best practice, best institutions (BPBI) yield.

As economies move further up the technology infrastructure scale, the institutions gap is further reduced and extension's role in reducing G(I) becomes small. However, G(R) is closed by national and international research programmes—and the BPBI yields rise, providing extension with more new practices to extend. The private sector grows in importance and markets are improved.

A Note on Statistical Methods and Issues for Economic Evaluation

The studies under review in this chapter sought to measure the impact of public agricultural extension programmes' activities in the following four areas: (1) farmer knowledge of technology and farm practices; (2) adoption or use of technology and practices; (3) farmer productivity and efficiency; and (4) farm output supply and factor demand.

Estimation of extension impact is subject to a number of problems which are also faced in the evaluation of other public sector investments. The approach commonly used is a statistical analysis relying on data measuring extension activities at the farm level. Alternatively, statistical analysis can be undertaken where observations refer to aggregate extension services supplied to a given region in a specific time period.

Studies assessing extension impact at the individual farm level that use a farm-level measure of extension may be affected by two basic estimation problems. The first is the problem of statistical "endogeneity" in extension-farmer interactions.8 Early studies seeking to measure the impact of agricultural extension by identifying the extension variable as some form of extension contact often treated the extension contact as being unrelated to the farmers' actions and characteristics. However, it is likely that one of the characteristics of more productive farmers is the desire to acquire information about changing farm conditions or new technologies. Such farmers may be inclined to attend more demonstration days, read more literature, and seek extension contact. Analogously, extension agents themselves may also seek contacts with better farmers who would be good performers even in the absence of extension contacts.

In such cases, the extension contact variable is endogenous, and the estimates of extension impact on farmers' performance are likely to be biased upward, because some of the better performance credited to extension would in fact be the result of the superior attributes of the group which interacts with extension. The problem of endogeneity can, in principle, be handled econometrically by using two-stage procedures or simultaneous equations approaches, but this has been done in only a few of the studies undertaken so far.

The second source of potential bias is the problem of indirect or secondary information flows where knowledge which originates from extension contacts is passed on to other farmers who do not directly interact with extension personnel. The extension of interfarm communications is substantial, as shown in Birkhaeuser, Evenson, and Feder (1991), where data on farmers' sources of information were reviewed. This review showed that most farmers in areas receiving extension services report that other farmers are the main source of information. Except for the contact farmers in T&V extension areas who are singled out for extension contact by the nature of the programme, direct contact with extension personnel was typically not the major source of information to farmers. Information may, of course, be diffused (to other farmers) from farmers who were informed by extension agents. In such cases, there may be little difference in performance between farmers interacting directly with extension and other farmers, and an estimate of extension impact based on individual extension contacts would erroneously indicate zero extension effect. Generally the presence of interfarmer communications tends to cause an understatement of extension effects when the approach of defining extension impact by the number of direct contacts is used.

In the studies reviewed below, data from farm surveys and secondary sources on farmer awareness, adaptation, and productivity were related to the provision of extension services in different regions and time periods.[9] Productivity is typically measured as production per unit of all inputs (including labour, land, and fertilizer), although in some studies an aggregate production function approach was used.[10]

Estimated coefficients measure the *marginal product* of extension—the added production due to a one-unit addition to extension services supplied. The extension variables also typically have a time dimension. The adoption of improved practices will typically occur at some rate in the absence of extension services, depending on schooling and infrastructure. Extension both accelerates practice adoption and affects the long-run level of practice adoption. For Type 1a and 1b economies, extension may have a strong level effect if it is effective. For more advanced economies, the extension impact is primarily a speed-up effect. Most studies find speed-up periods of three to five years. Recent studies for Africa (Kenya and Burkina Faso) find significant level effects, implying that extension impacts in these economies are long-term impacts.

Knowledge of the timing weights and the marginal product allows the calculation of a marginal product "stream" over future years associated with an investment in time (t). This stream can be discounted to find the marginal internal rate of return (r) to the investment.[11] This rate of return estimate is the measure used to compare extension studies in the next section. It can be interacted as the interest rate that investors (typically taxpayers) receive from investing in this activity.

Estimates of Economic Impacts: A Summary

Table 2 summarizes estimates of economic impact from 57 economic studies undertaken in seven African countries (Burkina Faso, Cote d'Ivoire, Botswana, Nigeria, Ethiopia, Kenya, Malawi), seven Asian countries (Bangladesh, Indonesia, Malaysia, Nepal, Philippines, South Korea, Thailand), three Latin American countries (Brazil, Paraguay, Peru), and the United States and Japan. The studies are grouped into several categories: the distribution of the estimates by level of statistical significance and, where reported, by level of rate of return to extension. (Note that some studies reported more than one estimated impact. Many studies did not calculate returns, and returns are reported only when the estimated coefficient had a high level of statistical significance.)

Of the 174 estimated impact coefficients, 59 (one-third) are reported to be not significant. Very few of these were actually negative. This set of studies, however, cannot be said to be fully representative of the regions or types of extension programmes. Quite possibly a number of studies that found little or no extension impact were not reported.

Awareness (Knowledge) Studies

Six studies of extension's impact on awareness and knowledge were undertaken. Three of these (India, Kenya, Burkina Faso) (see Table 3) examined the impact of T&V management on awareness of recommended practices. These studies find strong evidence that extension does create awareness and knowledge and that T&V management makes extension more effective in doing so.

Adoption Studies

Nine studies of adoption of farm practices were undertaken. All sought to determine the impact of extension in accelerating adoption. This evidence is somewha. less conclusive than the awareness evidence.

Most studies found that farm size and farmers' schooling also determined adoption rates. Most studies found evidence for some extension impact on adoption. The T&V studies found that T&V enhanced the extension effect. Two of the studies (Kenya and Burkina Faso) linked practice adoption to productivity. Both found that extension accelerated adoption and led to productivity change.

Productivity Studies

Forty-two studies reported estimates of extension impacts on farm productivity: 25 used farm survey data, and 17 used aggregated data, such as district-level data. Sixteen of the 25 farm survey studies used a farm-specific extension variable, usually a contact with extension. As noted earlier, these variables are highly vulnerable to the endogeneity problem. It is interesting to note that this category of studies actually had the highest proportion of insignificant estimates.

In contrast, the nine studies that relied on an extension supply variable such as the number of extension staff made available in a region or to a group of farms have a high proportion of highly significant estimates. The T&V studies were in this category, and they generally found a T&V management enhancement effect. Two of these studies used two-stage procedures to predict adoption or membership in T&V groups and found that the extension impact was in general realized via its effect on practice adoption and on T&V group participation.

The 17 studies based on aggregate data in most cases included variables measuring research, schooling, and infrastructure in addition to extension variables. Almost all found evidence for an extension impact. The studies that used interaction variables between extension and farmers' schooling generally found a net substitution relationship. Higher levels of farmer schooling reduced the impact of extension, and vice versa. The studies that examined the research-extension interaction generally did not find a significant interaction except in the U.S. studies.

Estimates by Period and by Country Group

There are no differences in the distribution of significant estimates or rates of return by period.

In an earlier section of this paper, I noted that the technological and institutional setting in which extension operates affects its design and impact. The estimates classified by country group show two things. First, they show considerable variation, with a substantial range of significance being reported. Second, they show a difference between Type 1 and Type 2 developing economies. For the Type 1 economies, 45 of 105 (43 percent) of reported estimates have a high level of statistical significance. For Type 2 countries, 38 of 56 (68 percent) of the estimates have a high level of statistical significance. Thus it appears, on the one hand, that it is possible to design effective extension programmes for the Type 1a and 1b economies where new technology is being developed at a slow rate. The T&V studies confirm this. But it also appears that the more dynamic technological environments of the Type 2 economies provide a setting for broader effectiveness of extension programmes.

Lessons

Perhaps the overriding lesson is associated with the range of results reported. Clearly, many extension programmes have been highly effective in aiding farmers to achieve higher productivity. It also appears that some programmes have not done so. Probably many programmes are underperforming and many lack the design and management discipline to be effective. Many extension programmes have large investments in monitoring and evaluation units. None of these, to my knowledge, has yet produced estimates of economic impact. Virtually all of the studies under review were conducted by economists in academic programmes. It clearly is important to extension policy makers that the competence of their monitoring and evaluation programmes be upgraded to enable more effective evaluation. At the same time, it continues to be important that independent evaluations of the type reviewed here be continued.

A second lesson appears to be available from the evidence by country type. Extension systems are most effective when researchers are effective. But research programmes are not effective everywhere. The evidence of effective extension programmes in the Type 1a and 1b countries, which are concentrated in sub-Saharan Africa, shows that properly designed and managed extension programmes can be effective in these environments. The two studies of T&V-managed programmes in the region stress that they have been effective.

A final lesson is that, in the long run, extension has its highest payoff in Type 2 economies where farmers have access to schooling, to new technology, and to extension. As conditions change, extension must change. The need for continuing evaluation is high at all levels.

Notes

1. A number of countries have moved from one country group to another in recent years. All of today's Type 2b countries were in the 2a or 1c categories twenty years ago
2. Agricultural extension programme development was also undertaken earlier than research investment in most countries (Judd, Boyce, & Evenson, 1986).

Table 2. Summary: 57 Economic Studies of Extension for Selected Countries[a]

Type or Category	Number of Studies	Distribution by Level of Statistical Significance			Distribution by Returns Estimates		
		Not Significant	Medium Significant	High Significant	Low	Medium	High
Awareness	6	7	2	27	nc	nc	nc
Adoption	9	16	8	17	nc	nc	nc
Productivity							
Farm Observation:							
Farm Contact	16	21	4	10	2	1	7
Extension Supply	9	11	3	21	1	1	12
Aggregate Observation	17	4	5	17	2	0	6
All Productivity	42	36	13	48	4	2	25
By Period							
Before 1980	17	12	3	13	2	2	7
After 1980	40	47	20	79	2	2	18
By Country Group							
1A-1B	9	16	6	24		1	9
1C	14	26	11	21	1	1	4
2A	12	5	3	28	2	2	3
2B-2C	13	8	2	10	1		4
Industrialized	9	4	1	9			5

Note: For statistical significance, the estimated "t" ratio is less then 1.5 for not significant, 1.5-2.0 for medium signifieanee, greater then 2 for high significance. For rates of returns, low is 5-25 per cent, medium 26-50 per cent, high 50 per cent or greater.

[a]. African countries: Burkina Faso, Cote D'Ivoire, Botswana, Nigeria, Ethiopia, Kenya, alawi; Asian countries: Bangladesh, Indonesia, Malaysia, Nepal, Philippines, South Korea, Thailand; Latin American countries: Brazil, Paraguay, Peru; the United States; Japan.

Table 3. Source of Awareness of Extension Recommended Practices

Study	Study of (Per cent)		
	Extension Stuff	Research Centres	Private Sector
Taiwan (Lionberger & Chang 1970)			
Shangfung	36	na	na
Liupao	24	na	na
Paraguay (Evenson 1988)			
Western Regions	21	1	10
India (Federal et al. 1986)			
T&V farmers	71	0	0
T&V group farmers	27	6	4
Non-T&V farmers	24	3	5
Burkina Faso (Bindlish et al. 1993)			
T&V group members	74	na	5
Non-T&V farmers	36	na	12

3. The perception that extension staff services were bargain-priced probably led to overstaffing of many extension programmes.
4. The recognition of the teaching component of extension has been growing in recent years, especially in the T&V-managed systems.
5. See Evenson (1986) for an earlier version of this gap analysis and a discussion of the economics of extension.
6. This gap-opening phenomenon is the source of new potential gains from extension and explains why extension programmes are demanded over long periods of time. If the BPBI yields remain stagnant in advanced countries, extension has a very limited role to play.
7. Schultz (1964), in his influential book *Transforming Traditional Agriculture*, stated that, in traditional agricultural economies where BPBI yields had been constant, farmers were "poor but efficient." In terms of Figure 1, he was saying that G(P) was actually not very large and that the potential for yield improvement from extension was also low. He effectively said that a rise in BPBI yields was required to create potential for extension to be effective.
8. The term *endogeneity* is a statistical term. Endogenous variables are chosen or controlled by he units being studied (for example, farmers). Exogenous variables are not chosen by the units. Exogenous variables can "cause" endogenous variables. Endogenous variables cannot be said to cause other endogenous variables.
9. The basic statistical model used in these studies is
   Z = a + bEXT + cSCH + d(EXT)(SCH) + eRES +f(EXT)(RES)
   In this expression, Z may alternatively be a measure of awareness, knowledge, testing activity, adoption, or farm productivity. Extension (EXT), schooling (SCH), research (RES), and other variables are the independent (exogenous) variables that determine the endogenous variable, Z. Interaction variables are often included to test for substitution-complementary relationships.
10. In the production function studies, inputs are included as independent variables:
   Output = a_1 + a_2LAND + a_3FERT + b'EXT + cSCH, etc.
   When a production function formulation is used, the interpretation of the coefficient b' measures the change in output, holding inputs (land, fertilizer, etc.) constant.
11. This internal rate of return is the rate r for which the following equation holds:

$$I_t = \sum W_{t+k} MP^*/(1+r)k$$

References

Benor, D., Harrison, J. Q., & Baxter, M. (1984). *Agricultural extension: The training and visit system.* Washington, DC: World Bank.

Bindlish, V., & Evenson, R. E. (1993). *Evaluation of the performance of T&V extension in Kenya.* World Bank Agricultural and Rural Development Series No. 7. Washington, DC: World Bank.

Bindlish, V., Gbetibouo, M., & Evenson, R. (1993). *Evaluation of T&V extension in Burkina Faso.* World Bank Technical Paper No. 226. Washington DC: African Technical Department.

Birkhaeuser, D., Evenson, R. E., & Feder, G. (1991). The economic impact of agricultural extension: A review. *Economic Development and Cultural Change, 39,* 607-650.

Evenson, R. E. (1986). The economics of extension. In G. Jones (Ed.), *Investing in rural extension: Strategies and goals* (p. 65-91). Amsterdam: Elsevier Applied Sciences Publishers.

Evenson, R. E. (1988). Estimated economic consequences of PIDAP I and PIDAP II programs for crop production. Unpublished manuscript, New Haven, CT, Yale University, Economic Growth Center.

Evenson, R. E., & Westphal, L. E. (1994). *Technological change and technology strategy.* UNU/INTECH Working Paper No. 12.

Feder, G., & Slade, R. (1986). The impact of agricultural extension: The training and visit system in India. *The World Bank Research Observer, 1,* 139-161.

Feder, G., Slade, R., & Sundaram, A. (1986). The training and visit extension system: An analysis of operations and effects. *Agricultural Administration, 21,* 48.

Judd, A., Boyce, J., & Evenson, R. E. (1986). Investing in agricultural supply: The determinants of agricultural research and extension investment. *Economic Development and Cultural Change, 35,* 77-113.

Lionberger, H., & Chang, H. C. (1970). *Farm information for modernizing agriculture: The Taiwan system* (p. 282-283). New York: Praeger.

Schultz, T. W. (1964). *Transforming traditional agriculture.* New Haven, CT: Yale University Press.

Swanson, B. E., & Claar, J. B. (1984). The history and development of agricultural extension. In *Agricultural extension: A reference manual.* Washington, DC: FAO.

Part II

Improving Extension Programmes and Processes

Chapter 5
Assessing Target Group Needs

N. L. McCaslin and Jovan P. Tibezinda

The assessment of target group needs–often called needs assessment–represents one of the first steps in planning and developing extension programmes. Programme planning and development is an ongoing and interrelated process that, in addition to assessing client needs, includes selecting appropriate content and methods in programme delivery, managing programme delivery, and evaluating programme processes and outcomes.

In the United States, needs assessment first emerged with the passage of the Administrative Procedures Act in 1946 (Summers, 1987). In the 1960s and 1970s, more than 30 of the 54 largest pieces of health and human services legislation mandated federal, state, or local needs assessment (Zangwill, 1977). Since then, there has been an increasing emphasis on involving citizens in the planning, conducting, and evaluating of programmes such as extension. On the international scene, an increasing emphasis has been placed on citizen involvement through bottom-up and grass-roots programme planning and development. This is in stark contrast to earlier times when needs were determined by outside consultants and programmes were then developed in response to these needs.

Needs assessment, broadly defined, is a systematic process for establishing priorities and making decisions regarding programme planning, development, and operations. In this chapter, needs assessment is defined as determining if gaps exist between "what is" and "what should be" in terms of the outcomes of extension programmes and then determining the priority of these needs (Kaufman, 1982). Emphasis will be placed on making decisions and setting priorities based on information gathered from the people likely to be affected by these extension programmes.

When needs are being determined, it is essential that distinctions are made between needs, wants, and interests. Needs refer to something considered necessary or required to accomplish a purpose. Wants, on the other hand, are considered desirable or useful, but not essential. Interests indicate an individual's concern or curiosity about something. It is not unusual for individuals to confuse needs, wants, and interests. Therefore, extension personnel undertaking efforts to assess target population needs should ensure that they understand the meaning of "needs."

Needs Assessment Techniques

This section describes different data-collecting techniques available for carrying out needs assessment. The techniques are discussed under four categories: individual, group, secondary source, and rapid rural appraisal.

Individual Techniques

Individual techniques involve collecting data from people one at a time. The people from whom the needs assessment data are collected do not interact with one another in the course of providing data. Individual techniques include face-to-face interviews, key informant interviews, questionnaires, informal personal observations, and formal personal observations.

Face-to-Face Interviews. This technique is appropriate when dealing with less literate audiences or complex issues about which there is little available information. Both structured and unstructured questions are appropriate for face-to-face interviews, depending on the issues involved and the time available for the interviews. Unstructured questions are useful when dealing with complex or sensitive issues which require probing in order to get accurate data. For example, small-scale farmers may not have a direct answer about how they budget their resources. However, if probed about what they do on a typical day or week, they may provide insights into their economic activities and therefore their needs.

Key Informant Interviews. Key informants are people who are considered experts in a given area because of their professional knowledge or their position of influence in the community or organization. Examples include teachers, religious leaders, grass-roots workers, and traditional and political leaders. There is evidence to show that interviewing several of these categories of informants yields fairly accurate information about the problems and needs of the community at large. Key informants are particularly useful if the needs assessment has to be done fast, using a limited budget.

Questionnaires. This needs assessment technique tends to be more structured than interview schedules and can be administered by phone, mail, or in group settings.

N. L. McCaslin is a Professor in the Department of Agricultural Education, The Ohio State University, Columbus, Ohio. **Jovan P. Tibezinda** is a Lecturer in the Department of Agricultural Extension/Education at Makerere University, Kampala, Uganda.

Questionnaires are commonly administered in developed countries by conducting telephone interviews; however, in developing countries they are rarely administered in this manner because of the limited availability of telephones.

When dealing with literate communities that have access to good mail services (public or private), needs assessment surveys can be conducted by mail. Dillman (1978) provides excellent tips on conducting mail surveys which apply to developed and developing countries alike.

Instead of relying on conventional mail services, questionnaires also can be hand-delivered to respondents and collected after they have been completed. Alternatively, one may take advantage of occasions such as annual club or association meetings when potential respondents might come together. In this case, the questionnaires are presented to group members, who are asked to complete and return them before they leave. For best results, the questionnaire should cover pertinent issues and be short enough to be completed in the time the respondents have. When used appropriately, this method may save both time and money necessary for collecting the information. However, one must be cognizant of the high possibility for biases in the information collected. This is because there is always a chance that the assembled group completing the questionnaire is not representative of the audience in mind.

Stevens (1980) indicated that extension workers are often assigned to their positions without proper equipment and adequate preparation. Maalouf and Contado (1983) indicated that there was a need to assess the training needs of extension personnel and establish priorities. When using a questionnaire to collect needs assessment information, let's assume that an agricultural officer has decided to identify the professional skills of extension agents most in need of improvement in their region. One of the first activities of the agricultural officer was to identify the professional skills that extension agents should possess. In reviewing the literature, the officer found that Hedges and Rawls (1988) had identified needs for professional skills of Ugandan extension personnel, such as programme planning and evaluation, utilization of formal and informal teaching techniques, and administration and supervision techniques. This list served as the basis for developing a questionnaire for determining the agents' needs.

Once the initial list of skills had been prepared, it was necessary to assess the agents' perceived current level of competence and the importance they placed on each of the skills. In order to assess the current level of competence, a Likert-type scale was developed that included the following five categories of competence: not at all competent, little competence, moderately competent, fairly competent, and very competent. Next, another Likert-type scale was developed to assess the agents' perceptions of the importance of each of the skills, using the following five categories: unimportant, little importance, moderately important, important, and very important. The two Likert-type scales and the list of skills were combined by rating the current competence on the left of the skill and the importance on the right of the skill (Figure 1).

Once the questionnaires had been completed, the agricultural officer was ready to analyse the data and establish priorities for developing the professional skills of his or her extension agents. Hershkowitz (1973) identified a criticality function that is helpful in establishing such priorities. A 2 x 2 matrix is created in order to establish the priorities (Figure 2). First, an overall mean score is calculated for ability and importance scores for all items on the questionnaire. Then, the mean ability score is plotted on the Y axis and the mean importance score is plotted on the X axis. Perpendicular lines are then drawn from each of these points, resulting in a 2 x 2 matrix. The matrix has four quadrants–high ability-high importance (HH), high ability-low importance (HL), low ability-high importance (LH), and low ability-low importance (LL).

Next, the mean ability and importance scores for each of the professional skills are computed and plotted in the matrix. Those falling in the low ability-high importance quadrant are those with the highest need for development. In this example, skill 1 (producing educational-teaching materials) and skill 5 (determining information needed for evaluations) were those most in need of development.

Informal Personal Observations. Valuable needs assessment data often are gathered through informal observation. Fieldworkers see or experience a lot as they travel and work with farmers in the field. If noted or remembered, this information can be used in needs assessment. People in the habit of keeping diaries or writing notes to themselves are more likely to provide more accurate observational data for assessing needs.

Formal Personal Observations. This needs assessment technique is based on using rating forms, checklists, or observation schedules for collecting information. Formal observations differ from informal personal observations in that the items to be observed are predetermined. This technique also can be used to collect both quantitative and qualitative data.

Group Techniques

Group techniques allow participants to interact with one another during needs assessment activities. Information can be collected in writing, as in the Delphi technique, or orally in a group setting such as a focus group. In both cases, successful needs assessment depends on competent leadership and on having participants who have both the knowledge and willingness to participate actively in the interactive group process (Caffarella, 1982). This section will discuss the Delphi, focus group interviews, the nominal group, and informal group methods.

Professional Skills Survey of Extension Personnel

INSTRUCTIONS: This questionnaire lists a number of professional skills that are used by Extension personnel in conducting educational programs. For each of the following skills we would like you to indicate: (1) your competence to perform the skill and (2) the importance of the skill. The following scales are used:

Competence to Perform the Skill	*Importance of the Skill*
1 = Not at all Competent	1 = Unimportant
2 = Little Competence	2 = Little Importance
3 = Moderately Competent	3 = Moderately Important
4 = Fairly Competent	4 = Important
5 = Very Competent	5 = Very Important

Please circle the number that best represents your response for each item.

EXAMPLE:

Competence	*Professional Skills*	*Importance*
1 2 3 4 5	1. Producing educational/teaching materials	1 2 3 4 5
1 2 3 4 5	2. Using audio-visual aids	1 2 3 4 5
1 2 3 4 5	3. Conducting a method demonstration	1 2 3 4 5
1 2 3 4 5	4. Assessing community needs	1 2 3 4 5
1 2 3 4 5	5. Determining information needed for evaluations	1 2 3 4 5
1 2 3 4 5	6. Developing program budgets	1 2 3 4 5

Figure 1. Sample of the partial questionnaire.

Delphi. In this technique, people with exceptional knowledge about a given subject area are involved in repeated questioning and feedback, using written questionnaires, until a consensus is reached on the subject (Miller & Hustedde, 1987). The process begins with specifying the needs to be assessed, who will be involved, and how the information collected will be used. Ten to fifteen people may be used if responses are not expected to vary a great deal. If major differences in opinion are expected, then larger samples are recommended.

Three to four rounds of questionnaires are used to collect the data. The first round contains open-ended questions asking the participants to write their responses.

After the first questionnaire is returned, the responses are summarized and developed into a second questionnaire. The second questionnaire asks respondents to rank their concerns, support for or disagreement with the various group responses, and explain their position. It identifies areas of agreement and disagreement on priorities. The responses are analysed by tallying and summarizing the comments made. The third questionnaire is developed from the summary of the second. The respondents are asked to review their prior responses in light of the group responses and make additional comments. Additionally, they are asked to vote on the order of importance of the items listed. The third questionnaire is often the last, but if there are major areas of disagreement, additional questionnaires may need to be administered.

Once the questionnaires have been completed, a final report based on the last questionnaire is prepared and sent to the participants. The report should summarize the goals, process, and conclusions or actions reached as a result of the study.

The Delphi has two advantages. First, it avoids the direct confrontation of people with opposing views. Participants do not face the pressure to conform to the majority position. Second, it eliminates the costs of having participants travel to a central place, a major limitation of focus group interviews and the nominal group technique.

The technique has limitations too. Several conditions must be met in order to use this technique (Miller & Hustedde, 1987). First, adequate time must be allotted. It is estimated that in the United States it takes a minimum of 45 days from the development of the questionnaire to the final report. In developing countries, the process could take longer because of relatively poor mail services. Second, the participants must have good writing skills because the technique is used in a written context. Lastly, the participants must be highly motivated because no one will be available to encourage them to complete the several rounds of questionnaires.

The Department of Agricultural Extension Education at Makerere University, Uganda, used this technique to identify and prioritize policy issues related to the conditions of service for women extension workers in the Ministry of Agriculture, Animal Industry and Fisheries.

Women extension workers were involved and three rounds of questionnaires used. Because of the poorly developed domestic postal system, the questionnaires were delivered and collected in person.

Focus Group Interviews. Krueger (1994) defines a focus group interview as a technique in which a group of people who possess certain characteristics provide data of a qualitative nature in a focussed discussion. Each interview involves a group of six to eight people who discuss a common topic for one to tow hours under the direction of a moderator and assistant moderator. The discussion is recorded on an audio tape and later transcribed and reported as qualitative data. Typically, several group interviews are conducted. In each case, the moderator raises various issues pertinent to the needs assessments, following an outline commonly known as the questioning route. The interviews are recorded and analysed for patterns and trends among participants in one interview, as well as across the entire set of groups involved in the needs assessment.

The following example indicates how the focus group technique could be used in conducting a needs assessment. Suppose a home economics extension agent noticed widespread signs of malnutrition (such as poor health and stunted growth) among children in a rural community of 2,000 households. The community produces a variety of food including poultry, livestock, fruits, and vegetables that are critical in diets for children. The community is served by one public health centre that functions fairly well in providing vaccinations and other basic medical services. Several drilled wells provide water for the residents.

The extension agent decided to carry out a needs assessment to explore the problem and to identify practical interventions to address it. The agent set the following objectives of the needs assessment: determining causes of malnutrition among children in the community and identifying measures to improve the situation. Focus group interviews were used because they provided an opportunity for the participants to engage in dialogue about the problem, its causes, and how to go about solving it (Krueger, 1994).

Five steps are critical in the use of focus group interviews (Archer & Layman, n.d.; Krueger, 1994), as seen in the example above:

1. *Developing a questioning route.* The questioning route was based on the objectives of the needs assessment. Brainstorming with colleagues and prospective users of the information helped to generate the questions. The questioning route, developed in the local language, consisted of four main questions and several questioning probes. The questions were: (a) What problems do par ents in this village face in feeding their children? Are the problems similar for all families? (b) What should be done to make sure children get proper nutrition? What can parents themselves do? Can all parents afford these measures? (c) What problems do parents in this village face in

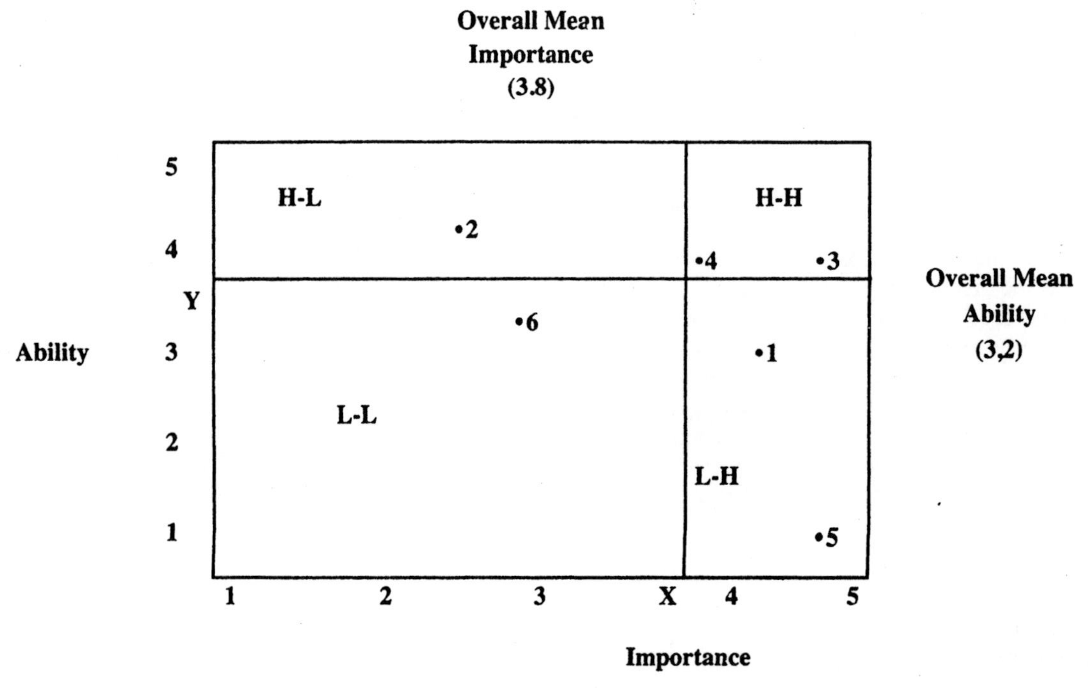

Figure 2. Matrix for determining needs.

keeping their children healthy? (d) What can parents do to improve the health of their children? Does the government have any role to play?

2. Recruiting the participants. The extension agent selected one of her assistant agents to help in conducting the interviews. Five interviews, one in each of five villages in the community, were conducted to help ensure representative data and to minimize travel costs on the part of participants. For each interview, eight participants were recruited to represent women's groups, local politicians, school teachers, religious leaders, health practitioners, and parents. The extension agent contacted the prospective particpant personally and explained to them the purpose and timing of the focus group interviews. The participants were selected on the basis of their knowledge of the village situation, ability to discuss freely in a group, and interest in child nutrition and health.
3. *Planning resources*. The extension agent developed the timetable for the sequence of steps involved, as well as the fiscal plan. She approached the administrators of local schools and secured rooms in which to conduct the interviews. A tape recorder was obtained from a local business person.
4. *Moderating interviews*. The extension agent moderated the interviews with the help of her assistant extension agent. Moderating is the process of keeping the discussion on track. It involved bringing the conversation back on target when irrelevant topics were introduced. This guidance had to be provided without reducing group enthusiasm and interest in the discussion. There are several personal attributes of a good moderator (Krueger, 1994), including (a) familiarity with group process either from previous experience in working with groups or through training in group dynamics, (b) good listening skills, © adequate background knowledge on the topic of discussion, (d) well-developed written and oral communication skills, and (e) a sense of humor.
5. *Data analysis and reporting*. Data analysis and reporting followed the interpretative summary format, whereby the data were not only described but also interpreted (Krueger, 1994). The analysis started with a debriefing immediately after the interview ended. The interview tape was played and a summary of the interview was written the next day. Each interview summary included key incidents, strong statements, and frequently occurring responses. Next, the key incidents, strong statements, and frequent responses were classified by question, coded, and grouped. The coding and grouping helped identify the general themes in the responses (Krueger, 1994; Miles & Huberman, 1994). The identified themes were then compared across interviews in order to develop a general picture for the whole community on a question by question basis. Lastly, specific concluding statements were made related to the two objectives of the needs assessment. Recommendations also were made regarding measures that extension and other organizations could take to improve child nutrition and health in the communities and region.

Krueger (1994) provides detailed guidelines and examples of focus group interviews as conducted in developed countries, specifically the United States. Focus group interviews have great potential for use in developing countries.

The Nominal Group. This technique can be used to generate possible items and set priorities in conducting a needs assessment. Although the data are generated in a group setting, verbal communication is minimized. For example, a chairperson of a farmers' association invited their extension agent to facilitate a meeting of selected members assembled to determine the activities for the following year. After learning that the farmers could read and write their native language, the extension agent suggested that the nominal group technique be used to assess needs for the association. The extension agent used the following six steps in conducting the nominal group process:

1. *Stating of question or problem*. This preceded the group session and involved clearly stating the question or problem to be addressed by the participants. The question addressed was: What activities should the association carry out next year?
2. *Generation of ideas*. The participants sat at a table facing each other and were asked to spend the first several minutes in silence, writing their ideas on a piece of paper. This silence allowed the participants to generate ideas uninterrupted and without being dominated by aggressive members.
3. *Presentation of ideas*. After they had finished writing down their ideas, each member presented one idea from his or her list in a round-robin fashion. The chairperson served as the recorder and wrote the ideas on a flip chart in front of the participants. The chairperson listed the ideas in concise phrases without attempting to change the wording or judge the ideas. The listing continued in order until all the participants had presented their ideas. The ideas suggested included organizing a farm fair, conducting an agricultural tour, acquiring an office for the association, increasing membership, and starting a farmers' market in a neighbouring town.
4. *Clarification of ideas*. The farmers were encouraged to seek clarification of any of the ideas listed. They were allowed to express their reasons for agreement or disagreement about each item, but argument was discouraged. This step ensured that the ideas listed were clarified without high status or aggressive members dominating the process.
5. *Rating of priorities*. At this stage, the farmers were asked to choose three of the most important

ideas from the list and rank them in order of priority. The rating was done on small cards which, after being collected, were shuffled to ensure anonymity. The votes were then tallied and the results disclosed on a flip chart in front of the group. The three highest rated items were: starting a farmers' market in a neighbouring town, conducting an agricultural tour, and increasing member ship in the association.

6. *Discussion and voting*. In the last step, the farmers discussed the vote, made additional clarifications, and voiced their agreements or disagreements. Following this discussion, the farmers decided that they would only start a farmers' market and work to increase their membership.

Delbecq (1975) is the original source on the nominal group technique. Miller and Hustedde (1987) have provided some guidelines on the use of the nominal group technique in community needs assessment in the United States. The technique is not widely used in developing countries because, like Delphi, it can only be used with participants who have developed writing skills.

Informal Group Methods. This category includes gathering information at group meetings and social gatherings. It is common for participants at meetings to talk about issues and problems in their family, community, or organization even when they are not part of the agenda. These side conversations may provide insights into the problems facing the organization or the individuals involved, as well as what can be done to address them. Tea-coffee breaks also provide another setting for members of an organization to talk about issues important to themselves, the organization, and the community at large. Important issues or needs of an organization may be identified by simply attending and listening to the conversations going on among its members during these short breaks.

Social gatherings such as recreational, cultural, and religious events also provide opportunities for collecting information. Valuable information may be obtained by listening actively and seeking selected individuals to clarify the issues overheard in the conversations.

Secondary Sources

Voss, Tordella, and Brown (1987) defined secondary data as "information gathered for the purpose other than the immediate or first application" (p. 156). Secondary data sources include census reports, previous studies, and administrative records and reports. Extension staff rarely use these data sources in needs assessment, probably because their application is not straightforward (Sofranko & Khan, 1988).

Census Reports. These data provide aggregate national and community statistics on such important sectors as population, housing, agriculture, education, health, and labour. When recent, these data can be used to identify the general areas of concern in the community and should serve as a good starting point in assessing needs. For example, small numbers of livestock, chicken, and fish would point to high chances of animal protein deficiency in the diets of community members. This might in turn imply poor health, especially among children.

Previous Studies. These reports include studies done by both groups and individuals. Although the issues studied may not be exactly the same as the needs assessment, there is always a chance that another development agency is, or has been, involved with similar kinds of issues. Local libraries and archives should be searched as a first step in needs assessment. In searching for previous studies, do not be restricted to school and alone. Also, do not forget the nonconventional sources of reading materials such as government departments, NGOs, and other local development agencies. In developing countries, local donor agencies (for example, the World Bank, Food and Agriculture Organization of the United Nations, United Nations Development Programme, United States Agency for International Development, and British Council) maintain valuable collections of (consultancy) studies and reports on past and present projects. Such reports are good for providing expert opinion regarding the problems and prospects upon which needs assessments could be based.

Administrative Records and Reports. Both government and nongovernment organizations maintain records on their activities. Most organizations also publish their activities quarterly or annually. When accessible, relevant organizational records and reports should be perused as one of the sources of needs assessment data.

Rapid Rural Appraisal

Rapid rural appraisal refers to the use of several data collection methods to gather practical information on development issues in local communities quickly (Freudenberger, 1994). These might include interviewing key informants, reviewing secondary data sources, mapping exercises, and conducting semistructured interviews with groups and individuals. Refer to chapter 6 for a detailed description of both rapid rural appraisal (RRA) and participatory rural appraisal (PRA).

Procedures

This section provides some guidelines to help in conducting a needs assessment. Extension personnel should be encouraged to be creative, efficient, and effective in designing their efforts. Consider the following guidelines:

1. Determine the purpose for conducting the needs assessment. Among these purposes are developing responsive programmes, generating awareness of programmes, satisfying official mandates, aiding in improved programme decisions, and promoting citizen participation and action.
2. Define the goals and objectives for the needs assessment. Show what it is you want to find out about and from whom. How are target clientele (including farmers of different resource levéls, genders, and ethnicities) involved in the setting of goals and objectives?
3. Select the approach you will take in collecting the information. Summers (1987) suggests four considerations: (a) the reasons for involving the public, (b) the decisions to be made using the information collected, (c) the need to generate representative information, and (d) the cost involved. Decide whether the needed information already exists, if a new data collection effort is needed, or if a combination of approaches is needed.
4. Design the instrumentation and procedures. When you design needs assessment instruments, it is usually best to keep the process simple. Long and complicated instruments discourage responses. Additionally, short instruments are less expensive to produce, distribute, collect, and analyse. Once a draft instrument has been prepared, it should be checked against the original purposes, goals, and objectives to make sure that nonessential information has not been included.
5. Prepare an estimated time line and budget for the needs assessment. The amount of resources available is likely to be one of the major determinants of the technique used for needs assessment. This is particularly critical in developing countries where resources are more limited. Although critical, budget constraints should not be used as an excuse for poor needs assessment.
6. Conduct a pilot test of the instrumentation and procedures. Special consideration should be given to collecting information from farmers of different resource levels, genders, and ethnicities. Many mistakes can be identified and eliminated in a pilot test with small groups of target clientele.
7. Collect the information. Limit the collection time in order to help develop a sense of urgency and keep the needs assessment targeted.
8. Analyse the data and information. If there is a large response, try to have access to a computer to analyse the data. There are also software packages to analyse qualitative data.
9. Prepare a report of the findings. Make it as user-friendly as possible. Do not feel confined to create one long document. It is probably better to divide the report into several brief documents for specific audiences. Consider using "white space" and figures to help communicate important points. Also consider developing audio-visual reports. Video tapes, transparencies, and slides also can be effective in communicating results.
10. Evaluate your efforts. Take time after the needs assessment has been completed to judge its merit and worth. What worked well? What problems were encountered? How could it have been done better? Once the evaluation is complete, share it with others interested in needs assessments. This will provide an opportunity for others to learn from what was done.
11. Use the needs assessment information. In order to have the information used, the following suggestions are offered: (a) issues that users perceive as important must be addressed; (b) the information must be communicated to the appropriate potential users; © groups must feel empowered to design and improve their programmes; and (d) the information must be available in a timely manner and in an understandable form.

References

Archer, T., & Layman, J. (n.d.). Focus group interview. *Evaluation Guide Sheet*. Columbus: The Ohio Cooperative Extension Service, The Ohio State University.

Caffarella, R. S. (1982). Identifying client needs. *Journal of Extension, 20*(July-August), 5-11.

Delbecq, A. (1975). *Group techniques for program planning: A guide to nominal group technique and Delphi process*. Glenview, IL: Scott Foresman.

Dillman, D. A. (1978). *Mail and telephone surveys: The total design method*. New York: John Wiley & Sons.

Freudenberger, K. S. (1994). *Tree and land tenure: Rapid appraisal tools*. Rome: FAO.

Hedges, L. E., & Rawls, W. (1988). *The status of agricultural education and extension in Uganda and the establishment of an agricultural education and extension department at Makerere University, Faculty of Agriculture and Forestry*. Kampala, Uganda: Makerere University, Faculty of Agriculture and Forestry.

Hershkowitz, M. (1973). A regional ETV network: Community needs and system structure. *Report for regional education service agency of Appalachian Maryland*. Silver Spring, MD: Operations Research Inc.

Kaufman, R. A. (1982). *Identifying and solving problems: A systems approach* (3rd ed.). San Diego, CA: University Associates.

Krueger, R. A. (1994). *Focus groups: A practical guide for applied research* (2nd ed.). Thousand Oaks, CA: Sage Publications.

Maalouf, W. D., & Contado, T. E. (1983). Basic and in-service training for agricultural extension. *Training for Agriculture and Rural Development, 31*, 39-47.

Miles, M. B., & Huberman, M. A. (1994). *Qualitative data analysis: An expanded sourcebook.* Thousand Oaks, CA: Sage Publications.

Miller, L. C., & Hustedde, R. J. (1987). Group approaches. In D. E. Johnson, L. R. Meiller, L. C. Miller, & G. F. Summers (Eds.), *Needs assessment: Theory and methods* (p. 91-125). Ames, IA: Iowa State University Press.

Sofranko, A., & Khan, A. (1988). It is not that simple. Improving needs assessment. *Journal of Extension, 26*(Winter), 14-16.

Stevens, H. (1980). Problems of agricultural extension in Africa. *Training for agriculture and rural development, 21*, 23-37.

Summers, G. F. (1987). Democratic governance. In D. E. Johnson, L. R. Meiller, L. C. Miller, & G. F. Summers (Eds.), *Needs assessment: Theory and methods* (p. 3-19). Ames, IA: Iowa State University Press.

Voss, P. R., Tordella, S. J., & Brown, D. L. (1987). Role of secondary data. In D. E. Johnson, L. R. Meiller, L. C. Miller, & G. F. Summers (Eds.), *Needs assessment: Theory and methods* (p. 156-170). Ames, IA: Iowa State University Press.

Zangwill, B. (1977, May). *A compendium of laws and regulations requiring needs assessment.* Washington, DC: Office of the Assistant Secretary for Planning and Evaluation, U.S. Department of Health, Education, and Welfare.

Chapter 6
Using Rapid or Participatory Rural Appraisal

Jules N. Pretty and Simplice D. Vodouhê

Conventional Methodologies for Learning

Agricultural institutions of all types have long relied on questionnaire surveys and quick rural visits to gather information on rural people and resources. Samples of people selected from a larger population are asked the same set of questions, and so it is assumed that the interviewers do not influence the process. Many informants are selected to account for all variation, and the resulting data are statistically analysed. Such surveys are used at practically all levels, from the large-scale census to small-scale, village-level research; by governments and NGOs; and for planning, research, and extension.

But there are problems with questionnaire surveys. The questionnaire designer has to determine the questions well in advance. Yet those who design these instruments cannot know which issues are important for local people. So they tend to increase the number of questions to ensure that all relevant issues are covered. This leads, in some cases, to forms of absurd length, with several hundred questions taking hours to administer. Rarely is attention paid to the nature of the interviewing process. In the structured survey, many of the contextual grounds for understanding are systematically removed or ignored, and all too often, the ill-trained enumerator further influences the process by prompting with answers. Despite many criticisms of this methodology (Chambers, 1983, 1992c; Fowler & Mangione, 1990; Rhoades, 1990; Gill, 1993), official surveys, such as sample censuses of agriculture or household expenditure surveys, remain remarkably popular.

At the other end of the spectrum are the brief field visits made by development professionals. But such "rural development tourism" is full of biases that misguide professionals into believing they have seen an accurate picture of rural life. Chambers (1983) characterized these biases into four main types: spatial biases, in which the better-off people living near roads and services are visited, with those who are remote and thus poorer being missed; time biases, in which visits are made during the seasons when roads are open and at times of day when people are busy in the fields; people biases, in which professionals speak only to rural leaders and articulate people who represent only the elite, dominant, and wealthy groups; and project biases, in which a showcase village or technology is repeatedly shown to outsiders, who get the impression that this is typical of all efforts.

What all this implies is that institutions come to believe that this selective information represents a comprehensive picture. Professionals are left with falsely favourable impressions of the impact of their work, and so they themselves have few reasons for initiating or encouraging change. Because of these flaws in conventional methods, development practitioners began in recent years to seek alternatives that avoided some of these problems.

Alternative Systems of Learning and Action

Partly because of these flaws in conventional approaches, there has been a recent rapid expansion in participatory methods and approaches. These began with the development of data-gathering methods which came to be known as rapid rural appraisal. During the late 1980s, this growing experience was supplemented by drawing upon long-established traditions that had put participation, action research, and adult education at the forefront of attempts to emancipate people. To the wider body of development programmes, these approaches represent a significant departure from standard practice. Some of the changes under way are remarkable. In a growing number of government and nongovernment institutions, extractive research is being superseded by investigation and analysis by local people themselves. Methods are being used not just for local people to inform outsiders, but also for people's own analysis of their own conditions (Chambers, 1992b, 1992c; Pretty & Chambers, 1993; Pretty, 1995).

The interactive involvement of many people in differing institutional contexts has promoted innovation and ownership, with many variations in the way that systems of learning have been put together. There are many different terms (Box 1), but they have the following important common principles (Pretty, 1994):

> *A defined methodology and systemic learning process.* The focus is on cumulative learning by all the participants and, given the nature of these approaches as systems of learning and interaction, their use has to be participative.
>
> *Multiple perspectives.* A central objective is to seek diversity, rather than to characterize complexity in terms of average values. The assumption is that different individuals and groups make different evaluations of situations, which lead to

---

Jules N. Pretty is the Director of the Sustainable Agriculture Programme, International Institute for Environment and Development, London. **Simplice D. Vodouhê** is a Lecturer in Extension Science, Faculty of Agriculture, National University of Benin, Cotonou, Benin.

different actions. All views of activity or purpose are heavy with interpretation, bias, and prejudice, and this implies that there are multiple possible descriptions of any real-world activity.

- *Group learning processes.* All involve the recognition that the complexity of the world will only be revealed through group inquiry and interaction. This implies three possible mixes of investigators, namely, those from different disciplines, from different sectors, and from outsiders (professionals) and insiders (local people).
- *Context specific.* The approaches are flexible enough to be adapted to suit each new set of conditions and actors, and so there are multiple variants.
- *Facilitating experts and stakeholders.*The methodology is concerned with the transformation of existing activities to try to bring about changes which people in the situation regard as improvements. The role of the "expert" is best thought of as helping people in their situation carry out their own study and so achieve something. These facilitating experts may be stakeholders themselves.
- *Leading to sustained action.* The learning process leads to debate about change, and debate changes the perceptions of the actors and their readiness to contemplate action. Action is agreed upon, and implementable changes will therefore represent an accommodation among the different conflicting views. The debate or analysis both defines changes which would bring about improvement and seeks to motivate people to take action to implement the defined changes. This action includes local institution building or strengthening, thus increasing the capacity of people to initiate action on their own.

These alternative systems of learning and action imply a process of learning leading to action. A more sustainable agriculture, with all its uncertainties and complexities, cannot be envisaged without all actors being involved in continuing processes of learning.

Box 1. A selection of terms and names for alternative systems of participatory learning and action.

AEA	Agroecosystems Analysis
BA	Beneficiary Assessment
DELTA	Development Education Leadership Team
DPR	Diagnóstico Rurale Participativo
FPR	Farmer Participatory Research
GRAAP	Groupe de Recherche et d'Appui pour l'Auto-Promotion Paysanne
MARP	Méthode Accélérée de Recherche Participative
PALM	Participatory Analysis and Learning Methods
PAR	Participatory Action Research
PRM	Participatory Research Methodology
PRAP	Participatory Rural Appraisal and Planning.
PTD	Participatory TechnologyDevelopment
PUA	Participatory Urban Appraisal
PfR	Planning for Real
PD	Process Documentation
RA	Rapid Appraisal
RAAKS	Rapid Assessment of Agricultural Knowledge Systems
RAP	Rapid Assessment Procedures
RAT	Rapid Assessment Techniques
RCA	Rapid Catchment Analysis
REA	Rapid Ethnographic Assessment
RFSA	Rapid Food Security Assessment
RMA	Rapid Multi-perspective Appraisal
ROA	Rapid Organizational Assessment
RRA	Rapid Rural Appraisal
SB	Samuhik Brahman (Joint trek)
SSM	Soft Systems Methodology
TfD	Theatre for Development
TfT	Training for Transformation
VIPP	Visualization in Participatory Programmes

The Different Interpretations of Participation

In recent years, an increasing number of analyses of projects have shown that participation by local people is one of the critical components of success in irrigation, livestock, water, and agriculture sectors (USAID, 1987; Reij, 1988; Finsterbusch & van Wicklen, 1989; Bagadion & Korten, 1991; Cernea, 1991; Guijt, 1991; Pretty & Sandbrook, 1991; Uphoff, 1992; Narayan, 1993; World Bank, 1994; Pretty, 1995).The terms "people's participation" and "popular participation" have now become part of the normal language of many development agencies (Adnan, Nurul Alam, & Brustinow, 1992; Bhatnagar & Williams, 1992). This has brought new dangers. The term "participation" has been used to justify the extension of state control and to build local capacity and self-reliance; it has been used for data collection and for interactive analysis. Participation has often centred on encouraging local people to sell their labour in return for food, cash, or materials. Yet these material incentives distort perceptions, create dependencies, and give the misleading impression that local people are supportive of externally driven initiatives (Reij, 1988; Bunch, 1991; Kerr, 1994). This means that *"more often than not, people are asked or dragged into participating in operations of no interest to them, in the very name of participation"* (Rahnema, 1992).

These many interpretations of the term participation can be arranged into seven clear types (Box 2). These range from passive participation, where people are involved merely by being told what is to happen, to

Box 2. A typology of participation: how people participate in development programmes and projects.

Typology	Characteristics of Each Type
1. Passive Participation	People participate by being told what is going to happen or has already happened. It is a unilateral announcement by an administration or project management without any listen ing to people's responses. The information being shared belongs only to external professionals.
2. Participation in Information Giving	People participate by answering questions posed by extractive researches using question naire surveys or similar approaches. People do not have the opportunity to influence proceedings, as the findings of the researche are neither shared nor checked for accuracy.
3. Participation by Consultation	People participate by being consulted, and external agents listen to views. These external agents define both problems and solutions and may modify these in the light of people's responses. Such a consultative process does not concede any share in decision making, and professionals are under no obligation to take on board people's views.
4. Participation for Material Incentive	People participate by providing resources, for example labour, in return for food, cash, or other material incentives. Much on-farm research falls in this category, as faermers provide the fields but are not invilved in the experimentation or the process of learning. It is very common to see this called participation, yet people have no stake in prolonging activities when the incentives end.
5. Funçional Participation	People participate by forming groups to meet predetermined objectives related to the project, wich can involve the development or promotion of externally initiated social organization. Such involvement does not tend to be at early stages of project cycles or planning, but rather after major decisions have been made. These instructions tend to be dependent on external initiators and facilitators, but may become self-dependent.
6. Interactive Participation	People participate in joint analysis, which leads to action plans and the formation of new local institutions or the strengthening of existing ones. It tends to involve interdisciplinary mothodologies that seek multiple perspectives and make use of systemic and structured learning processes. These groups take control over local decisions, and so people have a stake in maintainning structures or practices.
7. Self-Mobilization	People participate by taking initiative indipendent of external institution to change systems. They develop contacts with external institutions for resources and technical advice they need, but retain control over how resources are used. Such self-initiated mobilization and collective action may or may not challenge existing inequitable distribution of wealth and power.

Source: Pretty (1994), adapted from Adnan et al. (1992).

self-mobilization, where people take initiatives independent of external institutions. It is clear from this typology that the term participation should not be accepted without the term participation should not be accepted without appropriate qualification. If the objective of development is to achieve sustainable development, then nothing less than functional participation should suffice.

But the dominant applications of participation are rarely better than instrumental. A recent study of 230 rural development institutions.employing some 30,000 staff in 41 countries of Africa found that people participated in very different ways (Guijt, 1991). Participation was most likely to mean simply providing information to external agencies. Another study of 121 rural water supply projects in 49 countries of Africa, Asia, and Latin America found that participation was the significant factor contributing to project effectiveness, maintenance of water systems, and economic benefits (Narayan, 1993). Even though most of the projects referred to community participation or made it a specific project component, only 21 per cent scored high on interactive participation. Yet when people were involved in decision making during all stages of the project, from design to maintenance, the best results occurred. When they were just involved in information sharing and consultations, then results were much poorer.

Great care must therefore be taken over using and interpreting the term participation. It should always be qualified by reference to the type of participation. What is important is to ensure that those using the term define ways of shifting from the more common passive and incentive-driven participation towards the interactive end of the spectrum. A critical way to do this is to ensure the appropriate use of interactive and participatory methods.

Participatory Methods

The creative ingenuity of practitioners worldwide has greatly increased the range of participatory methods in use (see *RRA Notes*, 1988-1994; IDS/IIED, 1994; Pretty et al., 1995; Chambers, 1992a, 1992b; Mascarenhas et al., 1991; KKU, 1987; Conway, 1987). Many have been drawn from a wide range of nonagricultural contexts and were adapted to new needs. The methods can be put into four classes: for group and team dynamics, for sampling, for interviewing and dialogue, and for visualization and diagramming. These methods collected into unique approaches, or assemblages of methods, constitute systems of learning and action. In this section, some key methods are briefly described.

Participation calls for collective analysis and good rapport. Even a sole researcher must work closely with local people. Ideally, though, teams of investigators work together in interdisciplinary and intersectoral teams. By working as a group, the investigators can approach a situation from different perspectives, carefully monitor one another's work, and carry out a variety of tasks simultaneously. Groups can be powerful when they function well, because performance and output are likely to be greater than the sum of the individual members. But shared perceptions, essential for group or community action, have to be carefully negotiated. Yet the complexity of multidisciplinary team work is generally poorly understood. Various workshop and field methods are used to facilitate this process of group formation:

1. *Team contracts.* Team contracts developed by all the team members help to ensure good group dynamics and may include agreements to hold evening discussions and morning brainstorming sessions. One person may be elected to monitor team interactions to provide feedback. The monitor can make a note of each member's location and record interactions by drawing a circle around individuals' names when they speak or an arrow from talker to person being talked to, with duration of speech recorded in seconds. The results are used simply for showing team members how the discussion developed. It then becomes clear who has dominated and who was quiet.
2. *The night halt.* Rapport between outsiders and villagers is facilitated by staying in the village. Many have made this an essential part of participatory analysis and planning. It provokes change in outsiders' attitudes: they sleep and eat as villagers do; it allows for early morning and evening discussions when people are less busy; and it is an explicit com mitment by outsiders to village life.
3. *Work sharing.* When outsiders are taught some-thing by villagers, roles are reversed. Professionals soon learn how much skill is required, say, to plough a furrow, transplant rice, weed, lop tree fodder, cut firewood, dig compost, and wash clothes. Such activities prompt changes in attitude and help to build rapport.
4. *Rapid report writing, with self-correcting notes.* It is essential to record, as a team, the key findings before members disperse to their own organizations. Report writing is made easier by writing a brief summary of how diagrams were constructed and of the key findings. Individuals can be encouraged to keep a private diary or series of notes to focus on things they would like to improve the next time.
5. *Shared presentations.* The key findings should always be presented to villagers and outsiders. This is an important opportunity for cross-checking and feedback. Professionals present and invite comment and criticism. This is a fundamental reversal of roles and is crucial to establishing the trustworthiness of the findings.

To ensure that multiple perspectives are represented, practitioners must be clear about who is participating. Communities are rarely homogeneous, so there is always the danger of assuming that those participating are representative. Those missing, though, are usually the poorest and most disadvantaged. Sampling is an essential part of these participatory approaches, and several methods are available:

6. *Transect walks and direct observation.* These are systematic walks with key informants through the area of interest, observing, asking, listening, looking, and seeking problems and solutions. The findings can be mapped on a transect diagram. Most transect walks result in the outsiders discovering surprising local practices such as indigenous conservation practices, multiple uses of plants, and a great variety of crops. It has been instructive for many professionals to realize how much they do not see or do not think to ask about.
7. *Wealth rankings and social maps.* Wealth ranking is used to classify households according to relative wealth or well-being. Informants sort cards, each with one household name on it, into piles. The wealthiest are put at one end, the poorest at the other, and as many piles as desired are made. The process is repeated with at least three informants. Another method is to conduct the ranking directly on a social map. Villagers are then asked to indicate on the houses the relative wealth classes. Individual assets such as land ownership, animals, and tools can be marked for each household. Wealth rankings are useful for leading into other discussions on livelihoods and vulnerability; producing a baseline against which future intervention impact can be measured; providing a sample frame to cross-check the relative wealth of informants who have been or will be interviewed; and producing local indicators of welfare.

Sensitive interviewing and dialogue are a third element of these systems of participatory learning. For the reconstructions of reality to be revealed, the conventional dichotomy between the interviewer and respondent should not be permitted to develop. Interviewing is therefore structured around a series of techniques that promote a sensitive and mutually beneficial dialogue. This should appear more like a structured conversation than an interview:

8. *Semi-structured interviews (SSI).* This is guided interviewing and listening in which only some of the questions and topics are predetermined; other questions arise during the interview. The interviews appear informal and conversational, but are actually carefully controlled and structured. Using a guide or checklist, the multidisciplinary team poses open-ended questions and probes topics as they arise. New avenues of questioning are pursued as the interview develops. SSIs are a central part of all participatory methods.
9. *Types, sequencing, and chains of interviews.* Many types of interviews may be combined in sequences and chains. These include key informant interviews, by asking who the experts are and then putting together a series of interviews (e.g., men on ploughing, women on transplanting and weeding, shopkeepers for credit and inputs); and group interviews, which may be groups convened to discuss a particular topic (focused or specialist groups), groups comprising a mix of people whose different perceptions illuminate an issue (structured groups), casual groups, and community groups.

The fourth element is the emphasis on diagramming and visual construction. In formal surveys, information is taken by interviewers, who transform what people say into their own language. By contrast, diagramming by local people gives them a share in the creation and analysis of knowledge, providing a focus for dialogue which can be sequentially modified and extended. Local categories, criteria, and symbols are used during diagramming. Rather than answering questions which are directed by the values of the outside professional, local people can explore creatively their own versions of their worlds. Visualizations therefore help to balance dialogue and increase the depth and intensity of discussion:

10. *Participatory mapping and modelling.* This involves constructing, on the ground or on paper, maps or models, using materials such as sticks, stones, grasses, wood, cigarette packets, tree leaves, coloured sands and soils, rangoli powders, coloured chalk, pens, and paper. Great play is made of the issue of *who holds the stick or pen.* The person who holds the stick talks about what is most important to him or her. As maps take shape, more people become involved, and so want to contribute and make sequential changes. There are many types of maps: resource maps of catch ments, villages, forests, fields, farms, home gardens; social maps of residential areas of a village; wealth rankings and household assets surveys on social maps; health maps, where the health status of each family member is shown on each house, using coloured stickers or other markers (categories might include cases of malnutrition, ear infection, jaundice, and the like); topical maps such as aquifer maps drawn by the water diviner or soils maps by soils experts; impact monitoring maps, where villagers record or map pest incidence, input usage, weed distribution, soil quality, and so forth. Some of the most illuminating maps combine historical views with those of the present.
11. *Seasonal calendars and activity profiles.* Seasonal constraints and opportunities can be diagrammed month by month throughout the year. Ceremonies can be used as a cross-check so that names of months are agreed upon. People use pieces of stick, draw histograms in the dust or with chalk, or make piles of stones, seeds, or powders to represent relative quantities and patterns of rainfall, soil moisture, crops, labour, food consumption, illnesses, prices, animal fodder, fuel, migration, pests, income, expenditure, debt, children's games, and so on. Seasonal calendars can be drawn in linear fashion with twelve months to show a typical year or eighteen months to illustrate changes between years, or they can be drawn in a circle. Daily pat terns of activity can be similarly explored by charting typical activities for each hour of the day, amount of effort, time taken, and location of work. These can be compared for men, women, the old, the young, and others.
12. *Time lines and local histories.* Historical analyses have been found to be a good icebreaker for field exercises and include detailed accounts of the past, of how things have changed, particularly focussing on relationships and trends. These include technology histories and review, crop histories and biographies, livestock breed histories, labour availability, trees and forest histories, education change, and population change. Folklore and songs are valuable resources for exploring history.
13. *Venn and network diagrams.* Venn diagrams involve the use of circles of paper or card to represent people, groups, and institutions. These are arranged to represent real linkages and distance between individuals and institutions. Overlap indicates flows of information, and distance on the diagram represents lack of contact.
15. *Matrix scoring and pairwise ranking.* These methods are for learning about local people's categories, criteria, choices, and priorities. For pairwise ranking, items of interest are compared pair by pair;

Box 3. A framework for judging trustworthiness.

1. *Prolonged and/or Intense Engagement Between the Various Actors.* For building trust and rapport, learning the particulars of the context, and to keep the investigator(s) open to multiple influence.
2. *Persistent and Paralled Observation.* For understanding both a phenomenon and its context.
3. *Trangulation by Multiple Sources, Methods, and Investigators: la triangulation triangulée.* For cross-checking information and increasing the range of different people's realities encountered, including multiple copies of sources of information, comparing the results from a range of methods, and having teams with a diversity of personal, professional, and disciplinary backgrounds.
4. *Analysis and Expression of Difference.* For ensuring that a wide range of different actors are involved in the analysis and that their perspectives and realities are accurately represented.
5. *Peer or Colleague Checking.* Periodic review meetings with peers not directly involved in the original information was constructed and analyzed. Without participant checks, investigators can make no claims that they are representing participants' views.
8. *Reports with Working Hypotheses, Contextual Descriptions, and Visualizations.* These are "thick" descriptions of complex reality, with working hypotheses, visualization, and quotations capturing people's personal perspectives and experiences.
9. *Parallel Investigations and Team Communications.* If subgroups of the same team proceed with investigations in par allel using the same system of inquiry and come up with the same or similar findings, then we can depend on these findings.
10. *Reflexive Journals.* These are diaries individuals keep on a dialy basis to record a variety of information about themselves.
11. *Inquiry Audit.* The inquiry team should be able to provide sufficient information for a disinterested person to examine the processes produit in such a way as to confirm that the findings are not a figment of their imaginations.
12. *Impact on Stakeholders' Capacity to Know and Act.* For demonstrating that the investigation or study has had an impact, including participants having a heightened sense of their own realities, as well as an increased awareness and appreciation of those of other people; the report itself could also prompt action on the part of readers who have been directly involved.

Source: Pretty (1994).

informants are asked which of the two they prefer, and why. Matrix scoring takes criteria for the rows in a matrix and items for columns, and people complete the boxes row by row. The items may be ordered for each of the criteria (e.g., for six trees, indicate from best to worst for fuelwood, fodder, erosion control, and fruit supply); or participants may put stones, seeds, or berries into piles for relative scoring.

Matrices are also useful for ordering and structuring other types of information and include attributes matrices for technologies, problem-opportunity matrices, and manual discriminant technique matrices for contrast comparisons.

The Trustworthiness of Findings

Users who have presented findings arising from participatory learning are often asked a question along the lines of *"But how does it compare with the real data?"* (see Gill, 1991). It is commonly asserted that participatory methods involve only subjective observations and so respond just to selected members of communities. Terms like "informal" and "qualitative" are used to imply poorer quality or second-rate work. Rigour and accuracy are assumed, therefore, to be in contradiction with participatory methods.

This means that, unlike conventional investigators, the investigators relying on participatory methods are called upon to prove the utility of their approach. Conventional researchers use four criteria in order to persuade their audiences that the findings of an inquiry can be trusted (see Lincoln & Guba, 1985; Guba & Lincoln, 1989). How can we be confident about the "truth" of the findings (internal validity)? Can we apply these findings to other contexts or with other groups of people (external validity)? Would the findings be repeated if the inquiry were replicated with the same or similar subjects in the same or similar context (reliability)? How can we be certain that the findings have been determined by the subjects and context of the inquiry, rather than by the biases, motivations, and perspectives of the investigators (objectivity)? These four criteria, though, are dependent for their meaning on the core assumptions of the conventional research paradigm (Pretty, 1994; Lincoln & Guba, 1985; Kirk & Miller, 1986; Cook & Campbell, 1979).

Trustworthiness criteria were first developed by Guba (1981) to judge whether or not any given inquiry was methodologically sound. But these criteria *"had their*

Box 4. The institutional changes promoted by a course in "Formation pour l'Appui à l'autopromotion Rurale" (FAR) in Benin.

A one-month course was conducted for participants from eight countries of sub-Saharan Africa on the approaches and principles of PRA. Farmers and participants analysed the local resources and capacities; they investigated the past roles of actors in development; and they worked together on joint inventories for developing self-promoting capacities in villages. The process of learning involved regular rotation between workshop and the field so as to increase the opportunities for reflection and iteration.

During the fieldwork, many participatory methods were used, including visualizations, wealth rankings, mappings, transect walks, seasonal calendars, and so on. These enabled participants to learn from villagers. It also meant villagers and trainees got to know each other better, and so mutual confidence grew. After all the data gathering and joint discussions, it was the presentations that had the greatest impact on attitudes and linkages.

The process involved presenting the findings back to various groups in the community, discussing the range of conclusions, and seeking accommodation between differing and conflicting interests. This is very important for the participants and villagers: for the participants, it may be the first time they have asked farmers and rural people to comment on their findings; for the villagers, it may be the first time they have seriously been asked for their views. It is also an important opportunity for the various actors to learn from one another. In this case, it was the wealthy and more important people in the village who recognized their limited view on a range of issues, particularly problems raised by groups of women.

A second set of restitutions occurred between villagers and staff of local government agencies and NGOs working locally. Each group of trainees presented findings to the group, which then led to detailed negotiations between the farmers and various external professionals. They were able to reach agreement on a plan for the village, including the priorities for action.

The most significant impact was that the external agencies working in the area had not had any prior formal contact with each other. This was the first time they had shared perspectives. The production of the joint plan of action had the effect of strengthening the linkages among the various agencies, as well as involving farmers at the centre of the process. As a result, trust and understanding among the various actors has increased.

foundation in concerns indigenous to the conventional, or positivist, paradigm" (Lincoln, 1990). To distinguish between elements of inquiry that were not derived from the conventional paradigm, further "authenticity" criteria have been suggested to help in judging the impact of the process of inquiry on the people involved (Lincoln, 1990). Have people been changed by the process? Have they a heightened sense of their own constructed realities? Do they have an increased awareness and appreciation of the constructions of other stakeholders? To what extent did the investigation prompt action?

Drawing on these and other suggestions for "goodness" criteria (Marshall, 1990; Smith, 1990), a set of twelve criteria for establishing trustworthiness have been identified (Pretty, 1994) (Box 3). These criteria can be used to judge information, just as statistical analyses provide the grounds for judgement in positivist or conventional science. The use of a system of learning without, for example, triangulation of sources, methods, and investigators and without participant checking of the constructed outputs should be judged as untrustworthy (Olukossi, 1993).

Note, however, that it will never be possible to be certain about the trustworthiness criteria. We cannot say that x has a trustworthiness score of y points, but we can say that x is trustworthy because certain things happened during and after the process of joint investigation and analysis. The trustworthiness criteria should be used to identify what has been part of the process of gathering information and whether key elements have been omitted. Knowing this should make it possible for any observers, be they readers of a report or policy makers using the information to make a decision, also to make a judgement on whether they trust the findings.

Towards a New Professionalism in Extension

Extension has come to mean extending knowledge from a centre of learning to those presumed to be in need of that knowledge. But this model does not lead to enhanced capacity amongst extensionists and farmers. These participatory methods and approaches represent an opportunity to build better linkages between the various actors and to increase the learning from each other. When PRA and other participatory approaches are used, extensionists and researchers have the opportunity to work together on the same team. They exchange knowledge and experiences and reach some consensus with farmers on what is most needed. As a result, all parties come closer together. Farmers become more confident that professionals can help them, without imposing solutions on them. An example of how this can strengthen linkages is illustrated in Box 4.

Widespread and persistent change is more likely where all elements combine, namely, new systems of participatory learning, new learning environments for professionals and local people, and new institutional settings, including improved connections both within and between institutions (Roche, 1992; Pretty & Chambers, 1993; Pretty, 1995).

The wider challenge is for agricultural organizations to become learning organizations. To do this, they will have to promote experimentation, promote connectivity and group work based on roles rather than disciplines, and develop monitoring and self-evaluation systems to improve learning and awareness. The central concept of sustainable agriculture is that it must enshrine new ways of learning about the world. This has profound implications for agricultural development institutions. The focus is less on *what* we learn, and more on *how* we learn and *with whom*. This implies new roles for development professionals, leading to a whole new professionalism with new values, methods, and behaviour.

References

Adnan, S., Barrett, A., Nurul Alam, S. M., & Brustinow, A. (1992). *People's participation: NGOs and the flood action plan*. Dhaka, Bangladesh: Research and Advisory Services.

Bagadion, B. U., & Korten, F. F. (1991). Developing irrigators' organizations: A learning process approach. In M. M. Cernea (Ed.), *Putting people first* (2nd ed.). Oxford: Oxford University Press.

Bhatnagar, B., & Williams, A. (1992). *Participatory development and the World Bank: Potential directions for change*. Washington, DC: World Bank.

Bunch, R. (1991). People-centred agricultural improvement. In B. Haverkort, van der Kamp, & A. Waters-Bayer (Eds.). *Joining farmers' experiments: Experiences in participatory development*. London: IT Publications.

Cernea, M. M. (1991). *Putting people first* (2nd ed.). Oxford: Oxford University Press.

Chambers, R. (1983). *Rural development: Putting the last first*. London: Longmans.

Chambers, R. (1991, June). *Farmers' practices, professionals and participation: Challenges for soil and water management*. Paper presented at the 1991 workshop on Farmers' Practices and Soil and Water Conservation Programmes, ICRISAT.

Chambers, R. (1992a). The self-deceiving state: Psychosis and therapy. *IDS Bulletin, 23* (4), 31-42.

Chambers, R. (1992b, September). *Methods for analysis by farmers: The professional challenge*. Paper presented at the 12th Annual Symposium of the Association for FSR/E, Michigan State University, East Lansing, MI.

Chambers, R. (1992c). *Rural appraisal:'Rapid, relaxed and participatory*. IDS Discussion Paper 311. Brighton: IDS.

Chambers, R. (1993). *Challenging the professions: Frontiers for rural development*. London: Intermediate Technology Publications.

Conway, G. R. (1987). The properties of agroecosystems. *Agricultural Systems, 24*, 95-117.

Cook, T., & Campbell, D. (1979). *Quasi-experimentation: Design and analysis issues for field settings*. Chicago: Rand McNally.

Finsterbusch, K., & van Wicklen, W. A. (1989). Beneficiary participation in development projects: Empirical tests of popular theories. *Economic Development and Cultural Change, 37* (3), 573-593.

Fowler, F. J., & Mangione, T. W. (1990). *Standardized survey interviewing: Minimizing interviewer-related error*. Applied Social Research Methods Series Volume 18. Newbury Park, CA: Sage Publications.

Gill, G. (1991). But how does it compare with the "real" data? *RRA Notes, 14*, 5-13.

Gill, G. (1993). OK, the data's lousy, but it's all we've got (being a critique of conventional methods). Sustainable Agriculture Programme. *Gatekeeper Series* SA38. London: IIED.

Guba, E. G. (1981). Criteria for assessing the trustworthiness of naturalistic inquiries. *Educational Communication Technology Journal, 29*, 75-92.

Guba, E. G., & Lincoln, Y. (1989). *Fourth generation evaluation*. London: Sage.

Guijt, I. (1991). *Perspectives on participation: An inventory of institutions in Africa*. London: IIED.

IDS/IIED (1994). *PRA and PM&E annotated bibliography*. Sussex: IDS; and London: IIED.

Kerr, J. (1994, November). *How subsidies distort incentives and undermine watershed development projects in India*. Paper presented at the IIED Conference, New Horizons: The Social, Economic and Environmental Impacts of Participatory Watershed Development, Bangalore, India. IIED, London.

Kirk, V., & Miller, M. L. 1986. *Reliability and validity in qualitative research*. Qualitative Research Series 1. Beverly Hills, CA: Sage Publications.

KKU (1987). *Rapid rural appraisal: Proceedings of an international conference*. Thailand: Rural Systems Research Project, Khon Kaen University.

Lincoln, Y. S. (1990). The making of a constructivist: A remembrance of transformations past. In E. G. Guba (Ed.), *The paradigm dialog*. Newbury Park, CA: Sage.

Lincoln, Y. S., & Guba, E. (1985). *Naturalistic inquiry*. Newbury Park, CA: Sage Publications.

Marshall, C. (1990). Goodness criteria: Are they objective or judgment calls? In E. G. Guba (Ed.), *The paradigm dialog*. Newbury Park, CA: Sage Publications.

Mascarenhas, J., Shah, P., Joseph, S., Jayakaran, R., Devararam, J., Ramachandran, V., Fernandez, A., Chambers, R., & Pretty, J. N. (1991). Participatory rural appraisal. *RRA Notes, 13*. IIED, London.

Narayan, D. (1993). *Focus on participation: Evidence from 121 rural water supply projects*. UNDP-World Bank Water Supply and Sanitation Program. Washington, DC: World Bank.

Olukossi, J. O. (1993). Participatory research methodology: A few hints. *Newsletter National Farming Systems Research Network*, Zaria, Nigeria.

Pretty, J. N. (1994). Alternative systems of inquiry for sustainable agriculture. *IDS Bulletin, 25*(2), 37-48. University of Sussex: IDS.

Pretty, J. N. (1995). *Regenerating agriculture: Policies and practice for sustainability and self-reliance*. London: Earthscan Publications; and Washington DC: National Academy Press.

Pretty, J. N., & Chambers, R. (1993). *Towards a learning paradigm: New professionalism and institutions for sustainable agriculture*. IDS Discussion Paper DP 334. Brighton: IDS.

Pretty, J. N., Guijt, I., Scoones, I., & Thompson, J. (1995). *A trainers' guide to participatory learning and action*. IIED Training Materials Series No. 1. London: IIED.

Pretty, J. N., & Sandbrook, R. (1991, October). *Operationalizing sustainable development at the community level: Primary environmental care*. Paper presented to the DAC Working Party on Development Assistance and the Environment, OECD, Paris.

Rahnema, M. (1992). Participation. In W. Sachs (Ed.), *The development dictionary*. London: Zed Books.

Reij, C. (1988). The agroforestry project in Burkina Faso: An analysis of popular participation in soil and water conservation. In C. Conroy & M. Litvinoff (Eds.), *The greening of aid* (p. 74-77). London: Earthscan Publications.

Rhoades, R. (1990). *The coming revolution in methods for rural development research*. Mimeo. Manila, Philippines: User's Perspective Network International Potato Center.

Roche, C. (1992). *Operationality in turbulence: The need for change*. London: ACORD.

RRA Notes (1988-1995). Issues 1 to the present. London: Sustainable Agriculture Programme.

Smith, J. K. (1990). Alternative research paradigm and the problems of criteria. In E. Guba (Ed.), *The paradigm dialog*. Newbury Park, CA: Sage Publications.

Uphoff, N. (1992). *Learning from Gal Oya: Possibilities for participatory development and post-Newtonian science*. Ithaca, NY: Cornell University Press.

USAID (1987). *Women in development: A.I.D.'s experience, 1973-1985*. AID Program Evaluation Report No. 18. Washington, DC: Agency for International Development.

World Bank (1994, April). *The World Bank and participation*. Report of the Learning Group on Participatory Development. Washington, DC: World Bank.

Chapter 7
Developing and Delivering Extension Programmes

Artur Cristóvão, Timothy Koehnen, and José Portela

The Difficult Challenges Ahead

Extension planners throughout the world face the difficult challenges of being creative in their programme development efforts and responsive to the needs of rural communities and farmers. A way to overcome these challenges is to look at different programme development ideas and approaches, analysing how they function in practice, as well as their advantages and disadvantages.

This chapter presents an overview of the programme development process, with an emphasis on local-level extension work. First, it will look at basic concepts, discuss different ways of approaching the programme development process, and review current ideas. In the second part, a rationale for participatory planning will be discussed and a number of key aspects, such as priority setting, definition of objectives, and evaluation, will be stressed. The last part will summarize new roles, knowledge, and skills.

It is important to emphasize that, although in the text we usually use the word "planning," our discussion is centred on programme development globally, which we see as a set of dynamic cycles necessarily implying planning, implementation, and evaluation stages and activities, interacting and often overlapping, evolving along the programme process and according to the changing circumstances of the physical, socioeconomic, and political-institutional environment.

Extension Work: Increasingly Concerned with Responsive Planning

In today's world, the pace of change is accelerated, and people are continually involved in it, either as passive elements or as active citizens, more often as mere project recipients or targets. Development projects are delineated to help people adjust to change, for example to new agricultural policies or market demands. Also, at times, but not as frequently, they are directed to help grass-roots groups and rural communities to build change projects that are relevant to their own needs and aspirations. Both situations require planning and the preparation of different types of extension programmes.

Programme Development and Planned Change

In one way or another, when we talk of programme development, we talk of some kind of planned change, that is, deliberate efforts to change a given state of affairs. Social, economic, cultural, or technological changes are commonly assumed to be the purposes of planned and systematic extension actions. As change facilitators, extensionists should then be concerned with the preparation of programmes and projects that are responsive to the needs and interests of rural communities and farm families.

Some would argue that planning is a difficult mission, especially in the developing world, where the level of political and economic uncertainty is high. This leads to on-the-spot decisions or very short-term and incipient planning work. However, we strongly argue that the lack of adequate planning and continual evaluation is a major reason for the frequent failure of development projects and extension activities (Koehnen, Portela, & Cristóvão, 1992, p. 207): planners do not look at the diversity embedded in most situations; different clientele groups are not systematically involved, and some (surprisingly, major ones) are neglected; alternative solutions are not carefully compared; objectives are too rigid, not clearly defined, or not linked to activities; results are not duly studied; and the distribution of benefits is overlooked. So in many situations the challenge seems obvious: to allocate more time for planning and evaluation and to stop acting hastily and mechanically, without direction and purpose, like a clock lacking hands.

Extension authors and professionals strongly support planning. Forest and Baker (1994, p. 87), for instance, underline that "program planning helps justify budget appropriations and brings understanding among the public," adding that "the planning process offers opportunity to people who participate in it to learn, thus building leaders-hip skills in the community that will likely contribute to self-help, independence, and positive end results." In general, adequate planning and evaluation do the following:

Artur Cristóvão is an Associate Professor of Extension and Rural Development, University of Trás-os-Montes and Alto Douro, Vila Real, Portugal. **Timothy Koehnen** is a Guest Associate Professor of Extension and Rural Development, University of Trás-os-Montes and Alto Douro, Vila Real, Portugal. **José Portela** is an Associate Professor of Rural Sociology, University of Trás-os-Montes and Alto Douro, Vila Real, Portugal.

- Involve an integrated analysis of needs and interests, opening up new horizons for action
- Promote a concentration of efforts, channelling energies and resources in appropriate directions, and helping accomplish complementarities and synergies
- Strengthen programme resources and attract funds, thus allowing the sustainability or expansion of activities
- Improve team and community capacities, motivations, performance, and autonomy
- Show commitment to address and solve problems
- Strengthen the quality of projects and staff performance
- Serve as a means to open dialogue with other organizations involved in development

In this sense, it is understandable that "planning extension programs has become an increasingly accepted practice among national authorities" (Maalouf, in Rivera, 1987, p. 116).

Programme Development Assumptions

The world has a complex and uncertain nature. Change and the unexpected live with us and are important ingredients in the history of both humankind and the local society. Planning, then, is a risk-taking exercise, subject to the unexpected and to failure. Without it, however, life is even more uncertain and the task of reaching a balanced and sustainable development more difficult. So when talking about extension programme development, we assume that:

- Planned change may be an important factor for the social and economic progress of rural communities and families
- Extension services and agents must not act mechanically and without a vision
- Extension programmes can contribute significantly to learning, educational improvement, and development
- It is possible to select, organize, and manage programmes that contribute to change and development
- Extension educators, as change facilitators, can help individuals, families, and communities to reflect upon their realities and build relevant programmes, thus improving the quality of rural life

Comparing Major Approaches

An ideal or universal programme development model or approach does not exist. Different options are available, and choices depend upon a great number of variables. It is necessary, for example, to have a clear definition of who plans (the institutional scene), for whom (the potential beneficiaries), who takes the initiative, what the goals are, what the means are, what the time frame is, and what the sociopolitical environment is. Other variables may also be very important, such as the source of funds or the project's physical scope (Dusseldorp & Zijderveld, 1991, p. 4).

It is current in our days to contrast different programme development approaches by using the following dichotomies: centralized—decentralized, top-down—bot-tom-up, and blueprint—process. The two extremes represent, indeed, distinct ways of approaching programme development and correspond to somewhat opposite assumptions, theories, and practices. Dichotomies, however, are simple ways of representing an issue, and in between the two extremes may lie a continuum of possibilities which must not be disregarded. To facilitate the discussion, let us stick to this continuum for a while.

The first word in each of the pairs is centralized, top-down, and blueprint. In general terms, they correspond to the so-called conventional way of developing a programme. This is, in effect, what happens in many extension projects following the training and visit system or other conventional model and stressing the transfer of technology and information dissemination. For example, research stations develop technologies which are then transferred to the extension service through subject-matter specialists. Extension officers at the zone or district level plan the programmes, defining specific objectives and messages to be disseminated. At the village level, extension workers implement the activities according to fixed work schedules, under close supervision and leadership. Farmers' involvement is not, in general, a priority.

This type of approach is based on a number of key assumptions and principles: there are clearly defined and generally accepted objectives; there is a detailed and precise knowledge of the process to be implemented in order to reach the objectives; there is the political will to use the available power and resources; and there is a predetermined timetable and well-known resources (Dusseldorp & Zijderveld, 1991, p. 21).

It is generally accepted that the centralized, top-down, or blueprint approach facilitates management, monitoring, and evaluation tasks because activities and expected outcomes are defined and a chain of responsibilities and duties is well identified. In some cases, it can be the best choice, for example, in emergency interventions where a strong management style may be required to attain objectives in a timely and highly organized manner, or in situations where extension tasks are objective, that is, based on specific facts and knowledge rather than on feelings, beliefs, and values. In this last instance, as Boyle (1981, p. 100-101) stressed, "The programmer would be able to make objective decisions and probably have less need for clientele input."

Top-down or blueprint planning has been subject to strong criticism for various reasons. An important one is that it is too uniform, not taking into due account the sociocultural environment, the particular circumstances in which project implementation occurs, and the characteristics of the different clientele groups, for instance, planning the improvement of village irrigation schemes without taking into consideration the specific local needs, water rights, and rules of water allocation and distribution (Portela, 1990); or planning for the dissemination of a given technology package without an adequate understanding of the farming systems and the diversity of farmer's problems, potentials, rationales, and strategies.

Others also argue the following:

- This approach is agency centred and the programmes are planned from the inside to the outside; planners assess and define needs and problems and determine objectives and courses of action.
- Programmes are essentially based upon institutional policies and philosophy, not taking into account the diversity of perspectives about a given reality.
- The approach assumes a high degree of simplicity and order in the programme cycle, stressing the possibility of following a logical sequence of steps.
- The approach is rigid and assumes a high level of stability; problems will not change while the programme is being planned and new problems will not emerge.

Looking at the second word in each of the above pairs, we have a different set of key words: decentralized, bottom-up, and process. They correspond, in general terms, to what has been called participatory planning, currently proposed as a key element in farming systems development (FAO, 1994), farmer-first models (Chambers, Pacey, & Thrupp, 1989), participatory technology development (Farrington & Martin, 1993; Reijntjes, Haverkort, & Waters-Bayer, 1992), or local process facilitation activities (Röling, 1994).

The guiding principles are quite different from those of the top-down perspective (Bergdall, 1993; Dusseldorp & Zijderveld, 1991; Korten, 1991):

- Development is regarded as a long-term effort and process requiring continued commitment and collective responsibility.
- Programme personnel should act as partners and facilitators rather than experts.
- Participation of local actors is stressed.
- More time should be spent on needs identification and project preparation, with the active involvement of the intended beneficiaries.
- The programmes should grow step by step, securing close linkages to the felt needs and the local environment.
- The ultimate goal of the programme is to increase the power of the local actors to plan and implement their own improvements.

In general, this approach is said to be open and process centred, embracing error as a learning factor and leading to programmes and projects with an emergent nature.

The use of such principles and ideas is growing in small, local development projects with an integrated nature where citizen participation is highly valued and desired. For instance, in Sao Tome and Principe, a country in West Africa, such projects are currently being promoted by the Ministry of Agriculture and Rural Development as part of a major land reform programme. These projects are animated by interdisciplinary and interinstitutional teams, having as partners local farmers' associations and youth groups. The projects emerge out of a participatory rapid appraisal exercise, and activities draw funds from multiple sources (Cristóvão, Botelho, & Bom Jesus, 1994).

It is clear that this approach is not exempt from problems and criticisms. Some are quite evident; for instance, activities start without predefined objectives, making things more difficult for personnel and perhaps confusing for participants who often used to be recipients of interventions, not active partners; and the overall philosophy and practices contrast with the conventional ones with which most funders and official services are familiar, which may complicate relationships. The question of decentralization is critical in this last respect. In effect, even considering the efforts developed in the last decades in this direction, it is known that the capacities at the local level are still generally weak and "finding ways of building the capacity of local administrative units to implement development programs and of eliciting the support of central bureaucracies in that task offers an important challenge to governments of developing countries in the years to come" (Rondinelly, 1987, p. 54). Also important is the issue of participation. It can be argued that participation is not desirable in every situation and that it has advantages and disadvantages. For instance, it requires time, which in some instances is not available, and it may lead to social and political conflicts because it touches the question of power, and those who may lose it are likely to oppose and resist. Undoubtedly, each project situation requires a careful analysis regarding the purposes which might be accomplished through citizen participation, the ways to achieve it, and the costs and benefits deriving from it.

In most cases the choice is not at one extreme of the continuum, but somewhere in between. At the same time, a combination of different approaches may even be necessary and advantageous. Boyle (1981, p. 7), for example, distinguished three types of programmes: developmental, informational, and institutional. The first one is clearly process centred, and objectives derive primarily out of the needs and problems of the participants. The second is predominantly top-down oriented, and participants are primarly recipients of information. The third type corresponds basically to training

Box 1. Two Frameworks for Programme Development.

Developmental Framework

Although presenting a sequence of steps, Boyle (1981, p. 51-57) stresses the dynamic nature of the programming process, underlining that distinct steps and activities can happen simultaneously or in different sequences:

- Identification of the basis for programming: philosophies, policies, and procedures
- Situation analysis of community and clientele
- Identification of desired outcomes
- Identification of resources and support
- Design of instructional plan
- Programme of action: calendar of events and activities
- Accountability of resources, evaluation
- Communication of the value of the programme

Informational Framework

- Determine what content is available, needed, or desired.
- Provide information or knowledge.
- Determine the extent of the distribution of content.

and instruction and is developed from a field of knowledge and from the educator. However, Boyle stresses that in the course of extension work these types usually overlap; that is, a locally initiated participatory project can very well benefit from centralized efforts to disseminate information and from training activities (see Box 1).

Creating Extension Programmes with the People: A Rationale

Planning is currently recognized as a process and a social practice. As a process, it is seen as a dynamic effort evolving around problems and implying decisions and actions to achieve goals. As a social practice, it implies a negotiation of interests and the construction of some type of working platform involving different actors, such as researchers, subject-matter specialists, rural leaders, representatives of farmers' associations and groups, and so forth. "Negotiation" emerges as a key word in current views (Cervero & Wilson, 1994; Sadowske, 1991). That means that more and more programme development is seen as a democratic exercise, implying a dialogue around issues like the following:

- How can a given situation be described and analysed?
- What are the major problems, needs, and expectations?
- What are the alternatives to solve the problems? What kinds of resources, information, and technologies are required?
- What projects and activities should be implemented? When? How? Where? By whom?
- How should evaluation be seen? Who should do it and when?
- How and by whom will the programme be managed and controlled?

Democratic planning of extension work implies the open discussion of these questions, not necessarily in any exact order, but usually in some kind of iterative way, going back and forth in successive approximations.

Participation of men and women farmers, rural leaders and networks (both formal and informal), community groups, and other development institutions becomes a critical issue requiring creative approaches and continual efforts. In fact, as experience shows, the majority of farming and rural populations, often representing the largest share of production, although in words considered central to development, are not actively and systematically brought into the process. It is known that extension work overlooks women and is mostly dedicated to a minority of progressive and large farmers who tend to be relatively well-off. These farmers have more power to influence the system, serve many times as privileged linking elements between the village and government agencies, are actively involved in organizations, and are capable of attracting a good share of project benefits. On the other side, it is also known that often small and medium farmers lack the time, money, and motivation to participate. They feel powerless, and very often local history is not encouraging and supports the view that participation is politically risky.

A Broad View of People's Involvement

Creating programmes with the people implies a broad view of people's involvement. Participation should not be seen, as many times in the past, as the occasional presence of rural dwellers in an information meeting, the simple use of public services, the voluntary contribution (with labour, money, etc.) to a project, or as some kind of activity toincrease support to preplanned top-down projects.

Participation becomes, then, people's involvement in reflection and action, a process of empowerment and active involvement in decision making throughout a programme, and access and control over resources and institutions (Cristóvão, 1990).

A Different View about the People

A new view about the people and their roles is also essential. In fact, the view that planners, administrators, and others often hold about rural people, small and poor farmers in particular, is very influential. Still relatively-widespread is the idea that local society is practically undifferentiated and that farmers are passive and ignorant and lack the ability to understand situations, analyse them and proposed solutions, translate them into action, and evaluate the results. The direct consequence is the paternalistic and superior attitude of technicians who think they know best and have the right answers to development problems.

To summarize, democratic planning of extension work implies:

- A broad understanding of participation, implying the involvement of different groups and organizations in various programme development stages and activities at the decision-making level.
- A different view of rural people and an attitude of respect and humility regarding their knowledge and experience.
- A close attention to the way power is shared and distributed among different social groups, local and regional networks, and segments of the population, and between these and the national and regional power holders (political leaders and parties, for instance).
- An understanding of the way decision-making power is distributed in the official structures, or among others dealing with people and development, at different levels.
- A recognition of the fact that in any situation, even in small rural communities, the existing needs, interests, and aspirations are not homogeneous.
- A clear perspective about the advantages and problems associated with participation.

This rationale is not only politically and ethically sound, because people should be respected and participation is today seen as a basic need and a human right, but it is also economically justified, because experience shows that project success and long-term sustainable results require people-centred approaches.

Making Participation Happen

Participatory planning is not a linear, sequential activity. However, it implies particular attention to a number of key steps: describing and analysing the situation; identifying needs, problems, and aspirations; assessing opportunities and obstacles; generating programme and project ideas; establishing implementation plans; monitoring action; and evaluating process and results. Globally, this is a dynamic and continual process evolving in a succession of actions and cycles. In each step, a number of actors intervene, and planners act as facilitators, enabling active participation and ensuring the quality of the outcomes.

The adoption of a strategy to promote participation is a critical aspect. In this regard, experience shows that five major, interrelated elements are essential (Bryant & White, 1982; Cernea, 1992):

- Keep extension projects simple and manageable.
- Differentiate various groups in the project area.
- Work with different types of organizations.
- Take advantage of all possible methods and techniques.
- Improve people's capacities to participate.

Keep Extension Projects Simple and Manageable. It is known that small projects, developed at the local level, implying relatively simple skills, providing direct benefits to participants, and building on self-help arrangements, have better chances of mobilizing people's attention and active involvement (Bryant & White, 1982, p. 218). When projects are larger and more complex, for instance at the regional or provincial level, the task can be more difficult, but ways can be found to generate participatory dynamics, including decentralization mechanisms that allow a greater input to local extension units, and working with rural organizations and groups at different levels.

Differentiate Various Groups in the Project Area. How can we assure that the needs of particular groups such as landless farmers or poor women are reflected in the extension programming process? The answer to this question implies a clear identification of groups in the project area. In effect, "farmers" or "the rural population" are often taken as the "target group" without any further specification and assume a homogeneous group, which does not exist in reality. So identification of specific groups and subgroups to reach and work with becomes a critical element in a participatory planning process.

Identification can be accomplished in different ways. Albrecht et al. (1989, p. 48-49) suggest two common approaches:

- Classification of holdings according to size, socioeconomic features (e.g., full-time employment, extra income, a second job), and production structure (land-use system, management system, etc.). This approach usually requires extensive surveys, which are costly and time consuming.
- Building on local common distinctions: it is simpler and more practical to classify groups according to distinctions that already exist in a particular society. They are less artificial and a more faithful reflection of the problem situation from the angle of the population. The use of key-informant interviews or group discussions are usual procedures in this case.

A third approach, complementary to the previous ones, can be the identification of existing community organizaions, looking at membership and leadership, the roles they play, and assessing the kinds of interests they defend and the extent to which they serve particular segments of the local society (Verhagen, 1987, p. 1079). This leads us directly to the third element.

Work with Different Types of Organizations. It is widely recognized that some form of organization is essential to facilitate participation. For instance, Oakley and Marsden (1984, p. 69) summarized the methodological approaches to participation in various cases in Brazil, Senegal, Bangladesh, and India, and in all of them the formation and role of groups emerge as a crucial aspect. In this regard, four specific recommendations seem important:

- Working with a wide variety of organizations increases the chances of reaching a wider spectrum of people. The institutional map needs to be done, because in most situations different sorts of organizations exist: cooperatives, commodity groups, irrigation and credit associations, youth or women's clubs, village committees, leisure and cultural groups, development associations, and others with a variety of labels and structures representing diverse interests and objectives and working at the local, regional or national levels.
- It is important to consider and work with existing organizations. Indigenous ones in particular, as stressed by Bryant and White (1982, p. 129), are an important resource because of their cohesiveness and reliance on mutual trust.
- Creating new organizations is important in many circumstances. In this case, the process should be educational. The concerned people have to feel the need to be organized and have to devise their own objectives and forms of action. A process of reflection, discussion, and decision making is usually necessary (Korten, 1980). Extension workers and change agencies should be ready to encourage and help, providing counselling and training and facilitating contacts and access to similar experiences.
- In any situation, it is important to analyse the interests represented by existing organizations and their degree of power and influence. In many cases, they are led by better-off farmers or representatives of elite groups and are not at all a reliable support to a participatory strategy aiming at the interests of different groups and attempting to reach the disadvantaged.

Still, in relation to this element of the overall strategy, it must be stressed that the work with advisory committees or councils of different sorts has proven to be advantageous (Albrecht et al., 1990; Boyle, 1981; Hemp, Kaczor, & Zwilling, 1983). Such organizations involve a small group of selected individuals, usually between twelve and fifteen, and help extension in various ways regarding the development efforts within a given area, being a specific means to assure formal participation in programme development.

Take Advantage of All Possible Methods and Techniques. In any situation, with favourable or unfavourable institutional scenes, something can always be done so that extension programmes are responsive to local needs and concerns. Some alternatives are the following:

- Conducting community surveys, contacting specific groups, and administering questionnaires at extension meetings, training courses, or other types of public events.
- Collecting information from key informants and other knowledgeable individuals, as well as from formal and informal group discussions and meetings at the village or other levels.
- Deliberately contacting hard-to-reach groups and others with less power and visibility that are nor mally not involved in extension work.
- Obtaining public input through interactive radio or television programmes.
- Communicating with other services in such fields as education and health and exchanging information, experiences, and views.
- Keeping a permanent record of needs, concerns, views, values, and customs.

Improve People's Capacities to Participate. This implies, among other things, education and training. On the extension service side, agents and other officers frequently lack the skills to work with rural people in an interactive manner. Their training is often narrow and stresses the acquisition of technical knowledge and abilities. In-service training may be required so that extension people learn how to act as helpers and facilitators of participation in the different programme phases and moments.

Concerning rural people, and farmers in particular, it is important to act in such areas as education and leadership development. The acquisition of literacy skills, communication abilities, and organizational and leadership capacities is vital for establishing strong local institutions and for promoting conscientious and self-sustained problem-solving efforts. As Albrecht et al. (1989, p. 182) underlined, training can increase the efficiency of local organizations, make them assume more responsibilities, and acquire political weight, ensuring that their members' interests are taken into account at higher levels.

These and other ideas can form the basis of an action plan to create extension programmes with the people, including the following dimensions:

- Change or adjust extension's organizational structure and management in order to favour an internal climate of dialogue, cooperation, and creativity conducive to participatory planning.
- Maintain open communication channels within the extension service, with other related services, and with all involved institutions in order to facilitate a rapid exchange of ideas and joint decision making.
- Plan for the development of attitudes, knowledge, and capacities essential to promote participation, and provide training to extension people and rural citizens.
- Concentrate attention and efforts on small local extension programmes.
- Differentiate groups at the programme area, and understand the diversity of needs, concerns, ideas, potentials, and obstacles.
- Consider and work with different types of organizations, formal and informal, to ensure programme quality and responsiveness.
- Establish interlocking advisory committees at the necessary levels and programme areas, and work with them actively throughout the project cycle.
- Use all possible mechanisms to obtain citizen input.

Defining Objectives: Tools to Build Consensus on Major Directions

The definition of objectives is a key aspect in extension programme development, and objectives are seen as project guides: they provide direction, a basis for selecting activities, and a framework for evaluation and making decisions explicit. In conventional planning, it is said that objectives must be defined in a clear and explicit manner, sequentially, and preferably in quantitative and measurable terms.

In a process approach to programme construction that is open and dynamic, and in our times of rapid change, objectives should be open to negotiation, flexible, changeable, and adjustable. They can start simple and be less ambitious, but should allow for expansion as the capacities of the partners grow, as experience is acquired, and as confidence is gained. Continual programme evaluation might be quite influential in this regard.

Some tools may be used to facilitate the definition of objectives. Objective trees are a possibility. In simple terms, an objective tree consists of programme or project objectives defined in quantitative or qualitative terms, linked in some kind of hierarchical form in a tree graph. Objectives at a lower level are supposed to contribute to the attainment of an objective at a higher level, as shown in the example in Box 2. Several levels of objectives can be considered.

Objective trees, in summary, help to clarify the expected project outcomes and to make explicit the relationships between them. As such, they can be a useful tool in programme planning and evaluation.

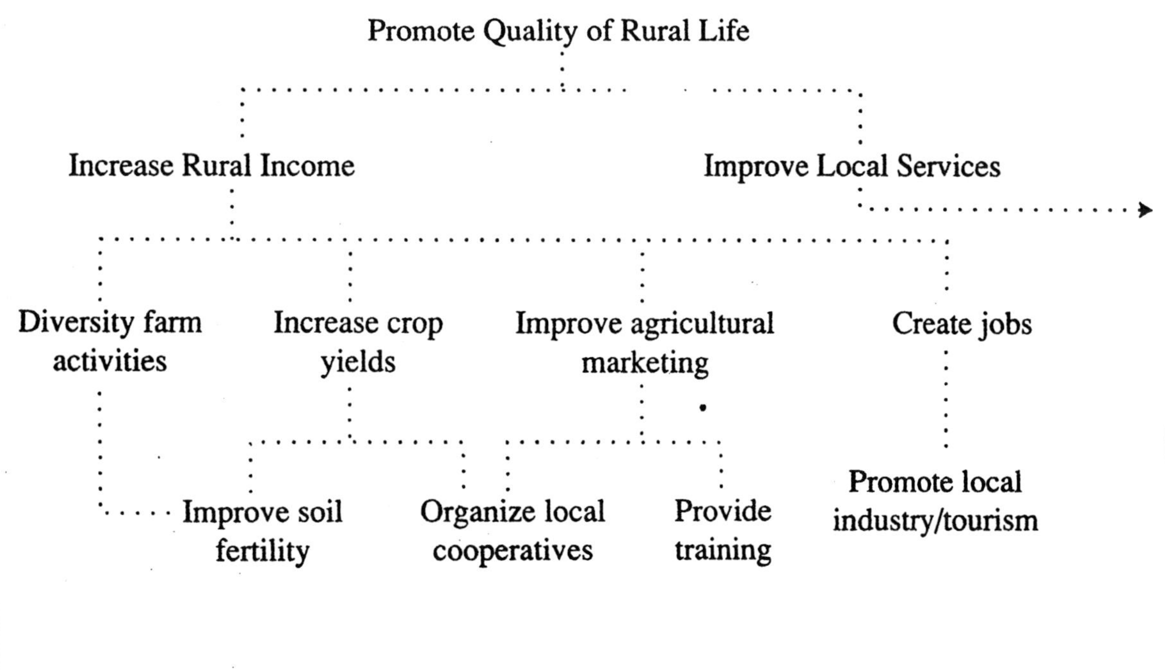

Box 2. Example of a partial objective tree for a rural development project. (Inspired by Delp, Thesen, Motiwalla, Seshadri, 1977).

Facing Broader Planners' Roles and New Skills

The views just presented imply new roles for extension planners, as well as for programme managers, implementers, and evaluators. These roles demand knowledge and skills in different areas. If the conventional programme development models stressed technical preparation, particularly in such aspects as designing surveys, analysing and reporting data, or preparing budgets, the emerging models require the same knowledge and skills and demand additional preparation, especially in the political and ethical fields:

- On the ethical side, extension people must be able to deal with values, which includes being aware of their own values, and have the capacity to assess the others' interests, values, and commitments regarding the programme being planned and development efforts at large. Critical in this sense are value considerations regarding who to serve, who to involve, and who to exclude; what kinds of change to stimulate; or what types of programme effects to avoid.
- On the political side, extension people must have the skills to analyse the institutional context and the respective power structures, formal and informal; improve their negotiation capacities, which includes being capable of communicating (especially listening) and working with different people and institutions, thereby developing mutual trust; and build capacities to evaluate permanently.

Moreover, a creative programme development practice requires permanent attention to the integration of planning, implementation, and evaluation stages and activities, demanding an attitude of cooperation from the concerned actors, and often the assumption of interchangeable roles and functions. Otherwise, blueprint or centralized approaches, as defined in the first section of this chapter, will determine extension practices.

A major challenge lies ahead: recognizing that extension work requires a definite move from "planning for" to "planning and creating with" and finding the ways that may help reach this change in each context.

References

Albrecht, H., Bergman, H., Diederich, G., Grober, E.,Hofman, V., Keller, P., Payr, G., & Sulzer, R. (1989). *Agricultural extension: Basic concepts and methods* (Vol. 1). Eschborn: BMZ/GTZ/CTA.

Albrecht, H., Bergman, H., Diederich, G., Grober, E., Hofman, V., Keller, P., Payr, G., & Sulzer, R. (1990). *Agricultural extension: Examples and background material* (Vol.2). Eschborn: BMZ/GTZ/CTA.

Bergdall, T. D. (1993). *Methods for active participation: Experiences in rural development from East and Central Africa*. Nairobi: Oxford University Press.

Boyle, P. G. (1981). *Planning better programs*. New York: McGraw-Hill.

Bryant, C., & White, L. G. (1982). *Managing development in the third world*. Boulder, CO: Westview Press.

Cernea, M. (1992). The building blocks of participation: Testing a social methodology. In B. Bhatnagar & A. C. Williams (Eds.), *Participatory planning and the World Bank: Potential directions for change* (p. 96-108). Washington, DC: World Bank.

Cervero, R. M., & Wilson, A. L. (1994). *Planning responsibly for adult education*. San Francisco: Jossey-Bass.

Chambers, R., Pacey, A., & Thrupp, L. A. (Eds.). (1989). *Farmer First: Farmer innovation and agricultural research*. London: Intermediate Technology Publications.

Compton, J. L. (1984). Extension programme development. In B. E. Swanson (Ed.), *Agricultural extension: A reference manual* (p. 108-119). Rome: FAO.

Cristóvão, A. (1990). Research in extension: Popular participation as a major issue. In M. Andersson & U. Nitsch (Eds.), *Proceedings of the 9th European seminar on extension education* (p. 204-229). Uppsala: Swedish University of Agricultural Sciences, Department of Extension Science, Rapport 2.

Cristóvão, A., Botelho, C., & Bom Jesus, A. (1994). *Diagnóstico rápido participado*. Plancas I. S. Tomé e Príncipe: CENFOPA, Centro de Formação Profissional Agrária.

Delp, P., Thesen, A., Motiwalla, J., & Seshadri, N. (1977). *System tools for project planning*. Bloomington, IN: PASITAM.

Dusseldorp, D. van, & Zijderveld, K. (1991). *Preparation and implementation of development projects in the third world*. Wageningen: Wageningen Agricultural University.

Farrington, J., & Martin, A. (1993). *Farmer participation in agricultural research: A review of concepts and practices*. London: ODI.

Forest, L. B., & Baker, H. R. (1994). In D. J. Blackburn, (Ed.), *Extension handbook: Processes and practices* (p. 86-99). Toronto: Canada: Thompson Educational Publishing.

Hemp, P., Kaczor, C., & Zwilling, A. (1983). *Course of study: Citizen's advisory councils in education*. Champaign, IL: University of Illinois, Department of Vocational and Technical Education.

Knowles, M. S. (1977). *The modern practice of adult education: Andragogy versus pedagogy*. New York: Association Press.

Koehnen, T. L., Portela, J., & Cristóvão, A. (1993). Bilateral curriculum development for assistance in food security in Lusophone Africa. In A. Bonano (Ed.), *The agricultural and food sector in the new global era* (p. 195-212). New Delhi: Concept Publishing.

Korten, D. C. (1980). Community organization and rural development: A learning process approach. *Public Administration Review, 40*(Sept.-Oct.), 480-512.

Korten, D. C. (1981). Rural development programming: The learning process approach. *Rural Development Participation Review, 2*(2), 1-8.

Maalouf, W. D. (1987). In W. M. Rivera (Ed.), *Planning adult learning: Issues, practices and directions* (p. 106-126). London: Croom Helm.

Oakley, P., & Marsden, M. (1984). *Approaches to participation in rural development*. Geneva: International Labor Office.

Portela, J. F. (1983). *Avaliação de projectos de desenvolvimento: Alguns problemas e conceitos*. Vila Real: UTAD.

Portela, J. F. (1990). Farmer-managed irrigation systems in the Trás-os-Montes Region of Portugal. In R. Yoder and J. Thurston (Eds.), *Design issues in farmer-managed irrigation systems: Proceedings of an international workshop of the farmer-managed irrigation systems network* (p. 295-299). Colombo, Sri Lanka: IRMI.

Reijntjes, C., Haverkort, B., & Waters-Bayer, A. (1992). *Farming for the future: An introduction to low input and sustainable agriculture*. London: Macmillan.

Röling, N. (1994). Agricultural knowledge and information systems. In D. Blackburn (Ed.), *Extension handbook: Processes and practices* (p. 56-67). Toronto: Thompson Educational Publishing.

Rondinelli, D. A. (1987). Administrative decentralization of agricultural and rural development programs in Asia: A comparative analysis. In W. Rivera and S. Schram (Eds.), *Agricultural extension worldwide* (p. 22-57). London: Croom Helm.

Sadowske, P. S. (1991). *Towards futuristic program planning in adult education*. Unpublished doctoral dissertation, University of Wisconsin-Madison.

Verhagen, V. (1987). *Self-help promotion: A challenge to the NGO community*. Amsterdam: CEBEMO/Royal Tropical Institute.

Chapter 8
Selecting Appropriate Content and Methods in Programme Delivery

Dunstan A. Campbell and St. Clair Barker

The issue of developing appropriate content is critical to the extension process; the performance of an extension system depends, in large part, on the appropriateness of its message. The more appropriate the message, the better will be the extension-clientele relationship and the more likely extension's programme will be supported.

Developing Appropriate Content

Appropriateness of content is situation driven: what is appropriate for one farm family may not be appropriate for another, even though both families operate within the same agroecological zone; or what is appropriate for one country may not be appropriate for another. Even at the national level, it has been shown that agricultural development can be facilitated by the development of appropriate technologies. Hayami and Ruttan (1985) pointed out that in Japan, where land is scarce, technology development was oriented towards biological technologies, for example to improve varieties. In the United States, however, where land is abundant, technology development was mainly in the form of mechanical innovations such as tractors. In the case of developing countries, their commercial relationship with the developed world predisposes them to accept or develop technologies that are not always appropriate for their farmers. For example, the research organization for banana production in the Windward Islands of the Eastern Caribbean because of market pressures and trade agreements have consistently produced technologies which require additional labour, although labour cost and availability are the biggest constraint to banana production in that subregion.

Given the arguments above, appropriateness should be defined within the scope of what is

1. Technically feasible
2. Economically feasible
3. Socially acceptable
4. Environmentally safe and sustainable

Technical Feasibility

Technical feasibility should be examined from two perspectives, the first being the ability of farmers to produce a commodity within their environment, and the second being what the farm family can achieve, not what can be achieved on research stations. In the first case, certain commodities do not fit into the production system of small farmers, while others do, for example scaleneutral technologies for crops like maize, rice, and bananas. In the second case, a distinction must be made between technological potential and technical feasibility. The technological potential is the tested output of a given technology and is dependent on controlled agronomic factors. Technical feasibility is what can be achieved within the holistic environment of a farm family; this is not limited to technical factors only, but also to social, economic, and political factors that affect the farm family. Small farmers, because they operate under less favourable conditions than those under which the technologies are developed, usually produce under the technological potential. Given this situation, what becomes important to farmers is whether or not that commodity can fit into their farming system. If the technology can fit, then it is technologically feasible.

Economic Feasibility

The ability of farmers to incorporate a technology or technological package into a farming system has economic implications. The farmers' resource base, both human and financial, must be considered. Do farmers have the financial resources to purchase the inputs to derive the benefits from the technology? Will this technology require the hiring of additional labour, and if so, is it available and affordable? Some Green Revolution technology failed to be adopted because some anticipated users of the technology could not purchase the fertilizer and other inputs which formed part of the package.

Economic feasibility can also be examined from the point of view of function within the farming system.

Dunstan A. Campbell and **St. Clair Barker** are outreach lecturers of the Faculty of Agriculture, University of the West Indies, St. Augustine, Trinidad.

Some small banana farmers in the Windward Islands of the Eastern Caribbean continue to grow bananas, although they have been advised against it. The rationale for this advice is that, among other things, soil erosion on the very steep slopes where most small farmers cultivate bananas and the resulting low yields make the cultivation of bananas economically unprofitable to them. However, they contend that banana cultivation allows them to have access to inputs which they can use to increase the profitability of other crops within their farming system, thus making the total farming system economically profitable. As such, banana cultivation within the system is seen by these marginal small farmers as a "facilitator" crop and not primarily for economic profitability. To these farmers, banana production is economically feasible within their farming system.

Social Acceptability

Innovations can be technically sound, but may conflict with the social norms of the end users or even cause societal disequilibrium. For instance, in Zaire, efforts to increase cassava production were constrained by the inability of the women to handle the increased yield. At the time the new varieties were introduced, the women in the targeted communities were already fully occupied. The task of processing cassava was their responsibility, but with the increase in yield, the women's workload increased correspondingly, thus making it physically impossible for them to handle. As a result, the farm families reverted to cultivating the local variety, which was lower yielding.

With regards to social disequilibrium, in some cases the adoption of technologies by small groups of individuals has resulted in a shift in power relationship within the society. For example, Campbell (1982) found that in the upland plateau of Ardèche, France, individuals from among the group of farmers who adopted tractors and mechanical harvesters were those who challenged the existing family-linked mayoral structure.

Within the social context, there is also what can be termed situational appropriateness, that is, what is most likely to be accepted or rejected by a group at any given time. For instance, it would be unwise to introduce programmes with a crop or livestock component which had recently experienced market failures. Farm families remember such failures and tend to be hostile to such programmes.

Environmental Safety

An extension programme must be cognizant of the effects it will have on the environment. The commercialization of agriculture results in farmers using increasing quantities of fertilizers and agrochemicals. These chemicals for the most part have negative effects on the environment because they contaminate groundwater and may disrupt food chains. For example, birds which feed on insects exposed to chemicals may become contaminated and may die.

Extension's Clientele

Extension's clientele has to be defined within the context of an agricultural policy framework. This framework should provide the boundaries for selecting, from among the broader user categories, the specific groups which are to be targeted. Given today's scope of agricultural development, that framework should not speak of generic small farmers, but should specifically state the groups—women, youth, landless farmers, or other categories that are specific to the situation under consideration. Swanson, Röling, and Jiggins (1984) have cited four major factors which seek to provide a framework for the development of appropriate technology among targeted groups of extension clientele. These factors are agroecological zones, access to resources, gender, and age. Ethnic groups can be added to this list.

Agroecological Zones

The first step in identifying appropriate technology for extension's clientele is to map the area into agroecological zones. Farming systems research (FSR) provides useful methodologies for this exercise (see Hilderbrand, 1986; Shaner, Philipp, & Schmehl, 1982). Mapping allows for the identification of agronomic variables such as soil type, rainfall, slope, and altitude, which will influence the development of location-specific technologies.

Access to Resources

Within each zone, there will be a wide variety of farmers because of socioeconomic factors. These factors explain the differences which exist in terms of access to the factors which facilitate production—land, labour, capital, markets, inputs, tenure, and information. These factors predispose farmers to adopt certain types of innovations. Swanson et al. (1984) provide a useful summary of these facilitating and impeding factors which should be considered in developing appropriate technology for extension's clientele:

1. *Land.* Size of holding: small, medium, large; type of tenure: owner operated, family land, renter-share-cropper
2. *Water.* Irrigated, nonirrigated
3. *Labour.* Family, hired (cost and availability), communal, or customary
4. *Inputs.* Availability of improved seeds, agricultural chemicals, fertilizers

5. *Markets.* Location, availability of storage and transport
6. *Capital.* Sources and cost of credit, type of collateral needed, and ease of obtaining credit
7. *Information.* Availability of extension service (worker to farmer ratio), appropriateness of technology
8. *Influence.* Ability to affect technology development, transfer to be appropriate to user needs such as user control, claim-making capacity

Gender

The role of women in agricultural production is now a topical issue. Several studies have documented women's contribution to both the economic and noneconomic sectors. However, there is need to move from this realization to its practical application in terms of the development and inclusion of appropriate technologies for women in extension programmes. In so doing, it must be taken into consideration that their access to resources and effective technologies is often constrained by gender barriers or blindness (Feldstein & Poats, 1989).

Recognition must also be given to the particular functions women perform within the farming system and their work schedule in the farm-household environment. Generally, women play a much greater role in the production of food crops than of export-oriented crops, and within cropping systems they perform certain tasks, for instance, weeding, fertilizing, and harvesting. These factors must be considered in developing appropriate technologies to address this group of players in the agricultural environment. These and other genderrelated issues are explored in more detail in chapter 9.

Age

In most developing countries, agriculture is an important sector, if not the most important sector, and a significant proportion of the agricultural activities takes place in rural areas where 20 to 80 per cent of the country's population may live. This resource is usually characterized by a low level of education, few skills, limited employment opportunities, and a strong desire of young people to leave the area.

Extension should consider age as an important characteristic for targeting not only from the point of view of youth but of other age categories. Agriculture usually has a low status because of the low level of technology it employs and low income-earning capacity. Extension programmes should aim at increasing both the level of technology used and income-earning capacity. If this can be done, then there is a good chance that the young people will remain. Other ways of achieving the above should also be explored. For instance, educational programmes which provide the skills to be self-employed may be a solution to the outmigration problem. This in turn might encourage the heads of households to invest more in agriculture, which will then tend to encourage the youth to remain. Campbell (1982), working among rural villagers in France, found that the adoption of new technologies by heads of household was a means of motivating the youth to remain in the villages. On farms where succession was guaranteed, there was almost always a higher level of technology adoption.

Ethnic Groups

In multiracial societies, ethnic groups should be separately targeted because of their different sociocultural characteristics. Language, food preferences, and religion are some of these characteristics. In cases where ethnic groups are found within the same extension district or zone, extension's programme should reflect these differences.

Incorporating Needs into Programmes

Rapid rural appraisal (RRA; see chapter 6) can be used by research and extension to provide agroecological, biological, and socioeconomic data for identifying problems and opportunities for extension's clientele. For effective utilization, these needs should then be categorized on the basis of type of needs; that is, whether they are material needs, knowledge-based needs, or infrastructural support needs (see chapter 5). Then the importance of the needs should be ascertained to prioritize them before incorporating them into extension programmes (see chapter 7).

Transferring appropriate technologies is not enough to ensure adoption. Incorporating them into the targeted farming systems may be constrained by other variables which needs assessment ought to identify. For instance, the use of nematicide can increase the yield of a commodity by destroying the harmful nematodes, thus increasing nutrient uptake and yield. However, because nematodes are microscopic organisms, their presence in the soil may not be so obvious to the farmers who have not had the education to understand microorganisms. Consequently, an education component will need to be added to the technology to increase its chances of being adopted.

There are also cases where appropriate material technologies are not accepted because inputs are unavailable. This underscores the importance of addressing the users' wide-ranging needs—knowledge, attitude, skills, economic needs, and institutional support—in order to achieve the adoption of material technology.

Program Delivery and Implementation

As noted in chapter 2, a number of different approaches can be used to organize and deliver extension programmes. The approach selected by a national extension organization should reflect its prescribed mission as articulated by policy makers, but it may also be influenced by donor agencies. In the final analysis, the type of approach being used will depend primarily on policy considerations, including the clientele to be served, the mission of extension, and financial considerations. In addition, the type of approach being followed will directly affect how programmes are delivered at the field level. Therefore, the reader is reminded that the process of implementing extension programmes is carried out in the context of and affected by the particular extension approach pursued by the national extension organization.

With regard to appropriateness of content used in these different approaches, Albrecht et al. (1989) provide a useful typology. They contend that all extension approaches can be classified as either production technology approaches or problem-solving approaches. Production technology approaches tend to emphasize the production targets more than the clientele; technologies used in these approaches are more concerned with addressing production issues than clientele-related problems. Timmer (1982) said that these approaches are geared to produce technoeconomic information and favour the larger, more commercial farmers. The small farmers because of their disadvantaged situation do not receive the full benefits from these approaches. The commodity-focussed approach and the technology-centred approach are examples of production technology approaches.

In problem-solving approaches, the clientele participate in defining their problem. Although these approaches use technoeconomic information, the socioeconomic considerations of the clientele stand out as important issues. For small farmers, this allows for the development of more appropriate content. The training and visit, community development-cum-extension, and animation rurale approaches fall into this category.

As stated earlier, extension's clientele is a heterogeneous group of people differentiated by resources, gender, age, and ethnicity. Communication theories indicate that the trickle-down effect of agricultural innovations rarely takes place among heterogeneous groups, and when it occurs, it does so at an extremely slow pace. Given this under-standing, it is therefore imperative to identify homogeneous categories within those groups of clientele in order to facilitate the effective transfer of technology.

The strategy is therefore to develop homogeneous groups of clientele to target extension programmes. The targeting of these groups should be done within the broader groups already defined, namely, agroecological zone, access to resources, women, youth, ethnicity, access to information, and age. Extension programmes should then be developed for each of the targeted categories (Swanson et al., 1984). For instance, special programmes should be developed for women which recognize their importance in the agricultural sector, take account of their informal communication network, and analyse their farm-home work schedule. In the case of programmes directed at ethnic groups, efforts should be targeted to the sociocultural factors which distinguish the group. Language, food preferences, and religious beliefs are among the important points that should be taken into consideration. In addition, extension officers in developing extension programmes should be guided by principles of learning and a knowledge of the diffusion process.

The Learning Process

Extensionists must be reminded that above all they are educators, and in pursuit of that function, they need to grasp the basic principles of learning and to understand the effectiveness of some teaching methods.

Learning is facilitated by the use of the senses. The more senses that are used, the more accelerated will be the learning. Use of this principle will assist extensionists in selecting methods which will provide an educational experience for their clientele. This will be achieved through setting up an environment and structuring the situation so as to stimulate the desired type of reaction.

Learning is also conditioned by the motivation and the abilities of the learner; thus it is important to create an environment conducive to learning, one in which the learners become more responsible for the outcome of their experience. This can be achieved through the selection of teaching methods and supporting material. The less abstract these methods, the more the learner will participate, and the more effective will be the educational experience. For example, if farmers are not performing a task effectively, then it will be much more effective to organize a method demonstration than a lecture or group meeting.

The Adoption Process

The other important consideration in selecting methods for the delivery of appropriate content is an understanding of the adoption process. Lionberger (1968) contended that the adoption process consisted of five distinct stages: awareness, interest, evaluation, trial, and adoption. According to Lionberger, the individual in the process goes through each stage within a definable time period. Later works by Lionberger and others have shown that these stages are not as distinct as he first proposed and that some of the stages may become condensed within the individual cognitive processes, thus making them unrecognizable as a behaviour which can be measured over time.

Although, in some instances, the steps in the adoption process may not be recognizable, yet the model provides a useful guideline for selecting extension methods in programme delivery. For instance, in the awareness stage, knowledge of the innovation is critical to the individual. Mass media and popular theatre are the preferred methods because they can reach many people at the same time. In using the mass media, extensionists must pay attention to the characteristics of the audience targeted. For instance, in multiracial societies, an ethnic group with a special language may require programming in that language. The other method, popular theatre, although reaching only smaller audiences, is a very effective means of building awareness because it uses the popular language and rhythms of the people in presenting content to its audiences.

In the interest stage of the adoption process, knowledge continues to be important, but building a positive attitude towards the innovation becomes the critical issue. For this reason, the desired methods should include information strengthening and attitude building as their goal. These methods should use the senses of hearing and sight, either individually or collectively. Group meetings, group discussions, and radio forums are recommended for strengthening knowledge, while field days and farm visits will allow individuals to see what they have been hearing, thus providing the opportunity for building the desired attitude towards the innovation.

Evaluation is the most critical stage in the adoption process, because the outcome usually determines whether or not individuals proceed to the trial and adoption stages. At this stage, people need to match knowledge against facts. Farmers need to be assured that what they heard and saw are indeed workable. Result demonstration, farmer exchange, and field days are recommended because they allow individuals to reinforce their interest by viewing tangible evidence. Within this group of methods, farmer exchange is an important method. Farmers selected for the exchange should be further advanced in the adoption process and within the same reference group as the visiting farmers. These types of experiences allow for the removal of doubts. Some skill training may be necessary at this stage to facilitate the individual's progression to the trial stage.

At the trial stage, the farmer's technical and management skill should be the main area to be targeted. The individual visit becomes the most preferred method at this stage, and the needs of individual farmers must be taken into consideration. This means that the extension officer will have to develop a plan for each individual farmer or group of farmers in similar situations. The extensionist has to remember that, although similar farmers are adopting similar techniques, the problems experienced are not always the same. At this stage, methods for reinforcing the farmers' interest by the use of farmer exchange and skill training can be useful in helping individuals to continue adopting.

Once the farmers start adopting, extension should continue to support their efforts. Recognition programmes and farmers' competition can be used to encourage farmers to continue adopting. The goals and criteria for these methods should be carefully developed so as not to bring out any negative effects because of poor planning and implementation.

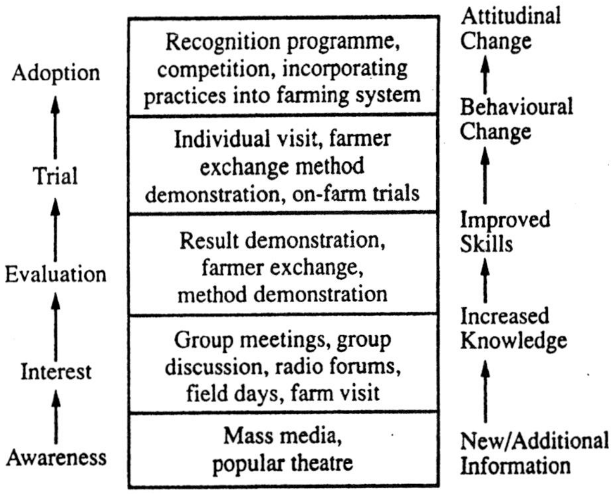

Figure 1. Recommended extension methods for use at different stages of adoption.

The recommended methods for the different stages of adoption are summarized in Figure 1. To achieve the maximum results from the methods selected at the different stages, extension officers must bring to bear their understanding of the learning process. For instance, in group meetings the extension officer can use a lecture format, supported with a video, and followed with group discussions.

In the final analysis, the method chosen will depend on the goal, resources, clientele relationship, and skills of the extension officers on the one hand, and on the size and educational level of the target group on the other hand. For instance, if extension officers lack the skills to organize and facilitate group meetings, then they will shy away from their use. Or if extension officers do not have vehicles, then they may not be able to conduct farm visits as frequently as might be desired or needed. Also, if the extension officer's constituency is very large, then it may become impractical to depend too much on individual visits.

As a guide in the selection of extension methods, Van Den Ban and Hawkins (1985) provide us with a useful set of criteria to judge whether the method is well chosen:

1. Is the chosen method adapted to whether we wish to change knowledge, skills, attitude, or behaviour?
2. Are the educational activities clearly specified so that we know what the farmer will see, hear, discuss, and carry out?
3. Are the different methods integrated in such a way that they reinforce each other?
4. Does the planned time scale make it possible to carry out all of these activities well?
5. When choosing learning activities, has the extensionist adequately considered the needs, skills, and means of the target group?

References

Albrecht, H., Bergmann, H., Diederich, G., Großer, E., Hoffmann, V., Keller, P., Payr, G., Sülzer, R. (1989). *Agricultural extension.* Vol. 1: Basic concepts and methods. GTZ Eschborn.

Campbell, D. (1982). *L'occidentalisation des campagnes francaises: Le cas du plateau ardechois.* Unpublished doctoral thesis, Université Paul Valery, Montpellier, France.

Feldstein, H. S., & Poats, S. V. (1989). *Working together: Gender analysis in agriculture.* West Hartford, CT: Kumarian Press.

Hayami, Y., & Ruttan, V. W. (1985). *Agricultural development: An international perspective.* Baltimore, MD: Johns Hopkins University Press.

Hilderbrand, P. E. (1986). *Perspectives on farming systems research and extension.* Boulder, CO: Lynne Rienner.

Lionberger, H. F. (1968). *Adoption of new ideas and practices.* Ames, IA: Iowa State University Press.

Morize, J., Dutilleul, J. P., & Beaulier, A. (1983). *Guide de l'agent du développement rural: Milieu physique, humain et agricole.* Presses Universitaires de France.

Rivera, W. M., & Schram, S. G. (1987). *Agricultural extension worldwide: Issues, practices and emerging priorities.* New York: Croom Helm.

Röling, N. (1988). *Extension science: Information systems in agricultural development.* Cambridge: Cambridge University Press.

Shaner, W. W., Philipp, P. E., & Schmehl, W. R. (Eds.). (1982). *Readings in farming systems research and development.* Boulder, CO: Westview Press.

Swanson, B. E., Röling, N., & Jiggins, J. (1984). Extension strategies for technology utilization. In B. E. Swanson (Ed.), *Agricultural extension: A reference manual.* Rome: FAO.

Timmer, W. J. (1982). *The human side of agriculture: Theory and practice of agricultural extension.* New York: Vantage Press.

Van Den Ban, A. W., & Hawkins, H. S. (1985). *Agricultural extension.* New York: Longman Scientific and Technical.

Chapter 9
Improving Women Farmers' Access to Extension Services

Janice Jiggins, R. K. Samanta, and Janice E. Olawoye

Over the last several decades, considerable effort has been made throughout the world to provide women farmers and women on the farm with efficient, effective, and appropriate technology, training, and information. The positive effects are beginning to show in agricultural production statistics and in indices of family welfare. Yet these successes still fall far short of what is needed at a time when public sector investments in agricultural research and extension are under pressure, when ever-greater demands are being placed on rural women in the face of rapid social transformation, and, in an increasing number of areas, when evidence of environmental degradation is mounting.

This chapter begins with a brief review of the need for gender analysis as the basis for agricultural sector policy making, extension programming, and agricultural project design and execution. The middle section examines some of the difficulties in providing women with extension services and what is being done to resolve them. Finally, the lessons of experience as the basis of an outline agenda for the coming decade are briefly considered.

The Need for Gender Analysis

The term *gender* describes the socially determined attributes of men and women, including male and female roles. (In comparison, *sex* denotes the physical and biological differences between males and females.) Gender has proven to be an essential variable for analysing the roles, responsibilities, constraints, opportunities, incentives, costs, and benefits in agriculture. Innumerable development projects, government programmes, research studies, and theoretical models have demonstrated that the improvement of women's access to agricultural research and extension services must begin with an analysis of men's and women's participation in the agricultural production process along two related dimensions: their role in agriculture and their role in the household.

Gender Analysis in Agriculture

It is now widely demonstrated that rural women, as well as men, throughout the world are engaged in a range of productive activities essential to household welfare, agricultural productivity, and economic growth. Yet women's substantial contribution continues to be systematically marginalized and undervalued in conventional agricultural and economic analyses and policies, while men's contribution remains the central, often the sole, focus of attention.

Women are typically, and wrongly, still characterized as "economically inactive" in statistical surveys of agriculture, a result that tells us more about survey methodology than about reality (Janelid, 1975). Agricultural extension services still do not attach much importance to reaching women farmers or women on the farm. Policy makers and administrators typically still assume (in the face of the empirical data) that men are the farmers and women play only a "supportive role" as farmers' wives (Samanta, 1994).

The official definition of a farmer in Nigeria in 1965, for example, was given as "an adult male . . . who has the right to the produce of a farm . . . women are not classified as farmers" (FOS, 1966, p. 3). Yet among many studies of rural women in Nigeria (WORDOC, 1988), Akor (1990) found that 92 per cent of the surveyed northern rural women gave farming as their primary or secondary occupation. Of these, 74 per cent owned or worked their own separate plots. While the official definition cf a farmer in Nigeria has been corrected to be gender neutral, as in most other countries, gender bias is prevalent in official agricultural circles and among field professionals. Similar investigations conducted in selected states in India show that more than 60 per cent of agricultural operations are performed by women farmers, yet the *fact* that "most farmers in India are women" (Shiva, 1991) is simply not reflected in extension provision or training.

Gender Analysis at the Household Level

The rural household typically is conceptualized in extension programmes and agricultural policies as a unit made up of individuals working in similar ways to meet common goals under the direction of a male head. In reality, the household is a more complex and dynamic social entity which may change its composition and goals over time as family members and dependents of varying age groups and sexes engage in various activities to meet the specific responsibilities assigned to each. However, while it is useful to draw attention to the fact that the division of labour along gender lines is a social constant and has profound implications for the organization of agriculture, men's and women's responsibilities

Janice Jiggins is a consultant in Andelst, Netherlands, and a Visiting Professor in the Department of Rural Extension Studies, University of Guelph, Ontario, Canada. **R. K. Samanta** is a faculty member and the Head of the TOT Systems and Policies Unit, National Academy of Agricultural Research Management, Rajendranagar, Hyderabad, India. **Janice E. Olawoye** is a rural sociologist and a Senior Lecturer in the Department of Agricultural Extension Services, University of Ibadan, Ibadan, Nigeria.

and privileges vary along sociocultural and socioeconomic lines specific to a particular time and place. It is thus misleading to make assumptions about the particular patterns in gender relationships to be found in any one household on the basis of data from elsewhere. Even within one country, sweeping generalizations are not advisable. Studies in Nigeria have revealed differences in gender relationships even in ethnically similar rural Nigerian communities just kilometres apart (Olawoye, 1985).

It is thus a mistake to view "rural women" as a homogeneous social classification or to derive policies and services for "women in agriculture" that are not based on empirical research which captures the diversity. As we will see later in the chapter, the consequence is that extension services need to be adapted to circumstance rather than designed on the basis of a single universal model. There is no one packaged extension model which can work for all women in all places (Olawoye, 1989; Berger, DeLancey, & Mellencamp, 1984).

The Changing Social Structure in Agriculture

As the composition and structure of rural households change (Snyder, 1990), gender responsibilities are under-going rapid change, typically with rural women becoming more responsible for household food security and children's welfare. One powerful indicator of these changes is the incidence of female-headed rural households, which is on the increase in most developing countries. In sub-Saharan Africa, women head an estimated 45 per cent of rural households in Kenya, 35 per cent in Malawi, 30 to 40 per cent in Zambia, and 15 per cent in Nigeria (ECA, 1973; Keller, 1986; World Bank, 1992a; FAO, 1993).

Typically, female-headed households are among the poorest, with the lowest level of food security (Heyzer, 1992), but in areas where female headship is the norm, as in the Caribbean, female headship can be a poor predictor of agricultural output, household welfare, or income status (Jiggins, 1994). In other cases, where women have had access to agricultural resources and services in their own right, as in parts of the Kenyan Highlands, women farming alone or with only sporadic assistance from migrant husbands have proved themselves more than capable of increasing farm productivity, efficiency, and profit (Jiggins, 1994; Saito & Weidemann, 1990).

Data summarized by the FAO (1993, Tables 3 and 4) serve as gross indicators of the potential overall size of extension's female clientele (covering *female farm operators* and *female-operated farms*) and their distribution among large, medium, and small farms. Case study data from twelve countries (Burkina Faso, Cameroon, Rwanda, Senegal, Bangladesh, China, Indonesia, Philippines, Colombia, Mexico, Egypt, Yemen) indicate the following in the study areas:

- 25 per cent of large farms were operated by women.
- 19 per cent of medium-sized farms were operated by women.
- China and Indonesia both had an exceptionally high proportion of women operators of large and medium-sized farms.
- The numbers and percentage of women operators were greatest on small farms.
- Except for Indonesia, Philippines, Senegal, and Yemen, the percentage of female-headed farms was above 15 per cent, and in Mexico (61.46 per cent), Egypt (28.10 per cent), and Kenya (27.09 per cent), they accounted for a quarter to two thirds of all farms studied.

Numerous instruments exist to guide agricultural professionals and policy makers in gender analysis relevant to research and extension programming. The instruments range from simple checklists and guidelines (FAO, 1982; Netherlands Ministry of Foreign Affairs, 1988), to more detailed methodological handbooks (Feldstein & Jiggins, 1994; Thomas-Slayter, Esser, & Dale Shields, 1994; World Bank, 1992a) and analysis of gender-sensitive agricultural programming (OIL, 1988; IFAD, 1985). Increasingly, both the donor community and national governments require gender analysis before making agricultural sector investments.

Constraints and Opportunities: Rural Women and Extension Strategies

The constraints affecting rural women's ability to improve yield, profit, and efficiency in agriculture include (1) women's legal and cultural status, which affects the degree of control women have over productive resources, inputs such as credit, and the benefits which flow from them (Olawoye, 1989); (2) property rights and inheritance laws, which govern access to and use of land and other natural resources (Jiggins, 1989a); (3) the relationship among ecological factors such as the seasonality of rainfall and availability of fuelwood, economic factors such as product market failures, and gender-determined responsibilities such as feeding the family, which trade off basic household self-provisioning goals and care of the family against production for the market (Jiggins, 1989b; Horenstein, 1989); and (4) the way that agricultural services are staffed, managed, and designed (FAO, 1993; Saito & Weidemann, 1990; Gittinger et al., 1990).

Improving Women's Access to Extension

Agricultural extension strategies traditionally have focussed on increasing production of cash crops by providing men with training, information, and access to inputs and services. This male bias is illustrated in farmer training centres, which have been established to provide residential training on technical subjects. Most do not provide separate washing and sleeping accommodations for men and women and do not provide facilities for the care of babies or young children, factors which may prevent women from attending the centres. Second, women's daily workloads do not usually allow them to be absent from home for residential training; even attending short courses may cause insuperable problems in arranging substitute care for children or the home. And third, even where attendance of women is quite high as a proportion of the total, women are given instruction mainly in home economics and craft subjects, not technical agriculture (Staudt, 1973; Perraton, Jamison, & Orival 1983).

Further, in the overwhelming majority of countries, extension services have been staffed predominantly by men. Only in countries such as the Philippines have women field staff been deployed in sufficient numbers and with sufficient resources to become effective agents of change among women farmers.

On the other hand, it is typical of ministries to assume that home economics services can substitute for agricultural training and information for women. Home economics and agriculture are both important, but they are not substitutes. Where home economics services have been provided, as Aidoo (1988) notes, female home economists worked almost exclusively with rural women, thus reinforcing the institutionalization of gender bias. Still, home economics services are far from universal and are poorly resourced, although some have struggled against the odds to provide farm women with technical information and training and with access to resources.

The introduction of the training and visit system emphasized the selection of *contact farmers* as a mechanism for passing on information to other ("follower") farmers in their area. The recommended selection criteria, such as title to land, literacy, or cooperative membership, as well as male extension staff's assumptions about women's roles in farming, have largely excluded women's involvement (see Aarnmink & Kingma, 1991 for a Tanzanian case study). In only a handful of countries (including China, Mexico, and Brazil) have women formed any significant percentage of contact farmers or follower farmers (Box 1).

In some countries, individual contact has been complemented by *group contact*, especially, but not only, where it may be difficult for male change agents to have any type of contact with individual women other than their own relatives. In many cultural settings, group extension significantly increases women's access (Berger, DeLancey, & Mellencamp, 1984; Ashby, 1981), because the group context calms the fears of male extension agents, husbands, and women about transgressing norms of approved social contact.

This may be particularly true in Islamic areas where women are in partial or total seclusion. Furthermore, in Islamic societies, there are probably not enough qualified adult females who are able to take up the post of change agent at the field level. However, in countries such as Bangladesh, the pioneering efforts of large-scale, non-government, rural development agencies such as BRAC and the Grameen Bank have demonstrated that religion and custom are not necessarily barriers to the hiring and field deployment of female staff, to the mobilization of women's groups and training of women leaders at the group and village levels by male staff, or to the development of efficient savings and credit services for rural women (Jiggins, 1994). Women-dominated and managed agroindustries, notably in poultry and silk, based on individual effort and rewards but mobilized through group mechanisms are becoming sizeable integrated enterprises backed by women's own savings and investments.

The group concept is being deepened and expanded in a number of ways, as agroindustry development opens up opportunities for developing functional linkages between women's groups and centralized processing facilities. The production and processing of herbs, medicinal plants, or perfume plants, for example, organized by TATA Industries in the private sector and BAIF in the nongovernment sector, provide examples from India of information and training linkages between women's groups and the agroindustry. The development of women-only cooperatives in the Dairy Development Movement in south India is a notable case (Chen et al., 1986; Jamal, 1994), but numerous other studies document the potential of this approach (Berger, DeLancey, & Mellencamp, 1984).

Other agencies have demonstrated innovative group approaches to overcoming women's illiteracy, which is a barrier to effective mass communication through written materials, and a restraint on women's ability to demand appropriate services. The rural wing of SEWA (the Self-Employed Women's Association, based in Ahmedabad, India), for example, has pioneered the use of video as the means by which groups of largely illiterate rural women can record and edit short films about their own farm and household environment, needs, problems, and solutions. Armed with these testimonials, they are demanding and receiving attention from district and provincial agricultural extension services, input

Box 1. Changing the Selection Criteria for Contact Farmers in Kenya.

The criteria laid down by the World Bank for selecting contact farmers specify that they should (1) represent the local range of farm size, cropping pattern, socioeconomic condition; (2) be regarded by other farmers as worthy of imitation; (3) be active, practising farmers; (4) be willing to adopt extension recommendations on at least part of their land, allow other farmers to observe the new practices, and be willing to explain these to other farmers; (5) to the extent possible, come from different families; and (6) be from geographically dispersed farms.

In practice, extension services commonly add other criteria such as a minimum landholding size, literacy, and ability to purchase inputs. Village chiefs and other formal leaders, who are typically men, and field extension agents, who are almost always men, usually make the selection, thus introducing other potential biases against women.

Three adjustments to selection criteria and the selection process have proven to be useful in Kenya in increasing the percentage of women selected:

- Encouraging chiefs, ministers, and other leaders to promote women's selection at local meetings and in the media
- Stressing the importance of selecting women farmers in extension training courses
- Emphasizing selection on merit from among those who are actually doing the work

A trial in three areas of Muranga District, Kenya, showed that in Makuyu, where selection criteria stressed active involvement in farming and the ability to meet face to face with the field agents, more than half of the contact farmers are women. In two other areas, where land ownership remained a criterion, about two fifths of the agents' selections resulted in only a quarter to a half of all contact farmers being women. In the two cases where the field agents were women, two thirds to 90 per cent of the contact farmers were women. In Meru District, where chiefs habitually select the contact farmers, far fewer women are selected.

(Source: Saito & Weidemann, 1990).

supply agencies, and rural banks. They are also using videos as tools for mobilizing and communicating their experiences with women in neigh-bouring villages.

Others are experimenting with the complementary methods of participatory rural appraisal (PRA), digital video, and picture-processing technology to explore further the possibilities for interactive dialogue between groups of women farmers and those more removed from agricultural reality, including male extension workers. Applications in agricultural research and technology development, as well as in environmental assessment and management, for example, are being developed in Tanzania with SIDA assistance. Other agencies such as ICLARM (International Centre for Living Aquatic Resource Management) are developing computer software that enables farmers and researchers to compare and contrast GIS information and maps with farmer-generated natural resource and on-farm resource maps as the basis for improved resource management and technological innovation.

Finally, we should mention the links between commercial extension and individual women as agricultural entre eneurs, particularly in emerging agroindustrial sectors such as ornamental plant and exotic flower production in countries such as Sri Lanka, Thailand, and Kenya. Where women do have access to land (homestead, family, or rented), they are as well placed as any small entrepreneur to take advantage of the input packages and marketing facilities offered by commercial companies in high-value agroindustries. Indeed, such companies often target women farmers as preferred clients because of the high standards of care and attention women bring to the management of the plants.

Changing Attitudes and Institutions

Experience has also shown that complementary strategies to bring about changes in attitude and behaviour within institutions are required (Poats, 1991). Gender sensitization training has been developed to initiate the task of attitude change within male-dominated extension and research bureaucracies and donor agencies (Rao et al., 1994). Training materials and methods for gender analysis in agriculture (Paris & Frio, 1994; Feldstein, 1994) have also been developed and are now in widespread use. Specialist material for training of trainers (Rao et al., 1994) has also been developed and is beginning to spread through agriculture training institutes, colleges, and universities.

But training needs to be complemented by other strategies to bring about change in institutional behaviours. Spring (1985, 1986) demonstrated in Malawi the range of often minor but critical adjustments which can increase women's access to and the relevance of extension significantly, even where most field agents are male.

Box 2. Farmer-to-Farmer Extension: Women's Participation in Networking among Farmer Groups in Tanzania.

The Farmers' Groups Network in Tanzania (MVIWATA) was formed by group representatives during a workshop hosted by Sokoine Agricultural University in 1993. The workshop was one of a series organized by farming systems researchers at Sokoine Agricultural University who have been working with farmer groups in technology development.

Researchers had learned the importance of structured, ongoing dialogue with farmers to identify priority problems, suggest and try out possible solutions, and disseminate technologies and information judged by both researchers and the farmer groups as useful. They distilled the following five principles from their experience:

- Multidisciplinarity, in recognition that farmers' problems are multifaceted
- The use of group approaches, in recognition that decision making is almost always built on group consensus
- On-farm development of technical innovations to ensure relevance
- Assistance with removal of critical production bottlenecks
- Empowerment of farmers through training, facilitation, and networking

The MVIWATA groups are very diverse. In Lushoto District, Tanga region, for example, nearly 100 women's groups are working with field officers in an agriculture and afforestation programme. In the Kilimanjaro region, small groups of five to twelve men and women members are working on activities such as rice or vegetable production to generate income. Each group member has his or her own plot, but they are trained as a group in crop production and farm management. The group also acts as the guarantor to enable members to gain access to credit.

Groups in a particular region form their own communication links to develop their own independent activities and build social solidarity, for example by composing songs about their activities and the network and passing these around local trading routes, or by holding farmer workshops on issues such as the new Cooperative Act. The national MVIWATA network publishes its own newsletter in Kiswahili and is linked to the newly created pan-Africa network, FAM-Africa (Farmers, Agriculture and Modernization in Africa).

(Source: Mattee & Lassalle, 1994).

For example, male extension agents were encouraged to ask their male farmer contacts to include their wives during visits, demonstrations, or farmers' meetings. Village leaders (typically male) asked to men identify women needing extension services. Field agents were required to devote a greater percentage of their time to working with women's groups. Women farmers' seminars were organized for women to share with researchers and field staff their solutions to the technical problems specific to women farmers' production systems, and women's field days were organized to celebrate and legitimate women farmers' successes and to promote farmer-to-farmer exchange among women (Box 2).

Participatory action learning (PAL) also has proved to be a powerful approach to institutional change. The Centre for International Agricultural Development (CIAD) at the University of Beijing and the All-China Women's Federation (ACWF) adopted a participatory learning approach to their work with farm women in Ningjin County (Huang Xiushen, 1995). They reversed the existing approach of "learning skills, implementing projects, and getting benefits" and started with "implementing projects" as the occasion and location for learning skills and ensuring that participants really did secure benefits that are meaningful to themselves and achievable in their locality. PAL began with joint diagnosis conducted by officers of ACWF-Ningjin County, men and women farmers, and members of CIAD. This phase included a case analysis about gender functions, a local workshop about gender-sensitive problem analysis to which local development agencies were invited, and a participatory appraisal of women's needs in the county. The next phase began with identification and prioritization of action plans based on the investigations, dialogue, and analyses. The plans enabled the swift approval of government and technical authorities and ready access to funding and technology. The Women's Federation (WF) formed a "socialized service network" to provide the backup training, drawing on the training expertise identified in earlier phases. For example, in Weijiazhuang, the WF selected a representative from each ten women and a group leader from among each three representatives selected. A scientific and technical team was then formed of the WF director and twelve group leaders. It is their responsibility to form a link with each participating household and with technical training resources for the purpose of training women to teach other women. At the end of the "learning skills" phase in Weijiazhuang, 80 per cent of 400 women have learned two practical skills from among such skills as preventing epidemics among pigs and chickens, trimming fruit trees, and ten other techniques. Over the following two years, WF cooperated with county, township, and village leaders to train 30,000 women in one of the practical skills.

Where other strategies are not possible, at least in the near term, it might also be useful to initiate extension services targeted specifically to and staffed by women field agents. The Tamil Nadu Women in Agriculture (TANWA) project, India, has been implemented by the state Department of Agriculture with DANIDA assistance since 1986 (Box 3). The project aims to provide small and marginal farm women with improved technical skills and technology through a training-cum-extension program. Over a seven-year period, a core team of thirty qualified women agricultural officers (AOs) has been recruited, and more than 14,000 women have received training, with significant improvements in food security, income, and productivity. The training is village based, given at times suited to women's workload to groups of twenty-five women, and focusses on practical skills and technologies selected and prioritized by the women in consultation with the AOs. The basic training stretches over a two-year schedule and is backed up by two-day training for the group leaders and short courses on special topics. In addition, farm women's conferences allow the women to interact with (male) research and extension officers and with groups from neighbouring villages. The appointment of the women AOs is a breakthrough in Tamil Nadu and follows the earlier success of appointing women dairy development officers in the dairy development movement. The success of women-to-women extension has built confidence among male AOs that farm women are worth contacting, while the group mechanism provides a culturally appropriate means for followup contact by male extension staff.

Scaling Up

Many of the examples cited in this chapter are relatively small scale. The challenge is to achieve impact on a scale that makes a difference. Three recent initiatives illustrate the ways in which this challenge is being addressed.

The Academy for Educational Development (based in Washington, D.C.), the government extension service, and relevant private sector agencies collaborated together in Honduras in adapting social marketing techniques to extension communication (UNIFEM & IWTC, 1990). They developed a model for communication for technology transfer in agriculture (CTTA), which is based on three principles: market research in order to identify user categories and needs, user participation in technology development and dissemination, and strong feedback mechanisms. Women farmers and farm women in various socioeconomic and spatial categories were identified explicitly as "users." The CTTA design focussed on the "four Ps" of social marketing—price, product, place, and promotion—as well as on two contextual factors—policy and politics—in devising more effective ways to involve users in the development, dissemination, and feedback of extension communications. During the pilot phase, the CTTA increased the outreach of extension services in Honduras from 3,000 to 16,000 farmers.

Box 3. Integrated Pest Management: The Farm Women's Field School, Kolumanivakkam, Tamil Nadu.

Since 1994, women farmers have been included in field training on integrated pest management (IPM) in Tamil Nadu. Of the thirty farmers' field schools (FFS) conducted in the 1994-95 rice season in Chengalpattu MGR District, for instance, four have been for women. The FFS runs one half-day a week for thirteen weeks for thirty farmers. They are taught by female agricultural officers to identify the pest and predator insects in rice, to monitor the number of each to ensure that predators are keeping the pests in check, and to observe the life cycles of the key insects and the fluctuations in populations through the production season. Weekly recording and analysis of the agroecosystem form an important part of the training.

The principles and practices of organic farming, integrated nutrient management, and the use of biofertilizers are included. Cultural, mechanical, and biological pest controls are practised, and participants are trained in the use of biopesticides and chemicals as a last resort if predator populations are insufficient to keep pest numbers within economic limits.

The women of Kolumanivakkam have composed a *mantra* to express their commitment to becoming experts in their own fields: "I can't lose. Why? I will tell you why . . . because I have Knowledge, Courage, and Enthusiasm." Prior to the introduction of IPM, farmers in the village regularly used pesticides, accounting for up to 20 per cent of the costs of cultivation. The women would work in the fields after the spraying and reported a range of symptoms such as nausea, headaches, eye troubles, skin rashes, and difficulties in breathing attributable to the chemicals. Now they have stopped spraying their homestead vegetables as well and have begun to monitor pests and predators in their home gardens. The men in the village are equally enthusiastic. They had heard about IPM from farmers in a neighbouring village who received training, and they encouraged their wives to attend the women's FFS when it was offered to them. IPM techniques save money, protect the family's health, and, through better crop management, also improve yields.

(Source: Farm Women Field School, Kolumanivakkam, Visit of Delegates, February 9, 1995, Directorate of Agriculture, Chengalpattu MGR District, Tamil Nadu).

Recent roundtable discussions of the position of women agricultural professionals in Africa (Doss, 1991) not only identified the all too familiar constraints and gaps, but moved to identify ways of promoting advocacy at both the popular and highest policy-making levels. Strategies for lobbying for changes in agricultural education and training provisions, for example, are beginning to be implemented, as well as strategies for changes in the legal status of women. In Benin, the Council of Ministries has instructed the Ministry of Rural Development to develop a policy for the advancement of rural women, to create quotas for the number of women offered professional positions in agriculture and rural development, and to establish a national network of women in the agricultural sector to keep track of policy implementation. A complementary initiative among women secretaries and ministers of education, vice-chancellors, and other leading educationalists has led to the formation of FAWE (Forum for African Women's Education). The promotion of girls' and women's education in science at all levels, especially the agricultural sciences, is a core part of FAWE's agenda.

The third example also comes from Africa. Over the last three decades, Nigeria has experimented with different agricultural development strategies with varying implications for rural women. In the 1970s, World Bank-supported Agricultural Development Projects (ADPs) were established in a number of Nigerian states. By the mid-1980s, ADPs were found in every state; technology development and extension were major components of their programmes. Toward the end of the decade, it became apparent that, while rural women had an important role in production, they were largely excluded from the ADP agenda. A Women in Agriculture (WIA) unit, with female extension staff, was established in every ADP throughout the country, with the goals of identifying the technical and information needs of rural women, assisting them to become more productive through training and technology dissemination, and meeting those needs through trained and qualified female agents working with women's groups. The WIA units today are fully integrated into the ADPs.

Despite gaps in the skills of many female agents, inaccessibility of women in many areas, and limited involvement of women in the selection and design of technology, the WIA experiment in Nigeria has integrated women's needs successfully into a national agricultural development strategy. To further improve the system in Nigeria, the following suggestions have been made:

1. Conduct additional gender-sensitizing programmes for policy makers and project implementers.
2. Collect gender-disaggregated data on rural and agricultural activities.
3. Encourage a more participatory approach, particularly involving more rural women.
4. Ensure equitable access to productive resources and extension services.
5. Design situation-specific implementation strategies, taking into account the unique sociocultural and ecological variations of each locality (Olawoye, 1994).

Recommendations for More Effective Extension Systems for Rural Women

Building on Present Potentials

It is a mistake to believe that rural women in developing countries do not possess skills and techniques which are an asset to the development process. Where groups already exist, capacity building of existing groups can be more successful than forming a new group to which members are less likely to be committed. Similarly, rural people are less likely to resist adoption of an innovation when the new technique is based upon a concept or procedure they are already familiar with or are currently using.

It is important for change agents to acquaint themselves with the organizations and knowledge systems available at the local level to determine how they can be improved, rather than assuming that nothing of significance is currently available. For example, instead of forming entirely new groups for women, local informal work exchange or savings groups could be strengthened through short training exercises on farm bookkeeping and record-keeping, leadership, and democratic procedures.

While groups have proven to be a highly successful mechanism, they are not a universal panacea or appropriate for meeting all women's needs in agriculture. The poorest women in particular can find the costs of participation (of time, for example) too high at times of seasonal stress or greatest need. Care is also needed to sort out situations where collective effort by groups is effective and where individual activity structured or accessed through group membership is more effective and efficient.

Institution Strengthening

Several aspects of institution-strengthening, including project integration and gender-sensitization of officials, have been presented. These efforts need to be pursued vigorously and extended. The World Bank (1992b) summarizes four organizational principles which might serve as broader guidelines to institutional development: (1) situation specificity, (2) project flexibility, (3) farmer participation, and (4) mainstreaming women's programmes. By incorporating these principles with the other factors considered earlier, agencies concerned with women's projects will be more likely to have a meaningful and sustainable impact (Box 4).

Interagency Cooperation

Many agencies currently focus on the integration of women in mainstream development efforts. Unfortunately, there seems to be little cooperation between projects administered by the same agency or between agencies, even where these agencies are under the umbrella of the same national government or development policy. In addition, international agencies and NGOs are attempting to achieve similar goals in the same localities, yet typically without any form of coordination or cooperation. The use of human and material resources could be managed more efficiently and with greater impact if development agencies would cooperate with each other (Olawoye, 1991).

Training Programmes

The following suggestions are made for improving and redesigning training programmes for farm women:

1. Adapt programmes to women's needs and skills.
2. Allow sufficient time to enable women to acquire new skills and adjust schedules to fit women's existing workloads.
3. Provide training in agricultural and other productive activities, not just home and family welfare topics.
4. Emphasize activities for which there is an actual income-generation potential.
5. Ensure the involvement and full participation of women from poorer and less educated backgrounds.
6. Use trainers who are not only technically competent and up-to-date, but who empathize with the needs and aspirations of rural women.
7. Provide practical field experience in the use of innovations.
8. Shift more resources to village-based training rather than residential training.

Retaining Benefits

Rural women seldom have autonomous control over the opportunities that may come their way or the benefits which flow from them. Many advantages won for rural women through development programmes are later lost, as illustrated in the following quotation: "When technological innovations do address women's tasks and make them more profitable, men often take them over. This was exactly what happened when pump irrigation was introduced for rice production in West Africa" (Gittinger et al., 1990, p. 10). For sustainable improvements, not only must benefits be targeted to rural women, but mechanisms must also be put into place to ensure that these benefits can be retained by the intended beneficiaries.

The Role of the NGOs and the Private Sector

NGOs have pioneered many of the initiatives subsequently incorporated in public sector extension services. NGOs continue to play a lead role in ensuring that women farmers and women on the farm receive training, information, and improved technologies. Their services often are increasing in scope and scale, either as complementary support to government efforts or to fill the gaps created as government expenditures and capabilities decline. An important emphasis which recently has been highlighted in NGO programmes is their support for membership-based community and farmer organizations. Women as well as men benefit from the expanding opportunities to develop farmer-to-farmer extension and training networks and to form partnerships with agricultural researchers and development agencies.

Box 4. Women, Agricultural Development, and Conservation in Honduras.

Many organizations are increasingly concerned with maintaining the productivity of the natural resources on which all agriculture ultimately depends. Yet once again institutions and structures are being created which either ignore women's natural resource management roles, or treat activities for women as "add-ons," outside the mainstream effort.

PRODESAI (Choluteca Integrated Agriculture Development Agency, a local world neighbours organization) in the Linaca region of Honduras is an exception. PRODESAI trains and incorporates local men and women as active agents of change in their communities. It has nurtured the formation and growth of a local organization, UCAMNE (Union Campesina en March, Nueva Esperanza—Peasants' Union on the Move, New Hope), which took over PRODESAI's work when it withdrew in mid-1993.

One of the founding members of UCAMNE is Dona Moncha, who first worked two mornings a week in four communities as a health promoter for community women's groups, receiving a small salary from PRODESAI. She and her husband and five children demonstrate in her own backyard garden the techniques she has learned. They have constructed stone walls and small contoured terraces to hold the land, conserve moisture, and prevent erosion and have planted live barriers of lemon grass and valeriana, which are used for medicine. They practise minimum tillage to grow maize, tomatoes, green peppers, local squash, basil, and other herbs and have planted leguminous trees to help maintain fertility and provide fuelwood.

(Source: Urban & Rojas, 1994).

Conclusion

Regardless of the rhetoric surrounding gender issues in agriculture and the countless research projects and "women's studies," the important question to be considered is: Have women benefited from these efforts? As Obermaier aptly concludes: "In the final analysis, women in developing countries are only interested in concrete actions—whether more and better projects for women are actually implemented and whether more women are supported in their efforts to solve their problems on the road to self determination" (1990, p. 7).

Women farmers' access to extension services must lead to concrete improvements for rural women themselves, as well as enhance the productivity of the agricultural sector and national food security through increasing marketed output.

References

Aarnink, N., & Kingma, K. (1991). *Female farmers and male extension workers: Women and agriculture in Tanzania II.* Leiden: Women and Autonomy Centre, Leiden University.

Aidoo, A. A. (1988). Women and food security: The opportunity for Africa. *Development, 2*(3), 56-61.

Akor, R. (1990). *The role of women in agriculture and constraints to their effective participation in agricultural development in Nigeria.* Paper presented at UNDP/ILO/DFRRI Training Workshop on Monitoring and Evaluation of Rural Women in Productive Skills Project.

Berger, M., DeLancey, V., & Mellencamp, A. (1984). *Bridging the gender gap in agricultural extension.* Washington, DC: ICRW.

Chen, M., Mitra, M., Athreya, G., Dholakia, A., Law, P., & Rao, A. (1986). *Indian women: A study of their role in the dairy movement.* New Delhi: Shakti Books.

Doss, C. R. (1991). African professional women in agriculture: An analysis of two roundtable discussions. Development Studies Paper Series. Morrilton, AR: Winrock International Institute for Agricultural Development.

Economic Commission for Africa (ECA) (1973). *Women and national development in African countries.* Position paper prepared by Human Resources Development Division, ECA.

FAO (1982). *Guidelines: Women in land and water development.* Rome: FAO, Land and Water Division.

FAO (1993). *Agricultural extension and farm women in the 1980s.* Rome: FAO.

Federal Office of Statistics (FOS) (1966). *Rural economic survey of Nigeria: Household enquiry—1963/64 and 1964/65.* Lagos: FOS.

Feldstein, H. S. (1994). Training in gender analysis in agricultural research: The use of case studies (chapter 32). In H. S. Feldstein & J. Jiggins (Eds.), *Tools for the field.* West Hartford, CT: Kumarian Press.

Feldstein, H. S., & Jiggins, J. (Eds.). (1994). *Tools for the field: Methodologies handbook for gender analysis in agriculture.* West Hartford, CT: Kumarian Press.

Gittinger, J. P., Chernick, S., Horenstein, N. R., & Saito, K. (1990). *Household food security and the role of women.* Washington, DC: World Bank.

Heyzer, N. (1992). Gender, economic growth and poverty. *Development, 1*, 50-53.

Horenstein, N. R. (1989). *Women and food security in Kenya.* Washington, DC: World Bank.

Huang Xiushen (1995, January). *Implement projects, learn skills and get benefits—Discussion about the new way of rural women's technical training.* Paper presented at international seminar on women in agriculture, Beijing, CECAT.

ICAR (1988, November-December). *International conference on appropriate agricultural technologies for farm women: Proceedings and recommendations,* New Delhi, ICAR.

IFAD (1985). *Rural women in agricultural investment projects 1977-1984: A report of findings based on an assessment study of IFAD experience relating to rural women.* Nairobi: IFAD.

Jamal, S. (1994). *Women in dairy development.* New Delhi: Concept Publishing.

Janelid, I. (1975). *The role of women in Nigerian agriculture.* Rome: FAO.

Jiggins, J. (1989). How poor women earn income in sub-Saharan Africa and what works against them. *World Development, 17*(7), 953-963.

Jiggins, J. (1989). Agricultural technology: Impact, issues and action (chapter 1, part I). In R. S. Gallin, M. Aronoff, & A. Ferguson (Eds.), *The women and international development annual: Vol. I.* Boulder, CO: Westview Press.

Jiggins, J. (1994). *Changing the boundaries. Women-centred perspectives on population and the environment.* Washington, DC: Island Press.

Keller, B. (1986). Development for rural Zambian women. In 1985: *Training for agriculture and rural development.* Rome: FAO.

Mattee, A. Z., & Lassalle, T. (1994). Diverse and linked: Farmers' organizations in Tanzania (p. 19-29). In *Linking farmers' organizations and researchers: Four case studies.* Network Paper 50. London: Overseas Development Institute.

Netherlands Ministry of Foreign Affairs (1988). *Women in agriculture.* Sector Papers No. 1. The Hague: Directorate General for Foreign Affairs.

Obermaier, D. (1990). Promotion of women remains controversial. *D + C: Development and Cooperation, 1*, 7-8.

Olawoye, J. E. (1985). Rural women's role in agricultural production: An occupational survey of women from six selected rural communities in Oyo State, Nigeria. *Nigerian Journal of Rural Sociology, 2*(1-2), 34-37.

Olawoye, J. E. (1989a). Difficulties for rural women in securing resources for agricultural production: Two case studies from Oyo State, Nigeria. *Rural Development in Nigeria, 3*(2), 77-81.

Olawoye, J. E. (1989b). Women in extension (chapter 14). In W. M. Rivera & T. O. Ogunfiditimi (Eds.), *Extension for development: Concepts and issues.* Ibadan: Lidato Press.

Olawoye, J. E. (1991). Harmonizing development efforts for rural women in Nigeria. In T. A. Olowu & J. A. Akinwumi (Eds.), *Development directions for 21st century Nigeria.* Ibadan: Ibadan Socio-Economic Group.

Olawoye, J. E. (1994a). *Women and forestry in Nigeria.* Technical Report to the Tropical Forest Action Programme, Ibadan.

Olawoye, J. E. (1994b, February-March). *Gender priorities and issues in agricultural extension delivery.* Paper presented at the inaugural conference of the Society for Agricultural Extension of Nigeria at ARMTI, Ilorin, Nigeria.

Olofin, Olutola Olayinka (OIL) (1988). *Occupational health hazards in traditional methods of processing gari and palm oil: A study of rural women in selected areas of Oyo State.* Unpublished bachelor's project, University of Ibadan.

Paris, T., & Frio, A. L. (1994). Training in integrating a gender perspective into farming systems research (chapter 33). In H. S. Feldstein & J. Jiggins (Eds.), *Tools for the field.* West Hartford, CT: Kumarian Press.

Perraton, H., Jamison, D., & Orival, F. (1983). Mass media for agricultural extension in Malawi (p. 147-202). In *Basic education and agricultural extension: Costs, effects and alternatives.* Staff Working Paper No. 564. Washington, DC: World Bank.

Poats, S. (1991). The role of gender in agricultural development. Issues in Agriculture No. 3. Washington, DC: CGIAR

Rao, A., Appel, M., Moore, K., & Feldstein, H. S. (Eds.). (1994). *Reflections and learnings: Gender trainers workshop report.* Amsterdam: Royal Tropical Institute; and New York: Population Council.

Saito, K., & Weidemann, C. J. (1990). *Agricultural extension for women farmers in Africa.* Washington, DC: World Bank.

Samanta, R. K. (1994). They reap less than they sow. *The Hindu* (April), No. 7. Madras: India.

Shiva, V. (1991). *Most farmers in India are women.* New Delhi: FAO.

Snyder, M. (1990). *Women: The key to ending hunger.* The Hunger Project Papers No. 8.

Spring, A. (1985). The women in agricultural development project in Malawi: Making gender free development work (p. 71-76). In R. S. Gallin & A. Spring (Eds.), *Women creating wealth: Transforming economic development.* Washington, DC: Association for Women in Development.

Spring, A. (1986). Reaching female farmers through male extension workers in Malawi. In *1985—Training for agriculture and rural development.* Rome: FAO.

Staudt, K. (1975). Women farmers and inequities in agricultural services. *Rural Africana, 29,* 81-93.

Thomas-Slayter, B., Esser, A. L., & Dale Shields, M. (1993). *Tools of gender analysis: A guide to field methods for bringing gender into sustainable resource management.* Worcester, MA: Clark University.

UNIFEM & IWTC (1990). *Reaching rural women with information on technology.* New York: UNIFEM.

Urban, A-M., & Rojas, M. H. (1994). *Shifting boundaries: Gender, migration and community resources in the foothills of Choluteca, Honduras.* ECOGEN Case Studies Series. Worcester, MA: Clark University.

WORDOC (1988). Women in agriculture. *African Notes, 3*(Special Number). Institute of African Studies, Ibadan.

World Bank (1992a). *Designing and implementing agricultural extension for women farmers.* Technical Note. Washington, DC: World Bank, Women and Development Division.

World Bank (1992b). *Towards a gender strategy for Nigeria: Integrating women's issues into the development agenda.* Washington, DC: World Bank.

Chapter 10
Implementing Strategic Extension Campaigns

Ronny Adhikarya

Strategic Extension Campaign: What and Why

A "strategic extension campaign" (SEC) methodology developed by FAO has been introduced in Africa, the Near East, Asia, and Latin America. This methodology emphasizes the importance of *people's participation* (i.e., intended beneficiaries such as small farmers) in strategic planning, systematic management, and field implementation of agricultural extension and training programmes. Its extension strategies and messages are specifically developed and tailored based on the results of a *participatory* problem identification process on the causes or reasons for farmers' nonadoption, or inappropriate practices, of a given recommended agricultural technology or innovation. The SEC technology transfer and application approach is needs based and demand driven and has a problem-solving orientation.

The SEC programme follows a systems approach, which starts with a farmers' Knowledge, Attitude, and Practice (KAP) survey whose results are used as planning inputs and benchmark-baseline for summative evaluation purposes. In addition, a series of practical and participatory approach workshops are conducted to train extension personnel, subject-matter specialists, trainers, and farmer leaders together on the skills of extension programme planning, strategy development, message design and positioning, multimedia materials development, pretesting and production, as well as management planning, implementation, monitoring, and evaluation. One of the strengths of this approach is in orienting and training relevant extension personnel to apply *a systematic, rational, and pragmatic approach to planning, implementing, managing, monitoring, and evaluating regular or routine programmes* of an agricultural extension service.

Empirical evaluation studies (using information recall and impact surveys, focus group interviews, and management monitoring surveys) of strategic extension campaign methods applied to specific FAO-supported extension activities conducted, for instance, in Bangladesh and Malaysia (on rat control), Thailand (on pest surveillance system), Malaysia (on weed management), Zambia (on maize production), Malawi, Jamaica, and Morocco (on population education), reported positive changes in farmers' knowledge, attitudes, and practices vis-à-vis the recommended technologies as well as significant economic benefits.

This SEC method has been replicated with FAO assistance in many countries in Asia, Africa, the Near East, and the Caribbean, with topics such as line-sowing method of rice cultivation, maize production, cocoa cultivation, tick-borne disease control, contour tillage, population education, and ploughing with draught-animal power. In addition to various SEC replications within a country, the multiplier effects of its method are felt beyond national boundaries. For example, extension specialists from Ghana, Malawi, Ethiopia, France, Malaysia, Thailand, and the Philippines who had been trained by FAO on this SEC method and implemented such programmes have now served as consultants and resource persons to train their counterparts, and/or have assisted in similar SEC replications in Sri Lanka, the Philippines, Malaysia, Thailand, China, Liberia, Zambia, Malawi, Kenya, Uganda, Morocco, Tunisia, Rwanda, Burundi, Guinea, Jamaica, and Honduras.

The Usefulness of SEC

The strategic extension campaign (SEC) is *not* an alternative to the conventional extension programme or activity. SEC is, and should be, an integral part of the programmes of an agricultural extension service. The effectiveness and efficiency of such a service could be increased because of SEC's emphasis on its problem-solving orientation, participatory planning approach, intensive extension personnel training, multimedia materials development, and extension management, monitoring, and evaluation procedures. Its activities should be carried out by extension personnel and to support the ministry of agriculture's policies, strategies, and priority programmes. The strategic extension campaign is useful and important to an agricultural extension service for the following reasons:

- It advocates a participatory planning approach.
- It has a needs-based and demand-driven orientation.
- It uses a strategic planning and integrated systems approach.
- It considers the human and behavioural dimensions.
- It has a problem-solving orientation.
- It employs a cost-effective multimedia approach.
- It provides specific extension support materials and training.

Ronny Adhikarya is a Senior Extension Education and Training Methodology Specialist, Agricultural Education and Extension Service, FAO/United Nations, Rome, Italy. The contents of this chapter are largely taken from several sections of R. Adhikarya, *Strategic extension campaign: A participatory-oriented method of agricultural extension,* Rome, Italy: FAO/United Nations, 1994.

- It has built-in process documentation and evaluation procedures.
- Its method is applicable to other extension programmes.

Participatory Planning Approach

This participatory approach extension method is responsive to intended beneficiaries' agricultural development problems and information needs because its extension objectives, strategies, methods, messages, and multimedia materials are specifically developed on the basis of survey results of their knowledge, attitudes, and practices (KAP) vis-â-vis the recommended agricultural technologies. Such a participatory approach in planning SEC activities increases the degree of relevance, and thus acceptability, of extension messages or recommendations among intended *beneficiaries who are consulted* during the planning process regarding their priority concerns and needs. It does not assume the target beneficiaries (i.e., farmers) to be ignorant or requiring all the information there is to know. Rather, it tries to understand and assess farmers' local indigenous knowledge, values, and belief system on farming practices, which may be good, need to be improved, or perhaps need to be discouraged. In short, it follows the well-known principles of rural reconstruction: "Start with what people already know," and "Build on what they already have."

Needs-Based and Demand-Driven Orientation

In order to make the best use of available extension resources, SEC activities concentrate on meeting the information, education, and training needs of intended target beneficiaries. Rather than providing them with the spectrum of information and skills related to a given recommended technology, SEC activities are geared to *narrowing the gaps* between knowledge, attitudes, and/or appropriate practice levels of the target beneficiaries vis -à-vis the technology recommendations. Furthermore, the focus of SEC activities is to create a demand (through information and motivation approaches) and/or to satisfy the demand (through education and training) among the intended target beneficiaries for the necessary relevant knowledge and skills for adopting the recommended technologies. Such a method needs to apply bottom-up and participatory planning procedures which will give *high priority in meeting the interests and needs of the target beneficiaries.* Tailoring the SEC messages and activities to the specific needs of the intended beneficiaries would not only increase the chances of success, but also would increase the efficiency in resources utilization.

Strategic Planning and Integrated Systems Approach

The SEC method advocates an integrated and holistic approach in extension strategy development, programme planning and management, training, media and materials development, and monitoring and evaluation. To ensure its relevance to audience needs and to utilize its resources efficiently, it relies heavily on both quantitative data and qualitative information obtained from target beneficiaries (i.e., farmers) to assist in problem analysis, objective formulation, strategy development, and management planning. It applies a *strategic planning* approach in programming and managing its activities to achieve *maximum* outputs or results using *minimal* inputs or resources in the *shortest time* possible. SEC activities such as surveys, strategy and management planning, multimedia materials design and development, training, field implementation, monitoring, and evaluation are integrated as a system, which is also an integrated part of a larger extension programme which has linkages with relevant agencies or units dealing with research, inputs-supplies, training, and marketing.

Human and Behavioural Dimensions

In order to minimize the heavy technology bias of many extension activities, the SEC method gives adequate consideration to *human behavioural* aspects, such as sociopsychological, sociocultural, and socioeconomic factors which may facilitate or impede adoption, or continued practice, of recommended technologies by farmers. Without sufficient understanding of their positive or negative attitudes and behaviour towards a given technology, the technology transfer process would be slow and ineffective, especially if the extension emphasis is on appropriate technology application by farmers. There is considerable evidence to suggest that nonadoption of a recommended agricultural technology or innovation is often related to, or caused by, *nontechnological* factors such as social, psychological, cultural, and economic problems.

The SEC method gives due attention to human and environmental factors which may influence the important decision-making process related to agricultural technology adoption and practices. It employs a behavioural science analysis, based on a participatory needs assessment and problem identification of the target audience, in developing appropriate strategies and tactics to overcome or minimize humanrelated constraints affecting the agricultural technology transfer and application process.

Problem-Solving Orientation

The SEC is particularly distinguished in that it normally focuses on *specific* issues related to a given agricultural technology recommendation. Its main aim is to solve or minimize problems which caused nonadoption of such a recommendation by intended target beneficiaries (i.e., farmers). Unlike more conventional extension programmes or activities, it does not "extend" the whole gamut of information on the recommended technology package. Instead, it *selects,* prioritizes, and utilizes only the most relevant and necessary information or facts which can maximize the effectiveness of extension efforts to minimize or solve the identified problems of nonadoption of a recommended technology. It stresses the need to provide strategic, critical, and "quality" information, which must also include nontechnological information as the reasons why nonadoption of agricultural technologies are often related to sociopsychological, sociocultural, and socioeconomic factors. Appropriate human behavioural science principles are thus applied to extension problem solving and in information positioning and utilization, which is *responsive* rather than prescriptive in nature.

The *segmentation* or classification of extension problems, objectives, strategies, and information needs according to a target audience's levels of knowledge, attitudes, and practices (KAP) in regards to a given recommended technology is not only conceptually important, but practical and useful as well. Problems related to low knowledge level require different solutions than those related to attitudinal problems. Similarly, strategies for changing negative attitudes on a recommended technology are likely to be different than those for solving incorrect practices in technology application or convincing people to try to practise a recommended technology. The implications of the different KAP levels would greatly influence the development of problem-solving strategies, message design, selection of multimedia mix (including when and how to utilize group and interpersonal communication channels, such as extension workers), and materials development, as shown in Figure 1. Application of a behavioural modification approach using information based on the KAP levels of the target audience for message development, media selection, and materials development alone could significantly increase the cost-effectiveness of extension activities.

Cost-Effective Multimedia Approach

One of the most common problems or constraints of a national extension service is the shortage of field extension personnel to reach large numbers of farmers in widely spread geographical areas with inadequate transportation. Moreover, extension workers are usually overburdened with an unnecessarily heavy workload which includes almost everything that has to do with farmers at the village level. Such an overreliance on extension workers is neither technically sound nor operationally efficient. Some extension functions for certain purposes such as awareness creation, information delivery, and motivational campaigns can be more effectively and efficiently performed by other means, channels, or nonextension groups under the coordination and supervision of extension workers.

Extension workers' workload could be reduced by mobilizing appropriate rural and community-based resources, including the increasingly accessible and low-cost mass communication channels (e.g., local radio stations, rural press, folk/traditional media, posters, flip charts, silk-screened printed materials, audio-cassettes, slide-tape presentations, leaflets, comics) to disseminate standardized and packaged extension messages, and by using local volunteers (e.g., school teachers, children, local and religious leaders) to serve as intermediaries in reaching farmers. Such an approach does *not* imply that extension workers can or will be replaced by these community resources. Rather, it is a rational approach of using available resources most efficiently for certain tasks, such as using extension workers for educational or instructional purposes, which require two-way interactions, field demonstrations, group discussion, and so forth that cannot be done as effectively by mass communication channels.

The SEC method employs a multimedia approach whereby a cost-effective *combination of mass, personal, and group communication channels* (including extension workers and trainers) and materials are efficiently utilized to reduce extension cost and efforts and to increase its effectiveness in dealing with a larger number of target audiences more rapidly.

Specific Extension Support Materials and Training

Most extension services in developing countries suffer from the lack of relevant and practical extension and training materials to support the field activities of their extension workers. Many extension workers rely primarily on their interpersonal communication skills, and thus their time during farmers' meetings may not be used as effectively as it should be.

Providing specifically designed and relevant training support materials to extension workers will not only facilitate their tasks and reduce their heavy workload. It will also ensure a certain degree of *quality control* in the delivery of technical information or extension message contents. Experience has also shown that extension workers' motivation, enthusiasm, confidence, and credibility increase when they are given relevant and attractive multi-media support materials which they can use to improve the effectiveness of their extension and training work.

FIGURE 2 - 5

General and Simplified Guidelines for Utilizing Results of Knowledge, Attitude and Practice (KAP) Survey for Planning and Development of Extension Campaign Strategies.

SITUATION	IF: Level of Audience's K Knowledge	A Attitude	P Practice	THEN: GENERAL/BROAD EXTENSION STRATEGY PRIORITY: Main Approach	Main Purpose	Information Positioning	Communication Channel Emphasis for Multi-Media Mix Usage: Mass	Inter-personal	Group	[illegible]
1	Low	Low	Low	INFORMATIONAL	CREATE AWARENESS AND INCREASE KNOWLEDGE	"WHAT"	High	Low	Low	DESIRABLE
2	Medium	Low	Low	INFORMATIONAL	INCREASE KNOWLEDGE. EXPLAIN NEEDS FOR AND BENEFITS OF ADOPTING THE IDEAS	"WHAT" AND "WHY"	High	Low	Low	DESIRABLE
3	High	Low	Low	MOTIVATIONAL	EXPLAIN BENEFITS, REWARDS AND USEFULNESS OF SUGGESTED IDEAS. INFORM CONSEQUENCES OF NON-ADOPTION	"WHY"	High	Low	Moderate	IMPORTANT
4	Medium	Medium	Low	MOTIVATIONAL AND EDUCATIONAL	AS ABOVE	"WHY"	High	Moderate	Moderate	IMPORTANT
5	Medium	Medium	Medium	MOTIVATIONAL AND ACTION	CHANGE UNFAVOURABLE PERCEPTIONS. NEUTRALIZE NEGATIVE ATTITUDES AND BELIEFS. COUNTER-ATTACK MISCONCEPTIONS	"WHAT" AND "WHY"	Moderate	Moderate	High	ESSENTIAL
6	High	Medium	Low	MOTIVATIONAL AND EDUCATIONAL	INFORM CONSEQUENCES OF NON-ADOPTION. CORRECT MISUNDERSTANDING	"WHY" AND "HOW"	Low	High	Moderate	VERY ESSENTIAL
7	High	Medium	Medium	MOTIVATIONAL AND EDUCATIONAL	AS ABOVE	"WHY" AND "HOW"	Low	High	Moderate	ESSENTIAL
8	High	High	Low	EDUCATIONAL AND ACTION	PROVIDE LOGISTICAL INFORMATION FOR ACTION. DEMONSTRATE & TEACH PROPER USE OF RECOMMENDED TECHNOLOGY	"HOW"	Low	High	High	VERY ESSENTIAL
9	High	High	Medium	EDUCATIONAL AND ACTION	DEMONSTRATE & TEACH PROPER USE OF RECOMMENDED TECHNOLOGY	"HOW"	Low	High	High	ESSENTIAL

Source: Adhikarya, (1985)

Note: The classification of KAP levels (high, medium, low) depends on the type of innovation/idea promoted, size of population, cost/effort required, etc. It's a rather arbitrary judgement as to what criteria to use to classity these levels, but normally low level is 0-30%, medium level is 31-60%, and high level is above 60% for knowledge or attitude, and 0-20% (low), 21-40% (medium), and 40% (high) for practice.

In SEC activities, extension workers are provided with *pretested* extension and training support materials whose messages are specially designed and developed on the basis of the extension programme's problem-solving strategy plan. Furthermore, these extension workers are also given *special training* to ensure their understanding of extension strategies, message contents, and management-implementation plans, as well as when, with whom, and how they should utilize the various multimedia support extension and training materials.

Built-in Process Documentation and Evaluation Procedures

The advantage of employing a Knowledge, Attitude, and Practice (KAP) survey, as one of the tools for participatory problem identification and information needs assessment, is not limited to obtaining specific baseline data and inputs for planning extension strategies and improving its management operations. It also provides benchmark information and data for the purpose of qualitative evaluation, in terms of changes in the levels of KAP over time. In addition, SEC activities have built-in evaluation procedures in the forms of *formative* evaluation (e.g., pretesting of materials and Management Monitoring Survey) and *summative* evaluation (e.g., Information Recall and Impact Survey), for which data and information from the target beneficiaries are essential. SEC uses various participatory-approach evaluation methodologies including, among others, the quantitative survey, focus group interview, pretesting, recall tests, content analysis, field monitoring, and cost-benefit analysis.

Another important aspect of SEC is that it not only provides empirical evaluation results, but it also usually includes a step-by-step documentation of its operational process through summary briefs as well as more detailed printed, audio, or visual reports and presentations. Such *process documentation* and evaluation results have proved to be instrumental for facilitating SEC replications and in obtaining necessary policy, institutional, and financial support.

Applicability to Other Extension Programmes

Most, if not all, of the important principles and techniques employed in planning, implementing, and managing SEC activities are applicable for developing and implementing any extension programme. The SEC's process, operational phases, and implementation steps (Figures 2 and 3) are essentially similar to that of a regular (but well-designed) extension programme. SEC could thus be considered a microcosm of an agricultural extension programme. It may be safe to assume that if SEC activities can be carried out successfully in a campaign context, which has a very short time period, then SEC processes, methods, and techniques, either partially or holistically, can be incorporated effectively into a regular and institutionalized extension programme which has a longer time span.

Given appropriate training in various SEC principles and techniques and through direct involvement in undertaking a planned SEC programme, trained extension staff could help in applying a systematic, rational, and pragmatic approach to planning, implementing, managing, monitoring, and evaluating regular/routine programmes of an agricultural extension service. As indicated in a number of SEC evaluation studies, many extension staff who had been trained and have implemented SEC activities have continued to use their skills in developing and implementing other institutionalized extension programmes for various agricultural technologies. Some have replicated the complete SEC process, while others have applied only certain SEC principles or techniques. These efforts have been appreciated and welcomed by many senior officials of the ministry of agriculture where SEC activities were undertaken because they could see its concrete outputs, results, and impact.

SEC Operationally Defined

In the context of agricultural extension, a campaign is one of the methods of extension which can reach a large number of target beneficiaries in a short time period. To complement and improve programmes of a national agricultural extension service, the SEC method gives special emphasis to the following:

- A campaign is purposive, problem-solving oriented, and focuses on a specific issue or recommended technology.
- Its goals are consistent with, and guided by, the overall agricultural development policies and extension programme objectives.
- Campaign objectives are specific and formulated based on intended beneficiaries' felt needs and problems identified through a baseline survey of their knowledge, attitudes, and practices (KAP) vis-à-vis the recommended technology.
- A specific campaign strategy is developed with the aim of solving problems that caused nonadoption, or inappropriate or discontinued practice, of the recommended technology.
- A strategic planning approach is applied in the process of target audience segmentation, multi media selection, message/information positioning and design, and extension/training materials packaging, development, and production, with a view to obtaining maximum output-impact with the least effort, time, and resources.

FIGURE 2 - 1

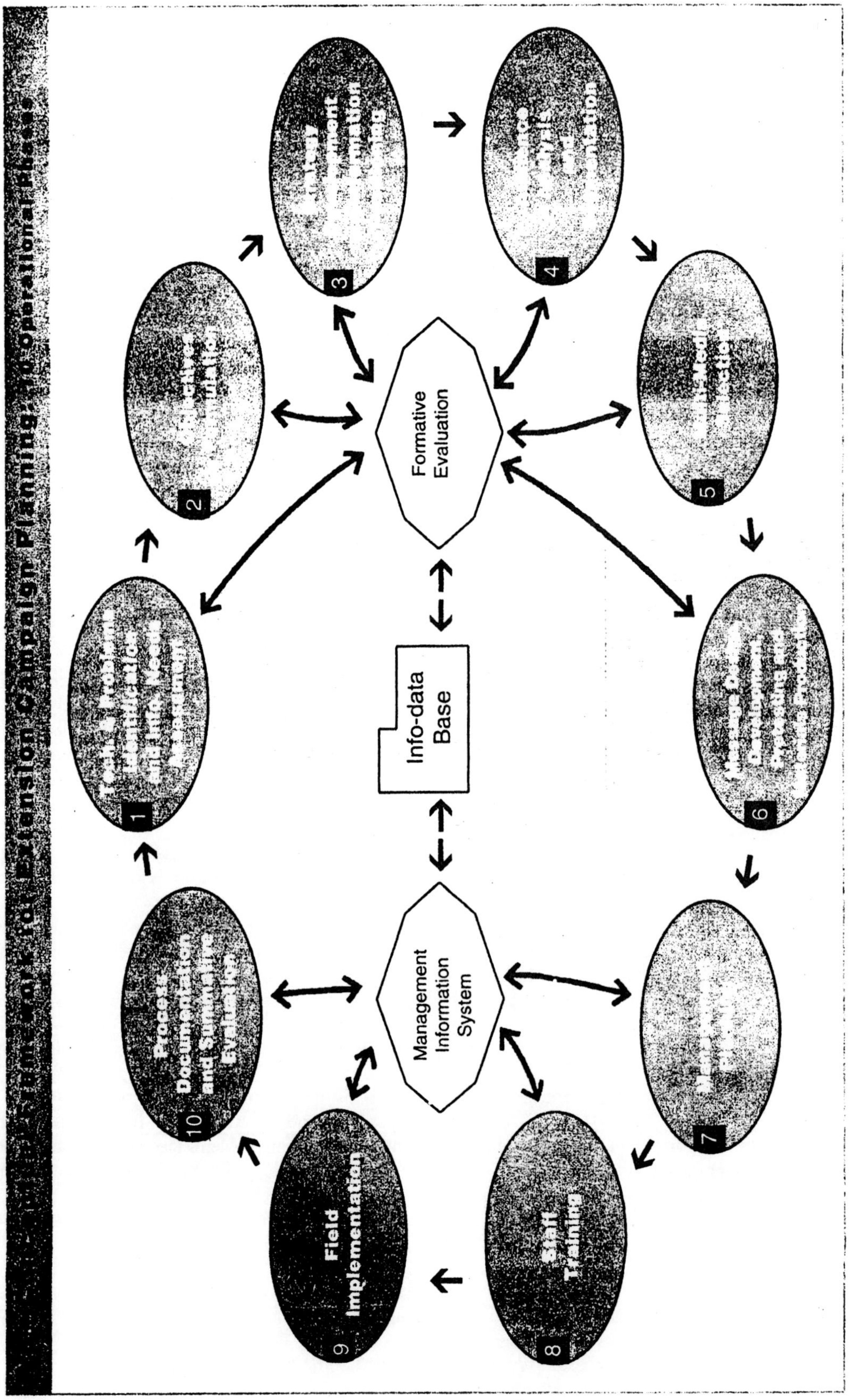

Adapted from: R. Adhikarya and H. Posamentier. *Motivating Farmers for Action: How Strategic Multi-Media Campaigns Can Help.* Eschborn, Frankfurt: GTZ, 1987
For more detailed information on similar frameworks, see: R. Adhikarya and J. Middleton, *Communication Planning at the Institutional Level: a Selected Annotated Bibliography*, Honolulu: The East West Center, 1979

FIGURE 2-2

Implementation Steps for Strategic Extension Campaign & Personnel Training

step	WORKSHOP ON	step	FOLLOW-UP ACTIVITIES
1	Purpose & method of Knowledge, Attitude, and Practice (KAP) survey (10 days)	2	Conduct K.A.P. survey, including Focus Group Interviews (through contractor, max. 3 months)
3	Campaign strategy planning, message design & prototype materials development (14 days)		
4	Methods for pretesting/formative evaluation of prototype multi-media campaign materials (6 days)	5	Field pretesting of prototype campaign materials (0.5-1 month)
		6	Revision and production of multi-media campaign materials (2-6 months)
7	Campaign management planning & training (6 days)	8	Briefing & training of extension workers and other field personnel involved in campaign activities (2-3 months)
9	Campaign evaluation methods and management monitoring survey procedures (5 days)	10	Field implementation of campaign (2-6 months)
		11	Conduct evaluation (max. 4 months) → Management Monitoring Survey (MMS) → Information Recall & Impact Survey (IRIS) → Focus Group Interviews (FGI)
12	Results demonstration & experience sharing on extension campaign planning, implementation & evaluation (5 days)	13	Publication/audio-visual documentation of campaign process, activities & results (max. 3 months)
		14	Replications & improvement of campaign process in other types of technology, locations or countries
		15	Utilization of good trainees as master trainers or resource persons to conduct similar workshops or follow-up activities in other parts of a country or in other countries

- Formative evaluation in the form of field pretesting of prototype multimedia campaign materials is conducted before they are mass produced.
- A comprehensive and detailed campaign management plan is an integral and vital part of the SEC process. It will not only spell out the implementation procedures and requirements, but will also be used to develop a management information system, including monitoring and supervision procedures.
- Special briefing and training for all personnel who are involved in SEC activities must be undertaken to ensure that they understand their specific tasks and responsibilities and have the necessary skills and support materials to perform such tasks effectively.
- Process documentation and summative evaluation on SEC implementation and impact are conducted, and those results are used to improve its ongoing performance (through Management Monitoring Survey) and to determine SEC's results and over all effectiveness (through Information Recall and Impact Survey, Focus Group Interviews, etc.), as well as to draw lessons learned from such experiences for future replications.

Integrated Process and Systems Approach

The conceptual framework of the strategic extension campaign (SEC) follows a generic model originally proposed by Adhikarya (1978) and described in detail in several of the author's other publications (Adhikarya, 1994; Adhikarya & Posamentier, 1987). The SEC programme planning framework and process are outlined in Figure 1, where all ten operational phases include participatory approach activities by soliciting relevant "feed-forward" and feedback from target beneficiaries. The SEC method advocates carrying out extension activities in a *systematic, sequential, and process-oriented manner,* rather than on an ad hoc basis. It is a *planned* extension programme with interrelated activities to be carried out following a management implementation plan by well-trained personnel within a given time schedule.

Staff Training as an Integral Part of SEC

Applying such a systems approach points out the need to train staff to master the whole extension process, rather than only some elements of the process or part of the activities. Thus, as can be seen in Figure 2, the suggested procedures in implementing a strategic extension campaign include training activities (through skills-oriented workshops) related to the operational phases or implementation steps which follow closely the conceptual framework and process.

It is, therefore, imperative that in the process of introducing or replicating the SEC method in a country, training a national group of extension planners, managers and trainers, subject-matter specialists, communication support staff, and field extension officers on relevant SEC methods and techniques must be an integral part of the SEC programme planning and implementation. Significantly higher cost and a longer time are required for the initial implementation of SEC activities, but such an investment in human resources development is perhaps one of the most cost-effective inputs which could significantly contribute towards institutionalization of an SEC approach in improving and strengthening a national agricultural extension system and service.

KAP Surveys, Evaluation Studies, and Followup Actions

In addition, operations research such as baseline Knowledge, Attitude, and Practice (KAP) surveys and other means of formative and summative evaluation must be built in to the extension process, programme, and methodology. The most vital part of the process is the actual field action and implementation activities which will be required for the preparation and followup of strategic extension campaign training and operations research activities. On the basis of the conceptual framework presented in Figure 1, we can identify three different but interrelated activity components and implementation steps which are part of the ten operational phases. These components, as shown in Figure 3, have various types of activities which are conducted in a process-wise sequence, as indicated by the various implementation steps (Figure 2).

Participatory Method of Planning and Implementation

The SEC method relies heavily on the participation of those who are involved in campaign activities, including campaign staff as well as target beneficiaries, in providing relevant inputs or suggestions throughout the campaign process. By so doing, SEC activities are more likely to address relevant problems and issues, as well as to provide practical solutions which are based on target beneficiaries' needs. Not only should the intended beneficiaries of a campaign be consulted, but potential campaign implementors such as concerned field staff, trainers, and community leaders should also be involved and given a role to play in different aspects of campaign planning, implementation, and management.

Through a careful and systematic method of soliciting fairly specific ideas and suggestions from a sample of concerned groups of people involved in SEC activities, a useful *campaign map* can be drawn to determine its direction, strategy, types of message contents, training needs, and field implementations requirements. A problem-solving and demand-driven strategic extension campaign also needs relevant "feed-forward" information from its intended beneficiaries and concerned campaign personnel. It is thus imperative that a participatory oriented method of planning be applied to ensure a needs-based extension campaign which can realistically be implemented and provide benefits to its intended clientele.

Suggested Conceptual Framework for Strategic Planning of Extension Campaigns

Planning is defined as a process of identifying or defining problems, formulating goals, thinking of ways to accomplish goals, and measuring progress towards goal achievements (Middleton & Hsu Lin, 1975). Such a plan must outline the management actions to be taken in implementing the strategy. Thus, in this context, campaign planning has to include both strategy planning (i.e., what to do) and management planning (i.e., how to make it happen).

Strategic planning can be defined as the best possible use of available resources (i.e., time, funds, and staff) to achieve the greatest returns (i.e., outcome, results, or impact). A strategic extension campaign plan should provide specific guidelines and directions in developing and making information, education, and communication activities operational. It must be constantly reviewed, especially at the implementation stage. Modification of the plan may be required because of specific local conditions and problems or alteration of the policies or objectives which guided the original plan. The plan should be flexible and ready for necessary modification as suggested by feedback results (e.g., through process and formative evaluation, including pretesting) in order to improve the strategy or management of campaign activities.

The process of developing a strategic extension plan can be divided into two major parts: strategy development planning and management planning. To provide a systematic approach in developing a strategic extension campaign plan, a generic conceptual framework (see Adhikarya, 1994; Adhikarya & Posementeir, 1987) is suggested based on a ten-phase circular model (Figure 1). The suggested process of developing a strategic extension campaign plan is described below, adapted from the ten phases of the conceptual framework originally proposed by Adhikarya (1978):

Part I: Campaign Strategy Development Planning

Phase 1: Technology and problems identification and information needs assessment
Phase 2: Campaign objectives formulation
Phase 3: Strategy development and information positioning
Phase 4: Audience analysis and segmentation
Phase 5: Multimedia selection
Phase 6: Message design, development, pretesting, and materials production

Formative evaluation should be a component in all of these phases, especially in phases 4 to 6. Formative evaluation in this context means the process of testing the suitability, appropriateness, or effectiveness of the campaign strategy and plan, including its multimedia messages and support materials, preferably before full implementation, in order to ensure good campaign performance or results.

When a plan for a campaign strategy is completed, it must be translated into action. At that stage, the task of an extension campaign planner shifts from strategy development to *management planning* which includes the following phases:

Part II: Campaign Management Planning

Phase 7: Management planning
Phase 8: Training of personnel
Phase 9: Field implementation
Phase 10: Process documentation and summative evaluation

These four phases should be supported by a management information system to provide planners with regular and up-to-date information for at least the basic components of the management objective: who will do what and when. There are three kinds of management activities for which such information is needed to make effective decisions: personnel, finance, and logistics.

KAP Survey as a Key Feature in SEC

Because the strategic extension campaign (SEC) method follows a participatory and demand-driven or needs-based approach, target beneficiaries need to be consulted in the process of identifying problems and needs regarding their requirements or acceptability of a given technology. While problem identification methods such as Participatory Rural Appraisal (PRA) and other macrolevel studies can provide good contextual information, a more specific and focused method of obtaining qualitative information and quantitative data should also be considered. A suggested procedure for conducting a participatory assessment of problems and needs is through a baseline survey of target beneficiaries' knowledge, attitudes, and practices (KAP) on specific and critical elements of a recommended technology. The KAP survey is problem-solving oriented and operates at a microlevel, with a focus on determining the knowledge, attitudes, and practice levels of target audiences vis-à-vis the critical elements of a given technology recommendation. The KAP survey seeks qualitative information from respondents, through focus group interviews, on the reasons and causes of their negative attitudes and nonadoption or inappropriate practice with regards to recommended technologies. KAP survey results are very useful for campaign objectives formulation and strategy development.

Results of a KAP survey can be used to analyse which specific elements of the technology package are not known to the majority of target beneficiaries, what are the reasons for their negative attitudes, how and why they have practised recommended technologies inappropriately, etc. KAP survey results can also be used for audience analysis and segmentation purposes to determine who needs which types of information through what combination of multimedia materials and channels. In addition, relevant findings from surveys on media consumption patterns and habits, media availability and reach, and other sociopsychological and anthropological research are useful inputs to the exercise of extension strategy planning and message development.

Lessons Learned

Experiences from various SEC applications have generated some important and useful lessons for increasing the effectiveness of agricultural extension systems and programmes. Details of the lessons learned are presented in Adhikarya and Posamentier (1987) and Adhikarya (1994). The main lessons are summarized below. A strategic extension campaign:

- Enhances the agricultural extension planning process
- Builds cadres of extension programme planners and trainers
- Helps in improving extension linkage with research
- Is needed most by small, resource-poor farmers
- Helps in improving extension linkage with training
- Reduces extension system's workload and increases its coverage
- Encourages partnership with, and participation of, community-based organizations
- Helps revitalize extension workers' professionalism
- Shows that extension programmes can be strategically planned, efficiently managed, and systematically monitored and evaluated
- Can contribute to improving and strengthening agricultural extension systems and programmes

References

Adhikarya, R. (1994). *trategic extension campaign: A participatory-oriented method of agricultural extension.* Rome: FAO/United Nations.

Adhikarya, R. (1985). Planning and development of rat control campaign objectives and strategies for the state of Penang, Malaysia. A summary report of a workshop on multimedia campaign planning for integrated pest control, organized by FAO/United Nations (projects GCP/RAS/101/NET and GCP/RAS/092/AUL).

Adhikarya, R. (1978). Guideline proposal for a communication support component in transmigration projects. A consultancy report prepared for FAO /United Nations (project 6/INS/01/T). Rome: FAO.

Adhikarya, R., & Middleton, J. (1979). *Communication planning at the institutional level: A selected annotated bibliography.* Honolulu, Hawaii: East-West Communication Institute, the East-West Center.

Adhikarya, R., & Posamentier, H. (1987). *Motivating farmers for action: How strategic multi-media campaigns can help.* Eschborn, Frankfurt, Germany: GTZ (Deutsche Gesellschaft für Technische Zusammenarbeit).

Middleton, J., & Hsu Lin, Y. (1975). Planning communication for family planning. Honolulu, Hawaii: East-West Communication Institute, the East West Center.

Chapter 11
Evaluating Extension Programmes

David Deshler

Getting Started with Evaluation: Avoiding a Passive Sabotage of Evaluation Efforts

Although extension educators, funders, and administrators are in favour of evaluating extension programmes, honesty requires us to admit that most of us are not overly enthusiastic about undertaking it. There are many reasons for our resistance. However, if we are to provide evaluation leadership, we must recognize forms of organizational and personal resistances, or these will sabotage our evaluation efforts. It is not uncommon to use the occasion of evaluation to mask our insecurity with distrust or criticism of others' intentions, motivations, competency, or adequacy of effort. We are sometimes caught in a bind between our own reluctance to ask ourselves hard evaluation questions and our resentment of outsiders asking questions.

The second form of resistance is fear, particularly the fear of change that evaluation might precipitate. It is the nature of organizations to be self-protective and defensive. Evaluation and organizational comfort appear to be somewhat incompatible because evaluation challenges organizations to change. We all resist change to some extent. Although evaluation is not revolutionary, it is a handmaiden to gradual change, and we have to recognize that our reluctance to participate in evaluation is partially a reluctance to embrace needed change.

Another resistance to evaluation comes from our need to avoid embarrassment about potential bad news. We need to know the quality of our efforts, but we also have a fear of finding out the truth about our achievements, particularly if we lack confidence. Most of us avoid tests for the same reason. We need to face up to our personal and organizational ambiguity regarding our need to know and our fear of knowing.

Still another real resistance results from the fact that evaluation is often an additional task to an already impossible workload. Those whose job description includes evaluation may need only to be reminded of this. However, potential benefits may have to be discussed and identified if collaboration is to become a reality among those who do not have a formal professional responsibility for under-taking it. Benefits include recognition for achievements, opportunities to improve practice, establishment of accountability, and learning new lessons about our efforts.

Creating Positive Evaluation Metaphors

The term "evaluation" can be threatening. If evaluations are to be positively embraced by extension organizations, it may be necessary for those initiating them to be positive in the language they use to describe the proposed effort. Negative metaphors that are sometimes used for evaluation, such as "under the gun," "a mine field," "an invasion of inspectors," "last rites before hanging," and "defend the fortress" reflect past negative experiences with evaluation. Why not create more positive metaphors for the evaluation task? For instance, we could describe evaluation in medical metaphors such as "complete programme physical" to identify physical soundness and potential programme health risks. The term "auto-diagnostico" has been used in the Dominican Republic and "self-strengthening" in Sri Lanka. Or we could use a business metaphor, such as "taking stock" of profits and liabilities. We could use sports metaphors, such as "watching video tapes" of yesterday's game to improve our "game plan." We can use "mirrors" to see ourselves as others see us. We could consider evaluation to be a "learning process" or a "sabbath" day of reflection, renewal, gratitude, and celebration. Being self-conscious and deliberate about the language that is used regarding evaluation may help reduce unnecessary resistance to the effort and may build positive support.

Exploding Myths about Evaluation

Several evaluation myths have often discouraged extension managers from engaging in useful evaluation.

Myth #1: Evaluate only when mandated. Many funded programmes require evaluation as a form of accountability. However, it is a myth that evaluation should occur only if it is mandated. On the contrary, evaluations that are self-initiated are more likely to be taken seriously for immediate programme improvement. Programmes become responsible and excellent just as often through self-initiated evaluation.

Myth #2: Evaluation is an add-on. It is a myth that evaluation is an add-on activity or at most a pretest with a posttest. It is most meaningful when it is integrated with decision making at every stage of programme planning and operation (Patton, 1991).

David Deshler is an Associate Professor of extension education, Cornell University, Ithaca, New York.

Myth #3: Evaluation is an activity for experts. It is a myth that evaluation should be undertaken only by technical experts. Yes, complex methods can be used; however, systematic evaluation can be undertaken by inexperienced managers, and specialists and educators themselves can be helped to critique their own work. This chapter is intended to help managers and specialists to do evaluation themselves.

Myth #4: Outside evaluators are best. It also is a myth that evaluation should be done only by external, outside, objective evaluators. Yes, external evaluators are often useful in challenging insiders to address what they have overlooked because of their nearsightedness. However, internal, self-initiated, and subjectively oriented evaluations also can be rigorous and valuable. In fact, because they often are participatory in generating, analysing, and interpreting data, they may result in greater acceptability of the findings and recommendations.

Myth #5: There is one best evaluation approach. Still another myth is that there is one best way to conduct an extension evaluation. Some approaches are probably better than others for addressing particular types of questions or concerns. However, the many types of evaluation approaches have their own strengths and limitations. Some situations require quantification and measurement, while others require qualitative, descriptive, and subjective data. Alternative approaches will be briefly described later in this chapter.

Myth #6: Quantitative data are best. A mixed-methods approach combining qualitative and quantitative methods can lead to better understanding and appreciation of phenomena under evaluation and provide triangulation, convergence, and corroboration of results from different methods. Qualitative methods are best for understanding the nature of something, while quantitative methods help in appreciating its extent. If we do not know the nature of something, we should conduct qualitative studies. After measuring something, we may still need to use qualitative methods to learn about variations and unique forms. The use of mixed methods can increase user responsiveness to evaluation information (Greene, Caracelli, & Graham, 1989). We should ask which combinations of methods are best for answering which evaluation questions, rather than deciding on a method and then forcing or changing our questions to fit that method.

Major Elements in Evaluation

There are at least five major elements in most evaluations: (1) focus questions, (2) objects or events to be evaluated, (3) data or evidence, (4) analysis and interpretation using judgement perspectives, and (5) judgements, conclusions, or findings. Purposes and approaches or models may vary, but these elements will be present in one form or another.

Selecting Evaluation Purposes: Unclear Purposes Ensure Unsatisfying Evaluations

There are many purposes for undertaking evaluations in any particular situation. It cannot be assumed that all stakeholders (farmers, extension staff, administrators, funders) share common purposes. Although people may never completely agree about the purposes, one of the tasks at the outset is to seek clarity about evaluation purposes through discussions with major stakeholders. Appreciating different purposes at the outset can reduce conflicts and disappointments down the road. Consider the following purposes.

Pseudo Self-Serving Purposes

Since organizations, including extension systems, have a self-serving tendency, it is not unreasonable to expect that some staff members, especially those in the highest places, may want a pseudo evaluation that will postpone, buy time, or avoid threatening change. In these cases, evaluators are not taken seriously, and the evaluation becomes a meaningless political diversion. In other cases, some members of organizations want evaluations as excuses for evading or avoiding administrative responsibility or to provide a scapegoat for criticism. Evaluations that are undertaken only to make the programme look good ("whitewash job") or to make someone or some aspect of a programme look bad ("hatchet job") are pseudo and illegitimate.

Enhance Accountability Purposes

It is quite common for external donors to expect that evaluation will provide accountability through evidence of impact, or to document cost-benefits, or to measure efficiency-effectiveness. In some cases, this evaluative evidence is considered in decisions to continue the programme; or to propose change, expansion, or reduction of a programme; or to change a policy, organizational structure, philosophy, or design. The potential for negative findings and the threat of discontinuing funding has led to "hiding the mistake," a dysfunctional practice. However, evaluations rarely provide a single basis for political decisions. They often are used by funders, administrators, or policy makers to justify their decisions even when the evidence of benefits is weak.

Improve Performance Purposes

This purpose of evaluation is sometimes called "formative" because the results are intended to help improve the programme during its formative stages. This is in contrast to "summative evaluations" when the purpose is to sum up or summarize the accomplishments at a point in time. When evaluations are to improve programmes, lessons learned about strengths and limitations of the programme are mined from the data so that changes can be made immediately. Sometimes the intent is to discover new approaches and alternatives or to adjust the programme to changing situations or client groups. Evaluation also is used to understand multiple reasons for apparent failure or to improve the management or operation of a programme.

External Evaluator **External Criteria**	**Internal Evaluator** **External Criteria**
External Evaluator **Internal Criteria**	**Internal Evaluator** **Internal Criteria**

Figure 1. Alternative locus of evaluators and criteria for planning extension evaluations.

Social Learning and Communication Purposes

Sometimes evaluations are intended to stimulate political dialogue or to resolve political conflicts intelligently. For example, an evaluation of extension in a country could provide an opportunity to debate the need to hire more women agents to respond to an increase of women in small-scale agriculture or to extend the extension network to subsistence farmers not being served. Often the most significant contribution of an evaluation is the creation of new expectations, new organizational arrangements, new linkages, and new purposes and goals. Evaluation may give visibility to a good idea and new language that can communicate new ways of viewing extension to others who also may want to share an experiment.

Evaluation purposes tend to vary, depending upon where one stands within a system like extension. External funders often want an accountability purpose, while field staff are more likely to favour a programme improvement purpose. Policy makers and programme administrators can often appreciate an evaluation that contributes to new ways of thinking about extension or new forms of extension. Farmers want an evaluation to improve the benefits they may receive from extension staff.

Recognizing the Politics of Evaluation: Know the Stakeholders or You Will Be Sorry

Evaluations are never value neutral. Political implications are always present because stakeholders have interests to defend. In this sense, all evaluations are political. Therefore, it is very important to know the values, expectations, and interests of the stakeholders at the outset. Failure to understand these interests may make the findings irrelevant or may cause unnecessary conflict and rejection of the findings. Knowing the interests also helps in the focus and design of the evaluation.

Values, expectations, and interests are reflected in stakeholders' visions regarding what "good" looks like in extension. Some may favour an extension system that empowers the poorest of farmers in their struggle for more land and a larger share of market benefits. Multinational agribusinesses may view the success of extension as increasing productivity for export. What a "good" or "excellent" extension system should look like will vary according to these different values and criteria. What different funders want may differ from what various groups of farmers may want in an extension system. Assumptions of stakeholders also vary regarding the relationship of extension education to larger social and political visions. For some, the extension system is intended to reproduce existing power relationships between government and farmers. For others the extension system is viewed as an instrument to transform those relationship. For some, extension is primarily a technology transfer system from researchers to farmers. For others, extension is a communication network among farmers, researchers, credit institutions, market organizers, consumers, and government policy makers. These expectations, values, and interests are reflected in the criteria that various stakeholders put forth as central for judging extension, regardless of the stated objectives in programme documents.

Political interests are usually behind the debate over external versus internal evaluations. Some external stakeholders favour external evaluators because they want an evaluation to serve the direction of change, and they suspect that internal evaluators will be self-protective and not face up to the necessary reforms. Local programme staff sometimes fear external evaluators who impose criteria or external visions through evaluation judgements without understanding the situation on the front lines. They do not want to be evaluated against unreasonable external criteria. The matrix in Figure 1 regarding locus of evaluators and criteria may help clarify conflicting viewpoints over external versus internal evaluation or external versus internal criteria for judging what is "good" or "excellent." One way to accommodate these interests is to employ evaluation teams of external and internal evaluators and to negotiate combinations of external and internal criteria.

Politics is about power and power relationships. Education, including extension education, also is about power, because one way of increasing power is through increasing knowledge. Extension is political in that it can either reproduce existing access to knowledge or privilege or provide a redistribution of power in a society through increasing knowledge access. Probably the most crucial political question that evaluation can raise is, "Who benefits most from extension programmes?" This question is political, because the answer often reflects class and economic privilege, ethnic or racial dominance, and gender interests and relationships between staff and various sectors of the population of farmers. How does the extension system privilege some and not others? Where are extension staff located in relationship to the political, economic, social, and cultural structures of a country? How is extension affected by relationships that governments have to various sectors of agriculture? These values, expectations, and interests provide the background for negotiating the purpose, focus, and design, as well as the interpretation of the findings of evaluations.

Selecting Alternative Approaches and Models: Which Model for Which Purpose?

What are several alternative approaches to programme evaluation? Choosing among these approaches is important because for each there are different assumptions about what data to collect, how to collect them, and how to make judgements about success. The following seven major approaches will provide a sufficient choice for most extension evaluation situations: (1) expert model, (2) goal-free model, (3) attainment of objectives model, (4) management decision model, (5) naturalistic model, (6) experimental model, and (7) participatory evaluation model.

Major Models for Programme Evaluation

Expert Model. This approach relies on expert judgement (Eisner, 1983). Usually, documentation is prepared in advance of experts' visits. The experts then interview, analyse documents, and make judgements using their own judgement perspectives or those set as standards by the outside organizations or stakeholders. Typically this type of evaluation brings in a team of experts from FAO or extension systems from several countries to make judgements and comparisons regarding strengths and limitations.

Goal-Free Model. This approach assumes that outside evaluators do not know, or need to know, what the programme has intended to accomplish, but that it is the task of the evaluators to uncover what is actually happening relative to farmers' interests regardless of stated goals and intentions. The focus point is to identify environmental and farming conditions and then to compare these needs with what people are actually experiencing as a result of the extension programme. The gap is then viewed as a starting point for making changes in the programme. An example is an evaluation that describes conditions of indigenous farming groups cultivating fragile hillside soils and comparing these conditions with access to and appropriate content of knowledge from existing extension services. This approach relies heavily on open-ended interviewing and observation by persons who do not have a vested interest in the programme (Scriven, 1972).

Attainment of Objectives Model. This approach assumes that the success of a programme can be determined by measuring a programme's outcomes against its own goals and objectives. This type of evaluation begins with clarifying measurable objectives and then gathering data that validate the extent to which these objectives have been met. For this model to be credible, an essential feature should be added, namely, the evaluation of the appropriateness of goals and objectives, given the circumstances and needs of farmers. For example, an extension system may have adequately met its objectives of increasing production of maize among large landholders, but at the same time it may have neglected to question its lack of commitment to small landholders or tenant farmers. If an attainment of objectives evaluation is anticipated, programmes are often tempted to set goals quite low so that outcomes will be met easily, thus appearing to be successful while ignoring major challenges. This model also has a "black box" limitation in that it tends to ignore the extension process, thereby failing to provide explanations for outcomes (Provus, 1971).

Management Decision Model. The purpose of this model is to provide relevant information as a management tool to decision makers. It assumes that evaluation should be geared to decisions during programme initiation and operation stages to make results more relevant at each particular stage. Participation of stakeholders is central to the process because evaluation should serve their decisions. Sometimes cost effectiveness and operations monitoring are included (Stufflebeam, 1971; Tripodi, Fellin, & Epstein, 1971; Gold, 1988). One limitation of this model is the tendency for the decisions of major stakeholders to be viewed as more important than those of various types of farmers, especially women in agriculture who may not benefit directly from such an evaluation unless care is taken.

Naturalistic Model. This model assumes that a programme is a natural experiment and that the purpose of evaluation is to understand how the programme is operating in its natural environment. There is an assumption that programmes are negotiated realities among the significant stakeholders and that evaluation serves this value-laden negotiation (Cronbach, 1981; Guba & Lincoln, 1989). Data should be collected and analysed from multiple perspectives. The outcome of the evaluation is dialogue concerning disagreements about objectives, expectations, problems, opportunities, policies, procedures, and suggested changes in methods or activities. Many positive collaborative changes can be made through this model

Level 5: Farmers conduct their own evaluation of extension independently of extension and report their findings to policy makers.

Level 4: Farmers carry out evaluation of extension in cooperation with extension managers and make decisions regarding changes in providing extension services.

Level 3: Farmers receive evaluation results and other information from extension staff and are asked to give reactions and recommendations for improving extension processes and resources.

Level 2: Farmers receive information, evaluation summaries, feedback on extension performance from extension staff, but are not asked to react.

Level 1: Farmers provide data and evidence of their achievements along with their reactions to extension without being involved in planning evaluation efforts.

Figure 2. Ladder of farmer participation in extension evaluation. Adapted from Arnstein (1969).

of evaluation if conflict resolution skills are combined with evaluation. Another purpose of this model is to diagnose or to identify the causes for certain behaviour on the part of some farmers, agency staff, or other development actors (Murphy & Marchant, 1988).

Experimental Model. The purpose of this approach is to determine whether changes in programme outcomes (learning accomplishments) were due to the contributions of the programme and not just to life's experiences or from other influences (Goldstein, 1986). This model asks the question, "Were differences in sustainable agriculture practice attributable to the programme?" The simplest way to determine causality between the programme inputs and comparable groups, a group that received the educational treatment and a group that did not. This means that programme accessibility, at least during the experiment, programme consequences is to compare at least two is withheld from those learners who serve as a control group. Because of the nature of human subjects, the ethics of withholding educational services, and the difficulty of controlling for external influences, it is extremely difficult and costly to operationalize this model. It is recommended that this model be used only when major changes are expected or when a major failure is anticipated in pilot efforts where causal claims are central to making major programme investments (Rossi & Freeman, 1982).

Participatory Evaluation Model. The purpose of this model is for extension educators and farmers themselves to initiate a critical reflection process focussed on their own activities. This is done through identifying a persistent major situation such as extension's neglect of women inagriculture; subject it to critical reflection, underlying assumptions, habits of mind, and cause and effect expectations; and then after creating new assumptions, change practices and validate or invalidate the results. The model assumes a democratic participatory process along with autonomy on the part of educators and learners at the local level (Brunner & Guzman, 1989; Greene, 1988). This is a form of what is usually called "participatory action research."

Benefits and Limitations of Participation in Evaluation

Why use participatory approaches to evaluation? Involving people who are on the receiving end is likely to assure the most efficient allocation of scarce resources and the early identification of ineffective or wasteful use of resources. People on the receiving end are ultimately the best judges of impact, whether benefits have been produced or not (Uphoff, 1989, 1992). Being included in planning, implementation, and evaluation will show farmers that they are regarded as responsible, capable individuals and not simply passive "beneficiaries" or a "target group." Participatory evaluation is self-educating. It can encourage the development of human capacities among farmers. Without participation in evaluation, sustainable agriculture is unlikely. Participatory evaluation also can decrease the paternalistic, directive, impatient, or insensitive relationships among officials, technicians, and farmers by improving staff attitudes toward working with persons having less status and education. Although an initial cost of time is required, participation in evaluation can speed implementation when participants take greater ownership of efforts.

Farmers' lack of experience is sometimes said to be an impediment to participatory evaluation, but this need not be an obstacle if the process whereby people gain experience can be planned for and invested in (Uphoff, 1992, p. 5). Lack of resources and organizational skills can be overcome. Power differentials, social stratification and cleavages, and personal conflicts do present limitations. However, these obstacles can be reduced by outside actors who can provide spaces for subgroups to express their views for the sake of broader community interests. Officials or NGO staff themselves may present the most formidable obstacle by their paternalism and preoccupation with control. They often feel threatened by people participation. Organizations tend to replicate in their environments the same attitudes, values, and social relations they exhibit internally. Practising participatory evaluation internally will provide a model for participatory evaluation that includes farmers. Positive experiences with participatory evaluation can help overcome staff anxiety.

How much participation should the farmers have in evaluating extension programmes and their own learning? Farmers, in many extension programmes, have participated in evaluation of extension through farmer associations and committees. However, as Arnstein (1969) has pointed out in a much quoted article regarding a ladder of participation, there are various levels of involvement and participation. In her model, the bottom rungs provide for merely token participation, while the top rungs allow for participant control. A ladder adapted from Arnstein that provides levels of farmer participation in evaluation of extension is depicted in Figure 2.The practice of levels 3 through 5 will increase authenticity of data, reduce paternalism in relationships, assure relevance, and make possible a collaborative commitment to positive change.

Levels 1 and 2 can be characterized as pseudo participation because they represent paternalism on the part of extension. Levels 3 and above can be characterized as genuine participation because they represent collaborative or empowering relationships.

Farmers are by nature involved in informal evaluation of their own practice. What is at issue is the degree to which extension managers will deliberately collaborate and share control with them in collecting, analysing, and reflecting on evidence of their achievements and judgements in regard to the helpfulness of events, technologies, inputs, processes, and learning experience.

Focusing the Evaluation Effort: You Can't Evaluate Everything, So Let's Set Limits

What should be the focus of programme evaluation? This question raises the spectre of evaluating everything, which is an impossibility. Choices and priorities among many possible questions have to be made. A full range of possible focusses can be represented at least in part by the levels represented in Figure 3 (modified from Summers, 1977), which shows eight major areas of focus for programme evaluation: (1) inputs, (2) activities, (3) participation, (4) reactions, (5) individual change, (6) organizational change, (7) community change, and (8) national change. Evaluations rarely cover all items of these areas. Most limit themselves to a combination of items that serve the evaluation model or the concerns of stakeholders.

Narrowing the focus usually begins during planning with the major stakeholders in a programme effort (farmers, extension staff, administrators, and funders). Interviews with these stakeholders are usually conducted at the outset of evaluation efforts, often during the programme-planning process, to identify the programme model or approach and to determine the questions that are central. Stakeholders often do not agree, so priorities, methods, and costs must be negotiated so that the task can be "reality based" and "doable." The key here is to determine the decisions that stakeholders intend to make based on the evaluation findings so that immediate use will be made of the information generated. Evaluation use is not based on the quantity of data, but on its timeliness and relevance to decisions and purpose.

8. *National impacts:* Political stability, economic fairness, agricultural environmental sustainability
7. *Community change*: Changes in administration of justice; health, welfare, and quality of life; fairness in the marketplace; change in human rights, status of women; change in economic and social indicators for poor; change of indicators of sustainable agriculture and natural resource management; change in communication patterns and access to education and news; public opinion change; fairer distribution of land and other resources; improved interorganization relations; evidence of conflict resolution; and cultural practice change
6. *Organization change:* Group operation and management; economic performance; technical operation and management; financial operation; group institutionalization and self-reliance, new groups of farmers included, new organization linkages; change in staff performance, new service delivery, new methods used, additional facilities and equipment; cost-benefits improved; new philosophy, purposes, and goals; improved organizational culture
5. *Individual change:* Changes in knowledge, attitudes, skills; sustainable agricultural practice; change in aspirations, selfimage, perspectives; expenditure of effort and money; use of methods, services; invention of appropriate technology; increased production or use of tools; compliance with or opposition to public policy; patterns of communication, career directions, and family relationships
4. *Reactions:* Testimonials; reactions to the relevance of content; appropriateness of technology, helpfulness, per ceived value of educational experience; reputation of the extension provider
3. *Participation:* Farmer access to extension services by social class, gender, and ethnic groups; intensity of face-to-face contacts; extent of media-assisted contacts; type of participation (volunteering, planning, recruiting, learning, experimenting, evaluating); indicators of commitment (attendance, continuity, frequency)
2. *Activities:* Participatory rural appraisal; planning; local knowledge documentation; farmer experimentation; farmer-to-farmer knowledge sharing; farm tours; farmer organizing; master farmer leadership training; farm demonstrations; exhibitions and fairs; residential workshops; marketing analysis; farm policy education
1. *Inputs-resources:* Organizational sponsorship and networks; funds; organizational design, facilities, equipment; philosophy, mission, goals, objectives; staff, resource people, volunteers; local and external research knowledge and relevance; cultural, economic, and political context

Figure 3. Focus for programme evaluation.

The seven evaluation models described above tend to focus on the different levels shown in Figure 3. For in-stance, the expert model most often focusses on data from inputs, activities, and participation, while the goal-free model tends to focus on individual change, organization change, or community change, ignoring the inputs and activities. The attainment of objectives model usually compares the philosophy, goals, and objectives of inputs to the extent of individual or organizational change outcomes. The naturalistic model emphasizes understanding activities, participation, and reactions as processes that occur within cultural, economic, and political contexts. The experimental model emphasizes causal relationships between inputs and individual or organizational change. The participatory evaluation model emphasizes activities and their relationship to benefits and values to farmers. It also emphasizes participation of farmers themselves in planning the focus, data collection, interpretation, and implementation of action that emerges from the evaluation process.

Selecting Methods for Programme Evaluation: Choosing the Right Tool for the Task

Which methods are right for the task? The basic rule here is that the selection of methods follows the selection of focus, not the other way around. Each evaluation question must be examined in relationship to what would constitute evidence for answering it. The following brief descriptions of data collection methods, although by no means exhaustive, can be used as a toolkit for a variety of circumstances. The list includes document analysis, observations, interviews, surveys, focus group committees, field visits and tours, village drama and role plays, maps, case studies, field trial documentation by farmers, remote sensing, and aerial photography.

Document Analysis. Examples include minutes of meetings, correspondence, budget records, workshop notes, participant papers, and newspaper reports, to name a few. These can be treated as data, analysed for content, and summarized in relation to questions, including extent of inputs into the programme; levels of participation, nature of goals and activities, and themes regarding problems, concerns, expectations, and new directions. Themes from documents can be a credible source of information. These documents usually do not reveal participants' motivations or subjective experiences. However, documents often reveal difficulties of programme operation.

Observations. Observers can be outsiders or persons who are involved in learning activities. Observers are usually given a short list of items that may include extent of participation and personal interaction, nonverbal indicators of interest or inattention, leadership roles, performance levels, and conflict indicators, to name a few. Both qualitative and quantitative data can be collected. Findings can be reported to the learners as a whole to start a reflective process about what may need to be changed, or they can be used as evidence of successful methods or learning outcomes (Worden & Neumaier, 1987). Observations of process and outcomes can be recorded by video or photo documentation. These data are very powerful graphic ways of communicating the nature of a programme and its outcomes to sponsors as well as to potential learners. Video records of learners' local knowledge also can be used to help them reflect on strengths and limitations of knowledge.

Interviews (Key Informants, Oral Histories, Storytelling). Interviews are probably the most widely used method for programme evaluation, including the evaluation of learning. Interviews with key informants and representative farmers are suited to indepth explorations of issues. If the questions are standardized, responses can be tabulated numerically to indicate item strength. If questions are open-ended, indepth unique responses can be generated, which in turn can provide information regarding reasons why the activities are viewed differently by diverse groups of participants. Individual oral histories can inform patterns of practice and the use of extension resources. Storytelling is one of the oldest forms of human social life. Evaluative judgements are usually embedded in stories. Interviews, oral histories, and storytelling with farmers who are not served is essential to a full picture of extension. Illiterate members of the community can participate fully through interviews, oral histories, and storytelling, demonstrating that illiteracy does not rule out participatory evaluation.

Group interviews (sometimes called focus groups) can be formed according to geographic locations or by farm type to discuss specific evaluation questions. Sometimes the groups may already exist. At other times new groups may be formed just for the evaluation. The purpose is not only to generate judgements using agreed upon criteria, but also to uncover unanticipated outcomes, applications, opportunities, and problems to inform future extension efforts. A community can reconstruct its history, chronology of events, crises, turning points, accomplishments, and so forth. Chalkboards, pictures, and drawings can represent milestones or decisions. The art of conducting group interviews can be learned. Many community members have a talent for this and are more likely than agency staff to evoke authentic responses. Subgroups are often necessary to listen to voices that are unlikely to be heard in groups dominated by those with status.

Surveys. The survey is a more standardized form of data collection that incorporates a prepared questionnaire. In most parts of the world, surveys of farmers in rural areas must be conducted by interview using interviewers who know the territory, language, and culture of potential respondents. Sometimes surveys can be administered at meetings or public gatherings; however, responses have to be treated as an "opportunity sample," rather than as a "random sample," and therefore generalization of the findings to a larger population is limited. Surveys are often used to evaluate the extent of practice, estimations of production yields, preferences for appropriate technology, and expectations regarding the future. They are best used with homogeneous populations rather than with quite diverse populations because standardized questionnaires are less likely to be sensitive to diversity. Evaluation of unique practices and adaptations of diverse farmer types is best done through interviews and observations.

Field Visits and Tours. There is no real substitute for field visits and tours to provide authenticity and reality to the conditions, limitations, and impacts of extension programmes. Evaluation teams composed of local farmers, extension staff, administrators, funders, and external evaluators provide balance and interactive learning regarding different perspectives. Team members can undertake observations and interviews and learn from each other about their findings during travel. Assignments can be distributed so that special knowledge regarding specific aspects of situations can be gathered. Some team members can focus on economic, social, and cultural aspects, while others focus on technical aspects. Comparing data, analysis and reflection on findings, and insights following field visits can generate a more holistic, balanced evaluation.

Documentation of Farmer Demonstrations. Farm visits, which have been used frequently by extension staff for teaching and the transfer of technology, also can be used for evaluation of appropriate technology. Farmers and local leaders can be taught to conduct their own field trials, thus fostering the pride and dignity of local people who can transfer appropriate technologies through their local languages. When farmers choose their own focus of inquiry, collect and analyse their own data, and come to their own evaluative judgement, they are more likely to adopt and pass on relevant and effective appropriate technologies. World Neighbors NGOs in Bolivia and Peru have developed a standardized means by which farmers engage in site-specific experimental documentation and reporting of crop yield comparisons, using simple calculators (Ruddell, 1994).

Village Drama and Role Plays. Asking farmers in farmer association meetings or in village gatherings to create a drama or role play that describes the interaction process of extension staff with the village on a specific practice will reveal a variety of evaluative data on social relationships, relevance of extension knowledge to local knowledge, and historical events that have affected the solutions to farmer problems.

Maps. The generation of maps can provide a basis for making judgements about access to extension resources by showing where farmer contacts have been made. Maps also can be made to show the location of sustainable agricultural practices. These facts can be used to evaluate the scope and effectiveness of extension efforts on practices in a region, watershed, or geographic area. When access maps are overlaid with social ranking maps, judgements can be made regarding benefits by social class. Concept maps created collaboratively among farmers and extension staff can provide explanations regarding success and failure of specific programme efforts. Reflections on these maps, which can be drawn on chalkboards or in the sand, can reveal contradictions in underlying assumptions and expectations, which in turn can lead to new experiments.

Maps showing before-and-after photographic and remote sensing images can be the basis for evaluative discussion regarding sustainability of farmer and extension staff practices. This is especially relevant to desertification, deforestation, soil erosion, and status of wetlands and wildlife habitat. These data can help direct extension activity to environments that are most at risk, as well as provide evidence regarding extension's positive impacts from collaboration with farmers on sustainable agricultural and natural resource practices.

Case Studies. In order to understand motivation of farmers or potential extension contacts, case studies of specific types of farmers or farming practices can be undertaken. Comparisons between farmers who have used extension technologies and those who have not are common types of case studies. The typology may be based on geographic regions, soil types, and cultural, age, gender, and economic differences. Case studies are best constructed through repeated interviews over time and often include, in addition to self-report, data from persons who know the subjects well. Oral histories, logs, and journals can also contribute to case study data if farmers collaborate in producing these case studies. Evaluators should guarantee the right to privacy and confidentiality of their sources. Case studies often reveal deterrents to participation, as well as ways participants have overcome deterrents to practice through extension contacts.

Selecting Methods for Evaluation of Teaching and Learning: How Do We Know That Learning Has Happened?

The Perspective of Learners

We have briefly considered methods for programme evaluation at the macro level. Let us now consider the evaluation of learning at the micro level from the perspective of the two most important stakeholders, the extension educators and the farmers or programme volunteers as learners in specific educational events. When learning is evaluated, there are many questions to answer. Central of course is how learners experience the learning process and what they actually learn (the outcome of learning), their knowledge, attitudes, skills, and aspirations, and their behaviour change. An evaluation also can focus on the extension educator as learner and the content, processes, and resources that are used. Because learning is always a social phenomenon, an evaluation can focus on the social environment, organizational context and the relevance of language, culture, and sometimes public policy to learning. These underlying cultural assumptions often explain resistance to learning, as well as the way learning either reproduces existing racial, gender, and economic power relationships or challenges these relationships. Not all evaluations include all of these questions. Educators tend to focus on questions that serve their own perspectives. Learners, including farmers, likewise may be interested in questions that serve their perspectives. Often adult learners are eager to reflect critically on their past and present learning contexts in order to overcome socially constructed deterrents to their learning.

When farmers consider evaluation of their own learning, they may ask themselves a broad range of questions. For example, have they, as learners,

- Gained knowledge or problem-solving skills that are useful to them?
- Increased their hopes and aspirations regarding the future?
- Learned how to learn better or gain access to more knowledge?
- Changed their assumptions, habits of mind, priorities?
- Gained confidence in taking leadership and presenting their ideas?
- Increased their commitment to experiment or take direct action?

The Perspective of Extension Educators

When extension educators consider evaluation of learning, they usually want to know how the learners perceive the process of learning, especially how they, as educators, have been helpful to the learning process. Educators ask themselves—and ask the learners to indicate—whether they have,

- Negotiated expectations and objectives?
- Introduced a variety of useful methods and materials?
- Encouraged the use of examples to illustrate concepts or practice?
- Given step-by-step instructions?
- Summarized the material presented?
- Related theory to practice?
- Showed concern about learners as human beings?
- Promoted discussion and learner interaction?
- Encouraged silent learners to participate?
- Used understandable vocabulary?
- Respected racial, ethnic, and gender differences and their unique contributions to learning?
- Appreciated learning handicaps and disabilities?
- Helped learners reflect critically on how they learn?
- Appreciated local knowledge of learners and made use of it?
- Helped learners learn from each other during learning activities?

Asking learners to reflect on educator behaviours also encourages critical reflection on their own learning.

The purpose of listing all of these questions from both the learners' and educators' perspectives is not that one should ask them all, but to stimulate discussion about which are essential items for a specific evaluation effort. The methods that follow can be used to gather evidence of learning through relationships.

Documentation of Local and External Knowledge. A basic principle of adult education is that learning should begin with prior knowledge. How can we appreciate what has been learned if educators do not know what participants already know? For centuries, farmers have passed on their indigenous or local knowledge. Extension educators can assist farmers in documenting this knowledge either in written, photographic, audio, or video forms. This is essential, since evaluation of new learning should be compared with what farmers already know. Also, farmer acknowledgment of the limitations of their local knowledge can form the basis for collaborative inquiry and the linkage of external knowledge with their local knowledge.

Rating Scales and Checklists. Educators can use these in checking the performance of learners. Learners can use these to check the performance of educators. They can be administered in group or field settings and can be easily revised (Worden & Neumaier, 1987). Learners can use them to judge their own performance, current knowledge, or educational expectations. Rating scales and checklists are not very useful in measuring attitudes or consequences of performance.

Feedback Committees. When residential training for extension staff and farmers takes place, participants can elect a feedback committee to provide evaluation observations to the leadership during the event. The feedback committee should be open to any complaints participants may have about the event, ranging from relevance of content, adequacy of facilities, or effectiveness of leadership, to the involvement of learners in discussion or activities. The committee can bring items to the attention of the instructor or to the group as a whole through written or oral form (Apps, 1991).

Group Discussion Assessment. This method can be incorporated into an ongoing group or meeting. It is relatively efficient in terms of costs and time use. The discussion is usually focussed on several open-ended questions, including those listed above. Groups can create their own questions and then make recommendations for changes.

Peer Review Panels. Farmers can become involved in evaluating one another's work through peer review panels. Panels can be taught to use standards and rating scales. Their evaluative judgements can be made with or without identification of reviewers. When peer review panels are used, it is important to establish a positive climate of constructive criticism.

End-of-Event Analysis. This can be done in several ways. The most frequently used is an evaluation form that is administered at the end of workshop sessions. Another way is to have these forms distributed, collected, and summarized by a feedback committee. They can then report these findings and conduct a discussion on the overall strengths of the workshop or training event. After discussing what should not be changed, they can then discuss what specific modifications should be made (Apps, 1991).

Testimonials and Stories. Testimonials and stories can provide subjective records of educational experiences and activities from the perspective of the learners. They are a form of results data and can qualitatively describe the nature and process of educational change. These stories also can be easily understood by others outside the programme as illustrations of types of outcomes and can lead to ideas for future programming. Disadvantages of this method include social desirability bias, nongeneralizability beyond the person giving the testimony, and difficulty in determining what happened as a result of the programme versus other influences on the person. Stories can be either written by the learners or created as a result of an audio-taped interview.

Pretests, Posttests, and Quizzes. In spite of the negative attitudes associated with tests and quizzes, they can be useful for diagnosing learner proficiencies, documenting prior knowledge, projecting learning achievements, and understanding learner attitudes. Repeating the quiz at the end of a learning event can document change that also can demonstrate to learners themselves that they have learned (Jacobs & Chase, 1992).

Interpreting Findings and Data: To What Do We Compare the Findings?

In some evaluations, the findings consist of narratives that describe the history of events. In other cases, evaluation findings are descriptions of the way things are without explanations or judgements. In still other cases, evaluations seek to answer questions of cause and effect or the relationships between methods and outcomes. As useful as these findings may be, they do not in themselves help us to judge the worth of programme efforts. How do we know that a programme deserves praise? This can occur only when the findings are judged in relationship to several judgement perspectives, five of which are briefly discussed below.

Standard-Referenced Judgement Perspective

Findings regarding a particular extension programme can be compared with examples of excellent achievement by using agreed upon criteria established by experts as "state of the art." For instance, such experts could describe excellence as (1) promotion of sustainable agricultural practice; (2) recognition of local knowledge and cultural practices; (3) high participation of farmers in appropriate technology generation; (4) high inclusion of farmers from all economic, social, and cultural groups; (5) access of women farmers to extension; and (6) promotion of justice in agricultural policies.

Cohort-Referenced Judgement Perspective

Findings from an extension programme evaluation can be compared with similar programmes. For this judgement, we ask how a particular programme or aspect of a programme would be ranked with extension programmes elsewhere. External evaluators are most prepared to do this since they may be knowledgeable about extension practices and impacts elsewhere. This approach is limited because few places are truly similar in extension resources, political and economic contexts, and other handicaps that can affect programme outcomes. However, a failure to make judgements using external cohort comparisons may result in a false sense of achievement and self-satisfaction. Such programmes, when compared, may be mediocre, outdated, poorly conceptualized, and wasteful while being considered excellent or acceptable by local practitioners and their funders.

Difficulty-Referenced Judgement Perspective

This perspective takes into consideration the difficulty of what is being attempted when making judgements regarding programme achievement. For example, an extension programme that is addressing sustainable agriculture in highland areas with few staff members in the midst of revolutionary unrest must be given extra credit for achievements under these difficult circumstances compared with programmes that have a large staff who are helping farmers with relatively rich soil increase their crop production. Difficulty-referenced judgement is the basis for giving higher scores for more difficult dives in Olympic competition. Achievement must be judged relative to difficulty and conditions.

Progress-Referenced Judgement Perspective

This perspective on the interpretation of findings gives credit and recognition to progress from past to present. Before-and-after descriptions are essential to making these judgements. Those who emphasize "base line" data usually want to use it to make progress-referenced judgements.

Alternative-Referenced Judgement Perspective

This perspective considers present descriptions of an extension programme in comparison with what could have been accomplished with the same resources used in alternative ways and places or with different people. This perspective asks, "How else could these resources have been spent with better or different results?" For example, would sustainable agricultural and environmental practices be more evident today if funds had been given directly to village groups or farmers' associations for conducting their own participatory action research? Is the present farm visit system better than a distance education system using radio and audio tapes? What would have been the results if all areas of extension programmes had worked more directly with women through their traditional organizations? Sometimes answers to these questions can be generated by projecting results based on pilot efforts or by alternative approaches that have already been tried in limited form in parts of an extension system.

These judgement perspectives also are helpful in evaluation of learning by individual participants themselves. Learners can judge their learning in comparison with expert standards, performance by their peers, their learning handicaps and external difficulties, their own progress from past to present, and what they would have found more important to have learned or done with their time. Often a combination of these judgements provides a balanced evaluation.

Managing the Work: Who Will Be Responsible for What?

All too often decisions are made to undertake an extensive evaluation without counting the cost and without people agreeing to do the work. These efforts are doomed to failure. Negotiating what is possible, given resource limitations and reasonable time expectations, is an essential collaborative task if people at all levels are to own the evaluation and act on the findings. Involving learners and educators together in critically reflecting on and interpreting the evidence will go a long way toward guaranteeing that evaluation itself is a major learning process.

Using Evaluation Findings for Improving Extension: Who Should Be Told What?

Typically, a single report provided to stakeholders neglects the interests of many other participant groups. One solution is to consider the findings, interpretations, and judgements as constituting a pool from which a variety of reports and styles of reporting can be fashioned to serve specific purposes and different users. Forms of reporting can include formal written reports, written executive summaries, letters to individuals and organizations, exhibits and pictorial displays, video narratives, news reports, news conferences, and public meetings. Stakeholders can then be given the information to which they are entitled in the form that best suits their purposes and best encourages learning.

Evaluation is a process of individual and collective learning (Choudhary & Tandon, n.d.) We can learn from our successes, but especially from our failures. Korton (1980) calls this "embracing error." Evaluation provides that occasion.

References

Apps, J. W. (1991). *Mastering the teaching of adults*. Malabar, FL: Krieger.

Arnstein, S. (1969). The ladder of citizen participation. *Journal of American Institute of Planners, 35*, 221.

Brunner, I., & Guzman, A. (1989). Participatory evaluation: A tool to assess projects and empower people. In R. F. Conner & M. Hendricks (Eds.), *International innovations in evaluation methodology* (p. 9-18). New Directions for Program Evaluation, No. 42. San Francisco: Jossey-Bass.

Choudhary, A., & Tandon, R. (n.d.). *Participatory evaluation: Issues and concerns*. New Delhi, India: Society for Participatory Research in Asia.

Cronbach, L. J., & Associates. (1981). *Toward reform of program evaluation: Aims, methods, and institutional arrangements*. San Francisco: Jossey-Bass.

Deshler, D. (1990). Concept mapping. In J. Mezirow & Associates (Eds.), *Fostering critical reflection in adulthood: A guide to transformative learning*. San Francisco: Jossey-Bass.

Eisner, E. (1983). Educational connoisseurship and criticism: Their form and function in educational evaluation. In G. F. Madaus, M. Scriven, & D. Stufflebeam (Eds.), *Evaluation models*. Boston: Kluwer-Nijhoff.

FAO (1990). *Participation in practice: Lessons learned from the FAO People's Participation Programme*. Rome: Food and Agriculture Organization, People's Participation Programme.

Gold, N. (1988). Stakeholders and program evaluation: Characterizations and reflections. In A. S. Bryk (Ed.), *Stakeholder-based evaluation*. New Directions for Program Evaluation, No. 17 (p. 62-63). San Francisco: Jossey-Bass.

Goldstein, I. L. (1986). Evaluation procedures. In I. L. Goldstein, *Training in organizations: Needs assessment, development and evaluation* (2nd ed.). Monterey, CA: Brooks/Cole.

Greene, J. C. (1988). Stakeholder participation and utilization in program evaluation. *Evaluation Review, 12*, 91-116.

Greene, J. C., Caracelli, B. J., & Graham, W. F. (1989). Toward a conceptual framework for mixed-method evaluation designs. *Educational Evaluation and Policy Analysis, 11*, 255-274.

Guba, E. G., & Lincoln, Y. S. (1989). *Fourth-generation evaluation.* Newbury Park, CA: Sage.

Jacobs, L. C., & Chase, C. I. (1992). *Developing and using tests effectively: A guide for faculty.* San Francisco: Jossey-Bass.

Korton, D. C. (1980). Community organizations and rural development: A learning process approach. *Public Administration Review, 40*(5), 480-511.

Murphy, J., & Marchant, T. (1988). *Monitoring and evaluation in extension agencies.* Washington, DC: The International Bank for Reconstruction and Development/The World Bank.

Patton, M. Q. (1991). Beyond evaluation myths. *Adult Learning* (October), 9-10, 28.

Provus, M. M. (1971). *Discrepancy evaluation for education program improvement and assessment.* Berkeley, CA: McCutchan.

Ruddell, E. D. (1994). *A simplified methodology for training peasant farmers how to conduct site-specific scientific field trials in deprived rural areas.* 1994 Conference Refereed Papers. Association for International Agricultural and Extension Education. Tenth Annual Conference, Arlington, VA.

Rossi, P. H., & Freeman, H. E. (1982). *Evaluation: A systematic approach.* Beverly Hills, CA: Sage.

Scriven, M. (1972). Pros and cons about goal-free evaluation. *Evaluation Comment, 3*, 1-5.

Stufflebeam, D. L. (1971). *Educational evaluation and decision making.* Itasca, IL: F. E. Peacock.

Summers, J. C. (1977). *Dimensions of program effectiveness and accountability.* Morgantown: West Virginia University, Office of Research and Development.

Tripodi, T., Fellin, P., & Epstein, I. (1971). *Social program evaluation: Guidelines for health education and welfare administration.* Itasca, IL: F. E. Peacock.

Uphoff, N. (1989). *A field methodology for participatory self-evaluation of P.P.P. group and inter-group association performance.* Ithaca, NY: People's Participation Programme of the U.N. ood and Agriculture Organization.

Uphoff, N. (1992). *Participatory evaluation of rural development.* New York: IFAD Monitoring and Evaluation Division for the Monitoring and Evaluation Panel.

Worden, P. E., & Neumaier, P. (1987). *A source book for program evaluation and accountability.* Fort Collins: Colorado State University, State Coopérative Extension.

Part III

Improving Extension Management

Chapter 12
Formulating Extension Policy

Tito E. Contado

In many countries, the problems of establishing or maintaining an effective agricultural extension service can be traced back to the lack of a realistic policy or an unstable policy framework for charting the mission of the extension system. Lack of agreement on the functions of extension, the clientele to be served, how extension will be financed, frequent changes in organizational structure and programme priorities, rapid turnover of the extension staff, and the proliferation and lack of coordination between different organizations that undertake extension work are some of the common problems that highlight the issue of extension policy. In addition, extension must be responsive to changes in the agricultural sector, the drive toward market reforms, and shrinking government budgets.

The importance of extension policy was recognized by the FAO's Global Consultation on Agricultural Extension when it recommended that "all national governments should develop and periodically review their agricultural extension policy. This policy should include the goals of agricultural extension, the responsible agencies and personnel, the clientele to be served, the broad programmatic areas to be addressed, and other relevant guidelines." The consultation further recommended that "the FAO, in cooperation with the donor community, should engage in policy dialogue with national governments to stress the importance of agricultural extension in national agricultural development and the need to have an explicit, formally enacted, agricultural extension policy" (Swanson, 1990, p. 11).

Formulating and enacting a sound, comprehensive, and useful extension policy is a difficult undertaking. Coutts (1994) attests to this observation when he undertook a case study of the introduction of a formal extension policy in Queensland, Australia. Therefore, the aim of this chapter is to outline the key dimensions of formulating an effective extension policy.

Role of Extension in SARD and Its Policy Implications

Farmers correctly view extension as a form of assistance to help improve their know-how, efficiency, productivity, profitability, and contribution to the good of their family, community, and society. At the same time, politicians, planners, and policy makers in many developing countries view extension as a policy instrument to increase agricultural production, to achieve national food security, and, at the same time, help alleviate rural poverty. In addition, some economists view extension as a policy instrument that will contribute to human capital development and economic growth; therefore, resources allocated to extension are viewed as an economic investment which must produce competitive economic returns. To the practitioner, agricultural extension enhances and accelerates the spread of useful know-how and technologies to rural people. These activities are expected to lead to increased and sustained productivity, increased income and well-being of farm people, and to the promotion of national food security and economic growth. These objectives are to be achieved through nonformal education and training programmes and two-way technology transfer and feedback systems where extension has an important contribution to make to agricultural and rural development.

At the same time, agricultural and rural development is no longer a matter of just increasing food and agriculture production; other factors must be addressed by policy makers and support service agencies in formulating and implementing agricultural and rural development policy. These issues include population and environmental concerns, and they have very strong implications for how key support services such as research and extension should be organized and financed.

The Population Factor

An important concern is the rapid population growth in many developing countries. This factor has a direct impact on the demand for food and other agricultural products, and it results in increased pressure on the land and other natural resources. In many countries, the population has more than doubled during the last three decades. During the last decade of the twentieth century, it is projected that more than 850 million people will be added to the world's population. Furthermore, world population growth is expected to increase 57 per cent from 5.3 billion in 1990 to 8.3 billion by the year 2025 (UN-World Population Prospects: The 1994 Revision, p. 24, Table A4). Because the bulk of these major population increases are occurring in developing countries, particularly in Africa and Asia, this implies not only an increase in the demand for food, but also more pressure on fragile and marginal lands, increased land fragmentation, and larger numbers of landless people in rural areas. These problems point to the need for more education and technical support to farm households, both to increase productivity and to preserve natural resources.

Natural Resources and Environmental Factors

Population pressure and the demand for increasing food output are now commonly associated with the degradation, depletion, and pollution of soil, water, and other natural resources (Alexandratos, 1995, p. 350). Numerous actions are required for a society to conserve, protect, rehabilitate, and manage its land, water, and other natural resources; therefore, extension has a central role to play in disseminating sustainable agricultural technology.

The Economics of Extension

Since the 1980s, funding of research and extension has become an increasingly important policy issue, one that has given rise to a progressive decline in financial support for extension. This decline is occurring in a situation where funding of extension has been chronically inadequate (Swanson, 1990). Many economists and development planners believe that public funding of extension should be higher (Wilson in Rivera & Gustafson, 1991, p. 13), and an FAO study on this issue, involving 114 member countries, showed that in most developing countries government support to extension is generally low when compared to agriculture gross domestic product (about 0.5 per cent of AGDP). The Global Consultation on Agricultural Extension recommended that "in countries where more than 60 % of the economically active population are engaged in agricultural production, approximately 1 to 2 % of the AGDP (depending on the size of the country and factor costs) should be considered the minimum level of financial investment to achieve both human resource development and technology transfer goals of a public sector agricultural extension system" (Swanson, 1990, p. 26-27).

This erosion in public support for extension may be explained, in part, by several factors, including the introduction of structural adjustment programmes in many developing countries. Also, many development economists look to the pattern found in industrially developed countries, where research and development (R&D) and technology transfer are predominantly in the hands of the private sector. In these countries, the communication and market infrastructure is well developed; therefore, the trend has been towards privatizing the extension function with a direct reduction of public funding for extension (LeGouis in Rivera & Gustafson, 1991, p. 31).

Privatization of extension in developing countries, where the circumstances of farming and farmers is quite different than in the developed countries (i.e., similar to conditions in developed countries some 50 to 100 years ago), is being promoted at the national and international level. This emerging trend is toward a more pluralistic conception of extension, as found in many developing countries today. For policy makers, the implication of this trend would imply the recognition and licensing of both public and private agencies or organizations (including NGOs) to become part of the national agricultural extension system or network. This approach implies the need for public and private sector cooperation to address the twin problems of poor extension coverage and resource limitations (Maalouf et al. in Rivera & Gustafson, 1991, p. 590). What is not fully apparent to some planners and policy makers is the relationship between the level of public funding for extension and the economic and social returns to a nation in the form of food security, affordable food for consumers, and the alleviation of widespread hunger and poverty.

Sustainable Agricultural and Rural Development

The integration of the food, population, and environmental nexus has led to a new platform of development, referred to by FAO as sustainable agricultural and rural development (SARD). This new concept of SARD has been defined by FAO as "the management and conservation of the natural resource base, and the orientation of technological and institutional change in such a manner as to ensure the attainment and continued satisfaction of human needs for present and future generations. Such sustainable development (in agriculture, forestry and fisheries sectors) conserves land, water, plant and animal genetic resources; is environmentally non-degrading; technically appropriate; economically viable; and socially acceptable" (FAO, den Bosch Declaration, 1991, p. 2).

The role of farmers is well recognized in SARD, particularly in the management and conservation of soil, water, and biological resources; maintaining an ecological balance; and applying environment friendly technologies such as integrated pest management (IPM). Farmers are the single largest group of users and managers of land, water, and other biological resources throughout the world. In 1970, about 790 million people were economically active in agriculture, and this number will increase to around 1.13 billion by the year 2000 and to 1.19 billion by 2010 (Alexandratos, 1995, p. 340). The majority of these men, women, and young farmers will need useful information, appropriate technology, and sound technical advice not only to increase their agricultural productivity and incomes, but also to make farming and rural life richer and more sustainable. Herein lies the important role of agricultural extension.

The Place of Extension Policy in SARD

The first principle to recognize is that extension is one of the most strategically important policy instruments for achieving the twin goals of food security and SARD in developing countries. In countries that already have an effective agricultural extension system, the cost of pursuing SARD with the farming population can be reduced by integrating environmental and sustainable themes into ongoing extension programmes that are relevant and useful to farm people (Stocking, 1994, p. 16).

Secondly, the new goals of pursuing SARD should be embodied into existing or new extension policies. The FAO-sponsored Expert Consultation on Integrating Environmental and Sustainable Development Themes into Agricultural Education and Extension Programmes noted that "without a persuasive, effective and unambiguous set of directives for good working practice, articulated in the form of high-level policy and agency mandate, the current ad hoc approach to environmental matters will likely continue" (Stocking, 1994, p. 18). In short, extension policy makers, with the participation of their various stakeholders, need to formulate extension policies that will address both farmers' needs and environmental concerns and that will guide extension managers during the first quarter of the twenty-first century.

Scope of Extension Policy

Agricultural extension policy is a part of national development policy in general and of agricultural and rural development policy in particular. Hence, agricultural extension is one of the policy instruments which governments can use to stimulate agricultural development (Van Den Van in Jones, 1986, p. 91). Extension is very much a part of what Roling refers to as the agricultural development mix. He notes that extension is a weak instrument when it stands alone, but that it becomes powerful when combined with price incentives, input supply, credit, seed multiplication, and so forth (Roling in Jones, 1986, p. 110). The Global Consultation on Agricultural Extension concluded that agricultural extension policy should be consistent with and supportive of national agricultural development policy and goals (Swanson, 1990, p. 11).

Each country should have a comprehensive agricultural extension policy which provides for coordination with research, education, input supply, and credit and marketing systems, as well as some flexibility to reflect the dynamic nature of the agricultural sector. The policy should include the mission and goals for agricultural extension, the responsible agencies and personnel, the clientele to be served, the broad programmatic areas to be addressed, and other relevant guidelines. In developing national agricultural extension policies, representatives of all major groups of farmers should be directly involved and other relevant agricultural organizations should be consulted. "By pursuing a comprehensive policy," the Global Consultation noted, "countries can expect the extension system to contribute to increasing agricultural productivity and farm income, and to improving the quality of life of most rural farm households in pursuit of the general goal of growth with equity. In addition, such a policy should help maintain and conserve the natural resource base for sustained agricultural development and enhance food security" (Swanson, 1990, p. 11).

Forms of Extension Policy

More research is needed in classifying extension policies both in developed and developing countries, as well as in those countries in transition. Three forms of extension policies will be discussed here.

Provisional Extension Policies

This is the most common form of extension policy in most developing countries. In the absence of more formalized extension policies, or at the time when the formally enacted policy has been suspended, a provisional or ad hoc policy comes into play. For example, Mozambique in the early 1980s did not have a national policy for agricultural extension. When the agricultural development policy shifted from a reliance on state farms to the involvement of small family farms, a provisional extension policy was formulated to provide farmers and the cooperative sector with improved training and technology. To develop and test this provisional policy, a UNDP/FAO-supported project assisted the government in defining a national agricultural extension policy and developing a programme of implementation.

Decrees and Proclamations

Decrees and proclamations are policies issued by the head of state or by the executive officer of government. Generally, this approach does not go through the process of consultation and debate involving various stakeholders and beneficiaries. An example of this form of policy was the Brazilian government decree abolishing the national agricultural extension authority and transferring some of its functions and staff to the national agricultural research authority. Also, it empowered the state-level rural extension authorities to continue their respective programmes.

In the Philippines, by a presidential proclamation, the name of the Bureau of Agricultural Extension was changed to Agricultural Extension Commission, and the title of "agricultural extension workers" was changed to "farm management technicians." The Bureau of Agricultural Extension, which was created by legislation, was subsequently restored until another proclamation abolished the Bureau of Agricultural Extension and created, in its place, the National Agricultural Training Institute. The agricultural extension function of the Department of Agriculture was then decentralized to the regions and provinces and put under the control of local governments. The decentralization aspect of the proclamation was later superseded by an enactment, in 1991, known as the Local Government Code (Serrano in APO, 1994, p. 319).

Legislated Extension Policies

Extension policies embodied by the country's highest law-making authority (e.g., congress or parliament) are common in many developing countries. Countries that have enacted extension policy through legislative action tend to have well-organized, financially stable extension systems that have sustained effectiveness and a cumulative impact. Examples of legislated extension policies which have worked well include the following:

1. The legislation that established the Cooperative Extension Service in the United States is known as the Smith-Lever Act of May 8, 1914. Important to policy makers from developing countries is that this policy stimulated the growth and efficiency of American agriculture from the 1920s to the present. Rogers (1995) summarized the worth of the Cooperative Extension System: "The U.S. agricultural extension model is one of the most widely recognized systems for the diffusion of innovations in the world today. Probably no other government or private agency can claim to be more successful in transferring technology."
2. The Japanese Agricultural Promotion Law of 1948 created and provided funding for Japan's Cooperative Agricultural Extension Service. The same extension policy has guided the Japanese extension system from 1948 to the present (Shinji Imai in APO 1994, p. 122). Under this law, the national government is responsible for two-thirds of the salary of extension workers, two-thirds of all operational expenditures for extension programmes and one-half of all expenditures for training extension workers and for rural youth work. The rest of the extension budget is the responsibility of the prefectural government (Agricultural Extension Service in Japan, 1978, p. 83).
3. Agricultural Extension policy in South Korea today is embodied in the 1957 Agricultural Extension Law and in the Rural Development Law of 1962. It is important to note that, because of "unhelpful interference from the administrative system," the 1962 Rural Development Law put together the Research Bureau, the Extension Bureau, and the Community Development Bureau under the new Rural Development Administration, freeing these two functions from the Ministry of Agriculture's administrative bureaucracy (Yong-Bok Chung & Youl-Mo Dong, 1984 p. 4, 5).
4. Thailand's agricultural extension policy was codified in the 1956 law that created the Department of Agricultural Extension as one of nine departments of the Ministry of Agricultur and Cooperatives. It outlines the functions, scope, organization, and mode of financial support for extension in Thailand.
5. Zimbabwe's Department of AgriculturalTechnical and Extension Services was established by law in 1981 and, although a relatively young institution, it is gradually building up its extension staff and its government funding.

Observations such as these, which demonstrate the relationship between formal extension policy and successful extension systems, led the Global Consultation on Agricultural Extension to recommend that "where possible, agricultural extension policy should be formally enacted through legislative action to provide a stable policy foundation, an explicit mandate, and clear direction for developing and executing programmes" (Swanson, 1990, p. 11).

Issues That Extension Policy Should Address

Extension Mission and Goals

Although extension has a generic and universal meaning, its mission and goals may need to be adjusted according to national objectives and the context and stage of agricultural and rural development in a given country. Should the mission of extension be to promote agricultural development through technology transfer? Should it give higher priority to human resource development in rural areas, or should it promote sustainable agricultural and rural development? The extension mission should be reflected in the name of the organization, and the preamble for extension policy should be included in the law governing the country's extension system. This mission then should be reflected in a statement of goals and objectives that are agreed upon and assigned to extension in a supporting policy document. This document should be periodically reviewed by policy makers and representatives from stakeholder groups.

Extension Approach and Functions

National extension systems can pursue one of several different extension approaches in implementing extension policy. Most extension systems in developing countries give primary attention to technology transfer, given national agricultural policies that emphasize increasing food production and achieving national food security. An example of a technology transfer approach would be the Training and Visit (T&V) Extension System that has been promoted by the World Bank through its lending programme. Although the U.S. extension system has been particularly effective in technology transfer, its main focus has been on increasing the skills and knowledge of rural farm families, who have become very effective consumers of agricultural technology. Therefore, the extension approach pursued by a country should reflect the mission of extension, and it will define the functions, programmes, and tasks that will be carried out by the extension staff.

Subject-Matter Coverage of Extension

Broadly speaking, the subject matter of extension is implied in the mission statement and even in the title of the extension service. What differentiates between agricultural and rural extension is the subject matter that the extension service will include in its programmes and the target groups to be served among the rural population. Very narrow subject-matter coverage such as promotion of food and cash crops and animal production may invite a costly proliferation of several specialized and uncoordinated extension initiatives. Broader subject-matter coverage such as promoting the entire farming system, sustainable agricultural, and rural development leads to a more unified agricultural extension system. Another issue is whether the extension system should include socioeconomic and sustainable development messages.

Geographical Coverage

Geographical coverage can be an important policy issue because of both political and cost implications. Most political leaders want their jurisdiction to be covered by an effective extension service; therefore, they must find a way to provide funds for extension programmes. If extension funding is to be provided by different levels of government (cost sharing), then the structure of extension must reflect these different sources of funding. Extension personnel will tend to be more responsible to those levels of government that provide extension funding. For example, if local governmental units provide some extension funding, then extension personnel will tend to be more responsive to the needs of farmers and political leaders within these local government units than they are if all funding comes from the national government. In short, having multiple sources of funding, especially from different levels of government, will increase the number of shareholders and result in an extension system that has a broader base of support and that is more responsive to stakeholders at the local level.

On purely economic grounds, some economists believe that agricultural extension should be concentrated in those agricultural areas that are well endowed in terms of both human and natural resources and where the rural infrastructure is already developed. However, to concentrate extension resources on larger, better educated, commercial farmers who frequently control the best land resources in a country will not lead to broad-based agricultural development. Furthermore, the use of only economic criteria in allocating extension resources may result in further degradation of soil and water resources as resource poor farmers continue to exploit marginal land without using appropriate farming practices. Finally, investing in resource poor farm families may increase their technical, management, and leadership skills, thereby enabling them or their children to move into higher paying, nonfarm jobs.

Clientele or Target Beneficiaries

A common criticism of extension services in developing countries is their neglect of the vast number of small-scale farmers in favour of fewer numbers of large farmers, or the very limited attention given to women farmers. This is a policy issue because of its implications for the mission and goals of extension, the priorities for technology generation by research, the cost-effectiveness of extension, and the sociopolitical goals of growth with equity and poverty alleviation.

The inclusion of women and rural youth in agricultural extension programmes is generally recognized in terms of their numbers and contribution to farming. Worldwide, an estimated 51 per cent of the active population in agriculture are women. In Africa, women's participation in food production is as high as 76 per cent (FAO, 1990, p. 5) in some areas. In 1995, there were an estimated 1.5 billion rural young people between 15 and 29 years of age, 1.3 billion of them in less developed countries. Given that rural youth may account for up to 60 per cent of the population in developing countries, should they be specially recognized for their crucial role in achieving sustainable agricultural and rural development across the coming generation of farmers?

Organizational Issues

The extension organization embodies different aspects of an extension system, and it provides the management framework for the extension service. This is a policy issue because it affects the scope, magnitude, and structure of the extension system, including factors such as control, cost-effectiveness, and the impact of the extension service. For purposes of illustration, four different forms of extension organization are discussed:

1. *Centralized organization.* Examples include the Department of Agricultural Extension in Thailand and Bangladesh, the Agricultural Extension Bureau of South Korea, and AGRITEX in Zimbabwe. In this form of organization, the national extension office manages and controls extension programme activities and resources at the regional, district, subdistrict, and village level. Clientele participation and feedback in programme planning are generally limited.
2. *Decentralized organization.* Examples of this form of extension organization are the agricultural extension systems in Brazil, Canada, India, Nigeria, and the Philippines. These systems have almost an invisible national or federal extension office, in that extension programming, management, and the control of activities and resources are vested with state or provincial governments.
3. *Cooperative type of extension organization and funding.* The distinguishing feature of this form of extension organization is the cooperation or partnership between the national, state or provincial, and local governments in funding, programming, and managing the activities and resources of extension. In the United States, extension is a joint undertaking of the U.S. Department of Agriculture (Federal Extension Service), the state land-grant universities, and the county governments. In Japan, extension is a joint undertaking of the national government and the prefectural government. In China, agricultural extension is a cooperative undertaking of the central, provincial, prefecture, and county governments. Cooperative programming, management, and support are demonstrated at the County Agro-Technical Extension Centre (CATEC), where normally 20 per cent of funding comes from the central government, 30 per cent from the provincial government, and 50 per cent from the county government.
4. *Pluralistic forms of a national extension system.* This is an emerging form of extension organization in many countries, but it is not yet reflected in national extension policy. This structure appears to occur in those countries where the need for extension services is widespread and/or where the public agricultural extension organization can no longer satisfy its clientele because of resource and management problems. As a consequence, many publicly and/or privately funded organizations, including nongovernmental organizations (NGOs), are beginning to conduct agricultural extension programmes. Publicly funded extension organizations may include the crop, livestock, and horticulture departments of the ministry of agriculture, state-funded agricultural colleges and universities, and commodity boards. Privately funded organizations may include rural development-oriented NGOs, agrobusiness firms (contract extension), and farmer organizations, including cooperatives and commodity associations. Generally, the geographical, subject-matter, and clientele coverage and the standard of work for each of these different organizations are not known. Also, these separate efforts are generally not well coordinated. For contemporary policy making in extension, it would be advisable that a roster or "map" of all the publicly and privately funded extension programmes be established and a national extension policy formulated that would recognize this multiplicity of extension funding and programmes, and then to study the feasibility of a policy that would promote integration of the agricultural extension system.

Extension Staffing Issues

By the nature of the mission and work that an extension system carries out, its worth to society is largely reflected by the quality and number of the technical and professional staff in the organization. For a national programme of extension, the human resource question that policy makers and extension managers are confronted with is: Given the mission, scope of the work, and available resources, what type of qualifications and how many extension staff should be employed by the extension system? Part of this staffingmatrix includes other questions: What should be the proportion of subject-matter specialists to field extension workers? What should be the proportion of field extension personnel to the number of farmers, farm households, or other target groups? How should extension staff be deployed, how often should they be transferred, and what incentives should be provided in order to ensure that they work closely with all groups of farmers?

Extension Funding

The most difficult and challenging policy issue facing extension today is to secure a stable source of funding. With the widespread trend to cut government budgets, including structural adjustment programmes, many policy makers have the impression that public extension is both expensive and a drain on the government's limited resources. At the same time, studies carried out in both developed and developing countries indicate that the returns to extension expenditures are high. Therefore, policy makers should examine this issue carefully in deciding what level of public funding is necessary to support extension in relation to the needs of farmers in the country.

While extension managers should look for ways to improve the efficiency and to reduce the cost of public extension (Wilson in Rivera & Gustafson, 1991, p. 13, and Gustafson in Rivera & Gustafson, 1991, p. 89), at the same time policy makers should look for ways to increase extension funding to adequate levels of support. In most developing countries, particularly the low-income, food-deficit countries, absolute levels of extension funding are very low. The Global Consultation on Agricultural Extension noted that "the current level of funding for extension in most developing countries is insufficient to provide adequate coverage for all groups of farmers, especially those who are resource poor and at the subsistence level. With a few notable exceptions, the needs of women and young farmers are largely neglected" (Swanson, 1990, p. 25). In countries where funding support to extension is low, the funding for extension should be increased to levels that reflect the anticipated economic rates of return and the social benefits when public funds are properly invested and managed (Alexandratos, 1995, p. 344; Evenson in Jones, 1987, p. 76-86; Pavelia in FES, 1974, p. 2).

The issue of funding extension continues to be the most difficult policy issue faced by extension. This issue is complicated by the increased demand for more extension services on the part of increasing numbers of farm households who have fewer land and water resources. Furthermore, extension is being called to integrate sustainable development messages into its extension programmes. This results in "working with less to do more."

Stability

A good extension policy promotes extension system stability, yet allows sufficient flexibility to reflect the dynamic nature of the agricultural sector. Extension should not be rigid; rather, "It should be responsive to all major groups of farm people and sufficiently inclusive to allow public, private, and non-governmental organizations to contribute fully to the agricultural development goals of the country" (Swanson, 1990, p. 11). Frequent organizational changes within extension, such as being transferred from one government agency to another, directly impact the organization's effectiveness. Such instability is costly in that trained staff are poorly utilized and opportunities for improved productivity are forgone. Extension policies in some countries have been successful in preventing disruptive and destabilizing change. For more than 80 years, the U.S. has followed, with some flexibility, the 1914 Cooperative Extension Service law. For almost 50 years, Japan has followed its extension policy, and Thailand has successfully followed its extension policy for the past 40 years. In these countries, agricultural extension is recognized as having contributed significantly to increased agricultural productivity and development.

Extension Policy Formulation

There is no standard formula to be used in formulating agricultural extension policies. It should be noted, however, that most existing laws and policies on extension have been formulated by planners and policy makers in the ministry of agriculture and agriculture committees in the legislative branch of government. Normally, agricultural extension professionals from agricultural universities or from abroad are called on to provide advice and to assist in drafting extension legislation. A congressional hearing is normally conducted before extension legislation is finally enacted into law.

To be more relevant to the needs of farmers and other clientele, extension policy should be reviewed and formulated through a participatory approach. This process could be initiated by dedicated professionals from the public and private sectors, with the active participation of farmers themselves, the private sector, and local government representatives. A proposed draft extension policy that results from this participatory approach would have to be legitimized by the ministry of agriculture and then enacted into law by the congress or parliament. The advantage of this approach would be greater relevance to local conditions and acceptance by stakeholders at the field level.

As the Global Consultation on Agricultural Extension noted: "There are sometimes contradictions between national development policy and the interests of the vast majority of the rural poor who are engaged in agricultural production. Representatives of all major farm groups should be involved, both through formal mechanisms and informal consultations, in the formulation and execution of agricultural extension policy. Farmer involvement in policy formulation and periodic review is the most effective means of creating a 'demand driven' national extension system" (Swanson, 1990, p. 11).

Conclusions

Lessons from the past can serve as a guide to the future in formulating relevant and useful extension policy in developing countries. Extension policy should include the following: (1) name of the extension system, (2) mission and goals, (3) intended clientele, (4) geographic coverage, (5) the dominant extension approach to be followed, (6) general subject-matter coverage, (7) institutional and organizational framework, (8) how extension will be financed, and (9) provisions for review and accountability within the extension system.

Finally, the ultimate test of extension policy is the impact that extension is having on the productivity of all major groups of farmers, including their incomes and quality of life. In addition, extension should be evaluated by its contribution to sustainable agricultural development. To extension policy makers, managers, specialists, and professional staff, the following checklist might prove useful: (1) Is extension policy developmental with a long-term vision? (2) Does it foster innovativeness and creativity on the part of the extension staff, and does it have more provisions for facilitating, rather than controlling, their work? (3) Does it foster stakeholder participation and confidence in the extension system? (4) Does it attract sustained financial support from government through the support of stakeholders and beneficiaries? (5) Does it follow appropriate procedures and methods to perform its responsibilities efficiently and effectively? and (6) Does it have reasonable provisions for accountability through periodic reviews?

References

Agricultural Extension Service. (1978). *Agricultural extension in Japan.* Tokyo: Japan Agricultural Extension and Development Association.

Alexandratos, N. (1995). *World agriculture: Towards 2010, an FAO study.* West Sussex, England: FAO and John Wiley & Sons.

Asian Productivity Organization (APO). (1994). *Agricultural extension systems in Asia and the Pacific.* Tokyo: Asian Productivity Organization.

Chung, Yong-Bok, & Dong, Youl-Mo. (1984). *A study of the agricultural extension experiences in the Republic of Korea.* Rome: FAO.

Chung, Yong-Bok, et al. (1985). *Agricultural extension services in Korea.* Seodun-dong, Suwon, Republic of Korea: Rural Development Administration.

Coutts, J. (1994). *Process, paper policy and practice: A case study of the introduction of a formal extension policy in Queensland, Australia, 1987-1994.* Den Haag: Koninklijke Bibliotheek.

FAO and Ministry of Agriculture, the Netherlands. (1991). FAO/Netherlands Conference on Agriculture and the Environment: The den Bosch Declaration and Agenda for Action on Sustainable Agriculture and Rural Development—Report of the Conference. Rome: FAO and the Ministry of Agriculture, Nature Management and Fisheries of the Netherlands.

FAO. (1991). *International directory of agricultural extension organizations.* Rome: FAO.

FAO. (1990). *Women in agricultural development: FAO's plan of action.* Rome: FAO.

Federal Extension Service (FES). (;.... 1974). *Extension international exchange.* Washington, DC: Federal Extension Service, U.S. Dept. of Agriculture.

Jones, G. (1986). *Investing in rural extension: Strategies and goals.* London: Elsevier Applied Science Publishers.

Kelsey, L., & Hearne, C. (1963). *Cooperative extension work.* Ithaca, NY: Comstock Publishing Associates of Cornell University Press.

Rivera, W., & Gustafson, D. (1991). *Agricultural extension: Worldwide institutional evolution and forces for change.* Amsterdam: Elsevier.

Rogers, E. M. (1995). *Diffusion of innovations (4th ed.).* New York: Free Press.

Stocking, M. (Ed.). (1994). *Report of the Expert Consultation on Integrating Environmental and Sustainable Development Themes into Agricultural Education and Extension Programmes.* Rome: FAO.

Swanson, B. (Ed.). (1990). *Global Consultation on Agricultural Extension: A report.* Rome: FAO.

Swanson, B. (Ed.). (1984). *Agricultural extension: A reference manual.* Rome: FAO.

Chapter 13
Improving the Organization and Management of Extension

M. W. Waldron, J. Vsanthakumar, and S. Arulraj

Managing people effectively in extension programmes is a skill that requires constant planning and development. An extension programme manager can be defined as the person who is vested with formal authority over an organization or one of its sub units. He or she has status that leads to various interpersonal relations, and from this comes access to information. Information, in turn, enables the manager to devise strategies, make decisions, and implement action (Mintzberg, 1988). Management is concerned with the optimum attainment of organizational goals and objectives with and through other people. Extension management organizations are characterized by many strategies, wide spans of control, democracy, and autonomy. Their management practices cannot be reduced to one standard set of operating guidelines that will work for all organizations continually. However, all managers of professional organizations face the same challenge: to manage one's time, objectives, and resources in order to accomplish tasks and implement ideas (Waldron, 1994).

Managers of extension programmes are painfully aware of the need for revision and development of the new skill sets held by today's high performers. If change is not handled correctly, it can be more devastating then ever before. High performers reflect, discover, assess, and act. They know that a new focus on connecting the heads, hearts, and hands of people in their organization is necessary. Astute managers know what needs to be done but struggle with how to do it. Quite often they prefer to consider themselves as teachers or communicators rather than managers. This results in underutilization of the increasing amount of literature on management theory and practice. The root of the problem is implementation. They must learn how to motivate others and build an efficient team.

More formally defined, management is the process by which people, technology, job tasks, and other resources are combined and coordinated so as to effectively achieve organizational objectives. A process or function is a group of related activities contributing to a larger action. Management functions are based on a common philosophy and approach. They centre around the following:

1. Developing and clarifying mission, policies, and objectives of the agency or organization
2. Establishing formal and informal organizational structures as a means of delegating authority and sharing responsibilities
3. Setting priorities and reviewing and revising objectives in terms of changing demands
4. Maintaining effective communications within the working group, with other groups, and with the larger community
5. Selecting, motivating, training, and appraising staff
6. Securing funds and managing budgets; evaluating accomplishments and
7. Being accountable to staff, the larger enterprise, and to the community at large (Waldron, 1994b).

The management functions listed above can be categorized by using the acronym POSDCORB (Bonoma & Slevin, 1978, from Gulick & Urwick, 1959):

- **Planning**: outlining philosophy, policy, objectives, and resultant things to be accomplished, and the techniques for accomplishment
- **Organizing**: establishing structures and systems through which activities are arranged, defined, and coordinated in terms of some specific objectives
- **Staffing**: fulfilling the personnel function, which includes selecting and training staff and maintaining favourable work conditions
- **Directing**: making decisions, embodying decisions in instructions, and serving as the leader of the enterprise
- **Coordinating**: interrelating the various parts of the work
- **Reporting**: keeping those to whom you are responsible, including both staff and public, informed
- **Budgeting**: making financial plans, maintaining accounting and management control of revenue, and keeping costs in line with objectives

Planning

Planning is the key management function of any extension worker. It is the process of determining in advance what should be accomplished, when, by whom, how, and at what cost. Regardless of whether it is planning long-term program priorities or planning a two-hour meeting, the planning aspect of management is the major contributor to success and productivity. Stated simply, "If you don't know where you are going, then you won't know when you have arrived!" Planning is the process of determining the organization's goals and objectives and

---

Mark W. Waldron is a Professor of Rural Extension Studies, University of Guelph, Guelph, Canada; **J. Vsanthakumar** is a Professor of Agricultural Extension, Annamalai University, India; and **S. Arulraj** is a Senior Scientist, Agricultural Extension, Sugarcane Breeding Institute, India.

The authors acknowledge the assistance of Michael Rzeznik, Ph.D. student in Rural Studies, University of Guelph, in preparing this chapter.

making the provisions for their achievement. It involves choosing a course of action from available alternatives.

Planning is the process of determining organizational aims, developing premises about the current environment, selecting the course of action, initiating activities required to transform plans into action, and evaluating the outcome. The types of planning that managers engage in will depend on their level in the organization and on the size and type of the organization. Generally there are four major types of planning exercises: strategic, tactical, contingency, and managerial. Strategic planning involves determining organizational goals and how to achieve them. This usually occurs at the top management level. Tactical planning is concerned with implementing the strategic plans and involves middle and lower management. Contingency planning anticipates possible problems or changes that may occur in the future and prepares to deal with them effectively as they arise (Marshall, 1992). Managerial planning is usually considered as microlevel planning. It helps in combining resources to fulfil the overall objectives of the extension organization.

A needs assessment may initiate a need for developing a plan. The planning process begins with the creation of a philosophy that consists of statements describing the values, beliefs, and attitudes of the organization. Its mission statement is a proclamation of its purpose or reason for being. After the philosophy and mission statements have been established, various goals and objectives are defined. Goals are usually general statements that project what is to be accomplished in the future. An objective is a concrete statement describing a specific action. Policies are predetermined guides to decision making; they establish boundaries or limits within which action may be taken. Managers are related to policy formation in two ways. First, they play a crucial role in implementing organizational policies that have been established by higher management. Second, they create policies within their departments as guides for their own work groups. Procedures outline the series of steps to be followed when carrying out a designed policy or taking a particular course of action. Rules are used to provide final and definite instruction. Usually they are inflexible.

Planning is designing the future, anticipating problems, and imagining success. In short, planning is essential for anyone who wants to survive. The functions of organizing, leading, staffing, and budgeting are means of carrying out the decisions of planning. Everyone is a planner—a planner of meals, of work time, of vacations, of families. Formal planning, however, distinguishes managers from non-managers, effective managers from ineffective managers. Formal planning forces managers to think of the future, to set priorities, to encourage creativity, to articulate clear objectives, and to forecast the future in terms of anticipated problems and political realities.

Long-Range Planning

Long-range planning is vitally important in that it focuses attention on crucial future issues which are vitally important to the organization. It involves studying societal trends and issues, surveying current and anticipated learners' needs, and being aware of long-term research directions and changes in technology. Many extension workers may think that such management is beyond their level of authority, control, or involvement. They may feel that such management is the prerogative of the director, the deputy minister, or the president. However, while senior levels of management must be involved, those who implement the objectives resulting from long-range planning should also be involved.

Strategic Planning

Strategic planning has been defined as that which has to do with determining the basic objectives of an organization and allocating resources to their accomplishment. A strategy determines the direction in which an organization needs to move to fulfil its mission. A strategic plan acts as a road map for carrying out the strategy and achieving long-term results. Occasionally a large gap exists between the strategic plan and real results. To boost organizational performance, people must be a key part of the strategy. A stronger, more capable and efficient organization can arise by defining how its members can support the overall strategy (Figure 1).

Strategic planning is different from long-term planning. Long-range planning builds on current goals and practices and proposes modifications for the future. Strategic planning, however, considers changes or anticipated changes in the environment that suggest more radical moves away from current practices. When doing strategic planning, the organization should emphasize team planning. By involving those affected by the plan, the manger builds an organizationwide understanding and commitment to the strategic plan (Flemming, 1989). The elements of strategic plans include:

- Organization mission statement—What
- Strategic analysis—Why
- Strategic formulation—Where
- Long-term objectives implementation—When and How
- Operational plans—When and How

Changes are essential to better position the extension organization and focus on client needs and moving forward in rural development and sustainability programmes. The strength and resilience of the traditional rural and farm population and the trend towards a decentralized society with more and more urbanites moving to the country suggest that successful rural communities will depend on people's ability to change, to adapt, and to work toward a better future. In the 1990s, facilitating farmer participation is a major extension activity (Chambers, 1993). Reorganization provides a framework for longer-term commitment to rural development. Organizations and sub units are being encouraged to put work teams in place to ensure that each sector integrates staff and services into a cohesive, focused business unit. Consultation and participation are believed to be essential for the successful development and implementation of organizational goals and objectives. Each work team is asked to develop an effective process for discussion of major challenges and opportunities facing the organization, if possible, over the next decade. Updated strategic plans are then developed. These plans form the framework for focusing organizational resources on the most strategic areas by using a staged approach. Updated plans are then implemented by work teams at all levels of management. Work-team objectives include:

1. Involving all levels of staff in consultation
2. Designing and implementing a process to develop-goals and objectives for the organization and unit; a strategic process for the next five to ten years
3. Defining and clarifying organizational structures and identifying functions, customers, and service delivery models
4. Identifying changes and staged approaches needed to move from the current situation to what will be required over the next three to five years
5. Identifying and recommending priorities for policy and programme development
6. Incorporating goals for expenditure reduction, service quality improvement, workforce management, accountability, technology, and business process improvement
7. Stating the start date and first report date

Managerial Planning

If long-range planning can be linked to "macro," then managerial planning can be linked to "micro." Managerial planning is the implementing of the strategic plan; it is the combining of resources to fulfil the overall objectives and missions of the organization. Managerial planning focuses on the activity of a specific unit and involves what needs to be done, by whom,

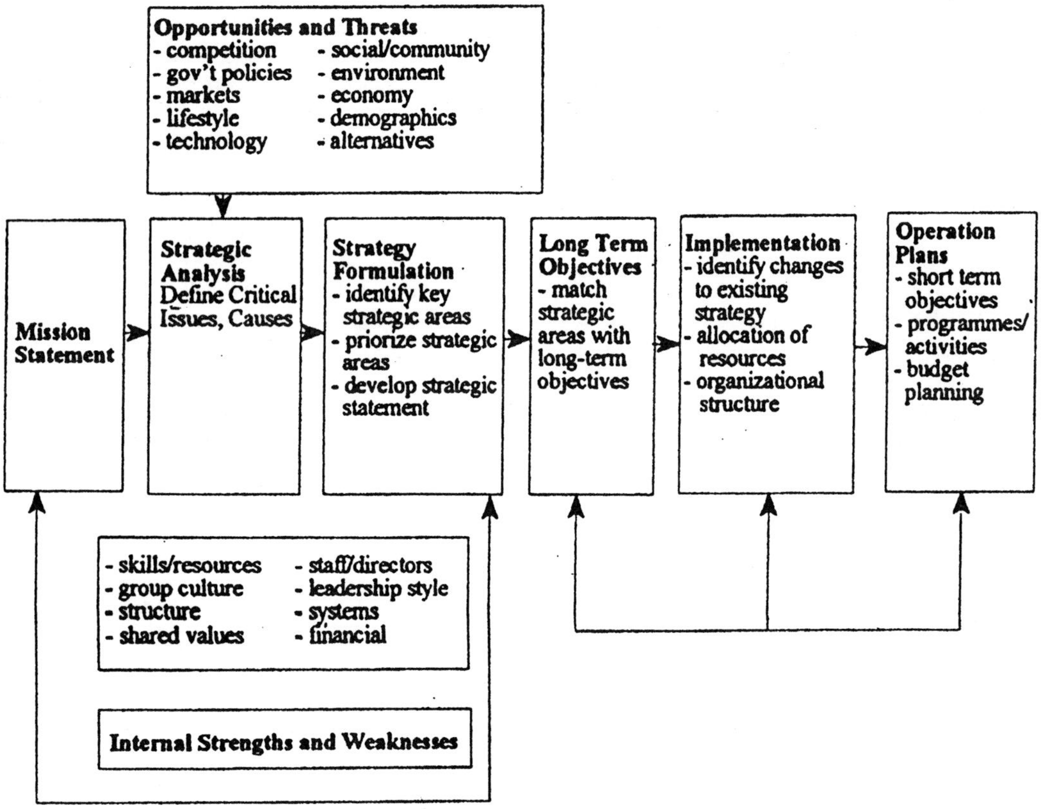

Figure 1. Strategic planning model (Source: OMAD Strategic Planning Factsheet, 1991).

when, and at what cost. The strategic planning process serves as an umbrella over the management planning process which deals with the following:

1. Establishing individual goals and objectives
2. Forecasting results and potential problems
3. Developing alternatives, selecting alternatives, and setting priorities
4. Developing associated budgets
5. Establishing personnel inputs
6. Establishing specific policies related to the unit
7. Allocating physical resources
8. Appraising how the management unit has succeeded in meeting its goals and objectives

Decision Making

Closely related to both strategic and managerial planning is the process of decision making. Decisions need to be made wisely under varying circumstances with different amounts of knowledge about alternatives and consequences. Decisions are concerned with the future and may be made under conditions of certainty, conditions of risk, or conditions of uncertainty. Under conditions of certainty, managers have sufficient or complete information and know exactly what the outcome of their decision will be. Managers are usually faced with a less certain environment. They may, however, know the probabilities and possible outcomes of their decisions, even though they cannot guarantee which particular outcome will actually occur. In such cases, there is a risk associated with the decision and there is a possibility of an adverse outcome. Most managerial decisions involve varying degrees of uncertainty. This is a key part of a manager's activities. They must decide what goals or opportunities will be pursued, what resources are available, and who will perform designated tasks. Decision making, in this context, is more than making up your mind. It consists of several steps:

Step 1: Identifying and defining the problem
Step 2: Developing various alternatives
Step 3: Evaluating alternatives
Step 4: Selecting an alternative
Step 5: Implementing the alternative
Step 6: Evaluating both the actual decision and the decision-making process

Managers have to vary their approach to decision making, depending on the particular situation and person or people involved. The above steps are not a fixed procedure, however; they are more a process, a system, or an approach. They force one to realize that there are usually alternatives and that one should not be pressured into making a quick decision without looking at the implications. This is especially true in the case of nonprogrammed decisions (complex and novel decisions) as contrasted to programmed decisions (those that are repetitive and routine).

One of the most difficult steps in the decision-making process is to develop the various alternatives. For example, if one is involved in planning a workshop, one of the most crucial decisions is the time, format, and location of the workshop. In this case, one's experience as well as one's understanding of the clientele group greatly influence the selecting of alternatives. Often decision trees can help a manager make a series of decisions involving uncertain events. A decision tree is a device that displays graphically the various actions that a manager can take and shows how those actions will relate to the attainment of future events. Each branch represents an alternative course of action. To make a decision tree it is necessary to: (1) identify the points of decision and alternatives available at each point, (2) identify the points of uncertainty and the type or range of alternative outcomes at each point, (3) estimate the probabilities of different events or results of action and the costs and gains associated with these actions, and (4) analyse the alternative values to choose the next course of action.

In extension, the decision-making process is often a group process. Consequently, the manager must apply principles of democratic decision making since those involved in the decision-making process will feel an interest in the results of the process. In such a case, the manager becomes more of a coach, knowing the mission, objectives, and the process, but involving those players who must help in actually achieving the goal. The effective manager thus perceives himself or herself as the controller of the decision-making process rather than as the maker of the organization's or agency's decision. As Drucker (1966) has pointed out, "The most common source of mistakes in management decision-making is the emphasis on finding the right answer rather than the right question. It is not enough to find the right answer; more important and more difficult is to make effective the course of action decided upon. Management is not concerned with knowledge for its own sake; it is concerned with performance."

Organizing

Once strategic planning and management planning are implemented, organizing to get the job done is next. Organizing is the process of establishing formal relationships among people and resources in order to reach specific goals and objectives. The process, according to Marshall (1992), is based on five organizing principles: unity of command, span of control, delegation of authority, homogeneous assignment, and flexibility. The organizing process involves five steps: determining the tasks to be accomplished, subdividing major tasks into individual activities, assigning specific activities to individuals, providing necessary resources, and designing the organizational relationships needed.

In any organizing effort, managers must choose an appropriate structure. Organizational structure is represented primarily by an organizational chart. It specifies who is to do what and how it will be accomplished. The organizing stage provides directions for achieving the planning results. There are several aspects to organizing—time, structures, chain of command, degree of centralization, and role specification.

Time Management

Managers must decide what to do, when, where, how, and by or with whom. Time management is the process of monitoring, analysing, and revising your plan until it works. Effective planning is a skill that takes time to acquire. It is difficult to implement because you have no one but yourself to monitor how effectively you are using your time. Everyone has the same amount of time—168 hours per week. How that time is managed is up to the discretion of each person. One extension agent joked that he was so busy taking time management courses, he had little time left to manage. Effective time management involves philosophy and common sense. Time is not a renewable resource—once it is gone, it is gone forever. To function effectively, managers have to be able to prioritize and replace less important tasks with more important ones. Most of us work for pay for only 1,800 hours per year. Effective and efficient time management encourages us to achieve and be productive while developing good employee relations.

Goals should be specific, measurable, attainable, realistic, and timely (SMART). Once the goals are known, it is important to think about how they can be achieved. Effective time managers facilitate planning by listing tasks that require their attention, estimating the amount of time each task will take to complete, and prioritizing them—deciding what tasks are most important to do first and numbering them in rank order. It is essential to know what is crucial and what is not. Some activities have relatively low levels of importance in completing a given task. By planning ahead, managers can decide what to do and take the time to come up with ideas on how to do it. They can make their own list of steps to eliminate or reduce time wasters. Maintaining a daily "To Do" list with priorities attached and maintaining a daily, weekly, monthly, and yearly diary is helpful. Managers should analyse their daily activities to see which are directed toward results and which are simply activities. They could learn how to manage meetings more effectively since considerable management time seems to be wasted in nondirectional formal meetings.

One of the methods that helps allocate time according to priorities is Pareto's Law or the 80/20 rule: if all items are arranged in order of value, 80 per cent of the value will come from 20 per cent of the items. For example, 80 per cent of the complaining in your department is likely to be done by 20 per cent of your staff. Four suggestions for better time management are (1) never handle the same piece of paper twice; (2) learn how to say "no" without feeling guilty about requests that do not contribute to the achievement of your goals; (3) when a visitor drops in to your office, stand up while you have your discussion to ensure that only a brief period of time will be consumed by the visitor's interruption; and (4) avoid being a slave to the telephone. By managing time well, managers are better able to solve problems quickly, make decisions, avoid frustration, keep from getting bogged down in day-to-day tasks, handle crises, work on their goals and priorities, and manage stress. Guidelines for scheduling time include:

1. Always put your schedule in writing.
2. Focus on the objectives you are trying to accomplish.
3. Continually review objectives, priorities, and scheduled actions to keep on track.
4. Schedule around key events and actions.
5. Get a productive start by scheduling early-day actions.
6. Group related items and actions whenever possible.
7. Do not hesitate to take large time blocks for important tasks.
8. Be sure to allow enough time for each task, but not too much time.
9. Build in flexibility for unexpected events.
10. Include some thinking time for yourself.
11. Consider how to make waiting and travel time useful or otherwise productive.
12. Try to match your work cycles to your body cycles.
13. Learn to control your unscheduled action impulses.
14. Prepare tomorrow's schedule before you get to the office in the morning.

Structures: Centralized versus Decentralized, Line versus Staff

Working productively and developing feelings of cooperation and effectiveness are related to having the right people doing the right jobs. Structure, then, can be defined as a system of interrelated jobs, groups of jobs, and authority. There is no standard organizational structure, but most organizations and agencies follow the "Christmas Tree" system with the star (e.g., president, minister) at the top, smaller branches at management levels, and bigger branches at the production levels. Some would claim that the lower branches support the upper branches, but as in the tree, the branches are supported by a single trunk, which can be thought of as the organizational mission and objectives. Each part of the tree has its specific function. When all parts work together, the system survives, functions productively, has balance, and is a pleasure to see! There are four primary elements in designing an organizational structure:

1. Job specifications—what each division/office/unit is responsible for
2. Departmentalization—the grouping of jobs and responsibilities in common sectors with the objective of achieving coordination
3. Span of control—a definition of how many job roles should be in each unit and which roles require coordination by a unit manager
4. Delegation of authority—assigning the right to make decisions without having to obtain approval from a supervisor

The resulting organizational structure will vary according to these four elements. An organization with decentralized authority and very heterogeneous departments will appear very different from one with centralized authority and a very homogeneous product.

Chain of Command

Once an organization starts delegating authority, then there is automatically a chain of command, "the formal channel which specifies the authority, responsibility and communication relationships from top to bottom in an organization" (Ivancevich, Donnelly, & Gibson, 1980). Thus authority flows from presidents to vice-presidents to divisional managers, from ministers to deputies to directors, from principals to vice-principals to deans, etc. In complex organizations, there may be bridges from one level to another and there will be complex procedures for maintaining the chain of command. Adult and extension educators, if working for an organization or agency, will be part of a structure and part of the chain of command. One cannot often make major changes in these two elements; it is wise, however, to be very aware of the organizational structure and chain of command if you wish to accomplish things efficiently.

Centralized organizations are those in which the key authority and decision-making role is focused on one or a very few individuals. Where authority is distributed among many managers, then one can see a decentralized structure. As the organization's various roles become more diverse in terms of programme, product, or geographical location, one can see a more decentralized organizational structure with authority being delegated to those who are closest to the action. Centralization refers to authority, whereas centrality refers to the proximity to the organization's stated mandate and objectives. One could have a very decentralized organization with each unit being responsible for programmes, staffing, and budget, and yet be very close to the main mission and objectives of the organization.

Another important point in terms of structure is the concept of line and staff functions. Line functions are those involved in creating, developing, and delivering a programme. Staff functions are those that are of an advisory and consultative order. Line functions contribute directly to the attainment of the organization's objectives, and staff functions contribute indirectly.

Staffing

A key aspect of managing an adult and extension enterprise is to find the right people for the right jobs. Much of one's success as a manager is related to appropriate human resource planning, regardless of whether it is the hiring of a secretary or an instructor for a particular work-shop. The staffing function consists of several elements:

1. Human resource planning—how many staff resources, with what backgrounds, and at what cost can be considered for objectives implementation?
2. Recruitment—how does one proceed to find the person with the appropriate mix of education, experience, human relations skills, communications skills, and motivation? An important component of the recruitment process is writing the job description. The description must be exact and specific but sufficiently general to solicit interest among potential candidates. The nature of the job, scope, authority, and responsibilities form the core of the job description. Indications of preferred educational background as well as salary range must also be included. In times of high unemployment, one can always expect several dozen applications for any one opportunity for employment. This leads to the next task of staff selection.

Staff Selection

The process of staff selection involves evaluating candidates through application forms, curriculum vitae, and interviews and choosing the best candidate for the specific job responsibility. One can even have a list of criteria and a score sheet for each individual. Even then, successful hiring is often a very intuitive act and involves some degree of risk.

As a means of giving some structure and design to the staffing process, the following guidelines are useful (Clark, 1973). Each job interview should be characterized by:

1. A clear definition of the purpose of the interview
2. The presence of a structure or general plan
3. The use of the interaction as a learning experience in a pleasant and stimulating atmosphere
4. The creation and maintenance of rapport between the interviewer and interviewee

5. The establishment of mutual confidence
6. Respect for the interviewee's interest and individuality by the interviewer
7. An effort to put the interviewee at ease
8. The establishment and maintenance of good communication
9. The willingness to treat what is being said in proper perspective
10. The just treatment of each interviewee

Staff Orientation

This is the process of formally introducing the selected individual to the particular unit, to colleagues, and to the organization. The selected person should be aware of the mission and objectives of the unit, the nature of responsibilities and level of authority, the degree of accountability, and the systems and procedures followed to accomplish the tasks associated with the job. A motivated individual will simply ask for such things as personnel manuals, administrative procedures handbooks, and aims and objectives statements. Such orientation tools should be available.

Directing

At one time there was a management emphasis on "directing" in the directorial (autocratic) sense, but in recent times, the concept of directing has become more congruent with leading than with pushing. Thus today, directing is more related to leading and leadership styles. Leadership in this context means the process whereby a work environment is created in which people can do their best work and feel a proprietary interest in producing a quality product or service.

McGregor (1960) proposed that managers might assume that employees are motivated in one of two ways. His dichotomy was labelled theory X and theory Y. Theory X relates to traditional management whereby managers assume that they must control, coerce, and threaten in order to motivate employees. Theory Y, the opposite of theory X, suggests that employees want to do challenging work, that they are interested in accepting responsibility, and that they are basically creative and want to be involved in policy development and objective setting. Today, theories X and Y don't really sound very revolutionary; the problem is that management styles and employee motivation do not fit easily into two theoretical labels. As a result, additional theoretical labels (e.g., theory Z) are being developed.

Coordinating

This important stage consists of interrelating the various parts of the work. It involves coordinating the various job roles and responsibilities of yourself and other staff, of your unit and other units within the same organization, and of your unit with the broader community.

There are two forms of coordination: (1) vertical reporting to your supervisor(s) and to your staff, and (2) horizontal reporting to your colleagues and your management team. Adult and extension educators are usually involved in very complex organizations such as governments, colleges and universities, and boards of education. Because of the size of the organization, the increasing demands for public accountability, the many government regulations and policies, the increasing competition among providers of adult education opportunities, and the changes in technology, it is essential that the coordinating role be given top priority.

How, then, can effective coordination be accomplished?

- Coordination needs professional, competent leadership, a democratic style that leads to trust, open communication, and ease of information flow.
- Coordination needs a constant definition and communication of mission and objectives that are understood by all managers.
- Coordination, to be effective, must have open, two-way channels of communication.
- Coordination involves a sharing atmosphere as well as commonly agreed on direction. Because effective coordination requires cooperation and communication, the meeting technique is still the most effective format for assuring the interrelationships among the various job responsibilities.

In recent years, formal systems of community coordination of adult and extension education activities have been developed. Such councils of continuing education provide more than just a network of workers but in fact lead to discussions of community needs, agency priorities, and an agreement as to who is going to look after what. Such voluntary coordination does not eliminate competition; it focuses on the multiway flow of information.

Reporting

This function, closely related to the coordinating function, consists of keeping those to whom you are responsible informed as to what is going on. It is essential that competent managers keep the information flowing, especially in this age when there is so much information being transmitted in so many forms. The reporting function is more than preparing an annual report, quoting statistics, and informing your staff of current developments. The reporting function is almost an evaluation function since it compares how you are doing with what you set out to do. It reviews your objectives and determines to what extent you are meeting your objectives. It consists of more than course numbers or annual statistics, but relates programme direction, policy changes, refinement in objectives, and changes in structures and priorities. It also uses the vertical and horizontal flows of information as presented previously.

One of the key elements of the reporting function is the annual report. Such a report gives you the opportunity to summarize programmes, projects, and activities and to provide statistics as well. Such a report can be used as a public information document by having it distributed to other adult education agencies in the community, to your senior levels of management, to your own managers, to your colleagues, and to the press. In addition, it will prove to be a valuable document to satisfy the requests you receive asking about your programme activities.

Budgeting

This management function includes fiscal planning, accounting and revenue, and expense controls. Budgeting requires specific planning, a thorough understanding of objectives and future programmes, a sixth sense of economic conditions and realities, and a hunch for predicting the unpredictable.

In many cases, an organization specifies the budget system being used. It could be based on (1) historical data (what you had last year with variations for the coming year); (2) 0-based data where the budget is created and justified on a line-item basis according to programmes and priorities; (3) an MBO system—management by objectives whereby specific objectives are funded; and (4) a PERT system—programme review and evaluation technique—where each programme is reviewed and assessed according to its contribution to specific goals. These are only a few of the budgeting systems in use. However, the key elements of any budget system consist of (1) determining what line items are necessary in terms of objectives; (2) in line with policies, determining the financial amounts for each line; (3) determining overhead, surplus, and/or profit margins; (4) determining anticipated revenue from fees, grants, gifts, contracts, etc.; (5) drafting a budget with specific amounts and justifications; and (6) discussing and making adjustments to produce a working budget.

The budget then becomes a guide which, however, may always be in a state of change. The budget process is not in a vertical something that one does only once a year; it is a continual process of regular review and possible revision. One should always be checking to see how one is doing compared with how one anticipated doing.

Budget management, then, consists of three parts: (1) budget determination—allocating revenue according to priorities and by line items; (2) budget accountability—how well the anticipated budget matches reality; and (3) using a +, 0 - notation in answering the questions and by placing the notations in the boxes on the chart. In this way, one can get a picture of the predominant types of management modes currently being used. While this may be useful in describing what is, it could be even more useful in describing what could be. It is also useful in providing some clues as to possible areas of role conflict—the scholarly research model would likely collide with the competent practitioner model (Waldron, 1994a).

Structure is the basis for many modern business organizations because we live in a structured society, although the concept of structural rigidity and hierarchy is now being challenged by a more educated, creative, and intrinsically motivated workforce. The structural approach shows graphically that the organization has a distinct physical shape or form provided by an internal form. A competent manager in this system is able to solve problems, to figure out what needs to be done, and then enlist whatever support is needed to get it done. This approach is favoured by traditional, hierarchical, job-specific, uncreative organizations.

A more organic management method is based on paradigms. A paradigm refers to a method of approaching a problem or situation and the kinds of assumptions, values, and attitudes associated with thinking about the situation (Ottaway & Terjeson, 1986). It connotes a pattern or structure that is dynamic, changeable, and responsive to the environment (Waldron, 1994). The most dramatic illustration of a paradigm shift was the shift from the Ptolemaic theory, which saw the earth as the centre of the universe, to the Copernican theory, which saw the sun as the centre of the universe. A paradigm shift results in a total restructuring in the ways we think about a situation and the kinds of assumptions we make about former observations. Covey (1992) speaks of paradigm shifts: things, people, and structure can and do change—nothing is constant. He shows how almost every significant breakthrough is first a break with tradition, with old patterns, with old ways of thinking, and with old paradigms. Senge (1990) states that a "shift of mind" is necessary because "the unhealthiness of our world today is in direct proportion to our inability to see it as a whole." In terms of management, extension managers should view people not as "helpless reactors, but as active participants in shaping their reality—from reacting to the present to creating the future" (p. 68-69).

The use of models is common practice in management thinking. Models are useful because various aspects of the structures can be viewed from different positions that can then lead to new perspectives. The goal of modelling is to achieve an accurate yet relatively simple representation of a system, complex entity, or reality usually on a smaller scale. It implies both structure and change. An important aspect of model building is collecting and preparing data. Information gained from the data is the foundation of the model.

The model should reflect the major aspects of the problem as simply as possible. Often, this requires trade-offs because simplicity and accuracy rarely go hand in hand. The constants, if any, within the model should be known with a high degree of precision. One mistake that

Model	Scholarly-Research	Professional	Entrepreneural	Missionary	Innovation	Competent Practitioner
1. Perceived Academic Status						
2. Mission, Mandate, and Objectives						
3. Reward System						
4. Decision Making Authority						
5. Market Orientation						
6. Account-ability						

Figure 2. Model classification comparison matrix (Source: Waldron, 1994a, p. 93).

inexperienced model builders often make is failing to take a broad perspective of the problem. They do not take into account other dimensions of reality that a solution may have an impact on. To accomplish this broad perspective, the extension manager should adopt a systems approach to model building and should focus not only on the immediate problem, but also on interrelationships that exist within and outside the organization and how these relationships will be affected (Stevenson, 1989). One model is usually dominant. To examine which model fits the needs of a particular extension manager's programme and his or her criteria, one can create a matrix similar to Figure 2. By answering questions similar to those presented in Figure 3, one can develop a good understanding of the specific model which would apply to extension programmes.

Systems Theory

Modern management is characterized by two approaches, the systems and the contingency approach. The systems approach views the organization as a total system comprised of interacting subsystems, all of which are in complex interaction with the relevant external environment (Lerman & Turner, 1992). Organizations are pictured as "input-transformation-output systems" that compete for resources. The survival and prosperity of an organization depend on effective adaptation to the environment, which means identifying a good strategy for marketing its outputs (products and services), obtaining necessary resources, and dealing with external threats.

Survival and prosperity also depend on the efficiency of the transformation process used by the organization to produce its goods and services, on worker motivation, and on cooperation. Efficiency of the transformation process is increased by finding more rational ways to organize and perform the work and by deciding how to make the best use of available technology, resources, and personnel. Top management has primary responsibility for designing an appropriate organizational structure, determining authority relationships, and coordinating operations across specialized subunits of the organization (Yukl, 1994). A system can survive only when it delivers an output that can be exchanged for new inputs as well as for maintaining the system. Similarly, an extension service is expected to produce some beneficial output.

Contingency Theory

Theories that explain management effectiveness in terms of situational moderator variables are called contingency theories. The contingency or situational approach recognizes that neither the democratic nor the autocratic extreme is effective in all extension

Perceived status

1. Is the unit headed by a dean?
2. Does the chief manager have tenure?
3. Is the boss a professor?
4. Does the unit have representation on senate, governing council, or the key academic decision-making body
5. Are the employees expected to do research and publish in journals?
6. Is the dean or director expected to have a doctoral degree?

Mission, mandate, objectives

1. What model would best apply to your mission, mandate, and objectives?
2. Is your unit perceived as a moneymaker?
3. Do your objectives give priority to research, innovation, product development, or product delivery?

Reward system

Ways of rewarding you for your effort:

1. Increments negotiated by a union or association
2. Salary negotiated with a senior administrator
3. Bonuses based on meeting targets
4. Ratings by colleagues

Decision-making authority

1. Does your unit decide on what courses and programmes need to be delivered?
2. Does your unit determine course content?
3. Can your unit allocate and reallocate budgets?
4. Does your unit have a unit-controlled development fund?
5. How independent do you think the unit is in terms of decision making?

Market orientation

1. Is your unit market driven?
2. Do you publish a market-oriented course calendar?
3. Do you use charge cards for tuition fee payment?
4. Do you publish regular press releases and media information?
5. Do you do market research and needs assessments?

Accountability

1. Do you report to a president or vice president-academic?
2. Do you present proposed policy to a senate or governing council?
3. Do you make use of a community advisory council?
4. To whom would you send an annual report?

Figure 3. Criteria for model classification (Source: Waldron, 1994a, p. 93-94).

management situations. Different traits are required in different situations. Table 1 describes the major features of five contingency theories and the Vroom and Yetton (1973) normative decision model. The table makes it easier to compare the theories with respect to content and validation. A synopsis of each is presented below:

- The path-goal theory examines how four aspects of behaviour influence subordinate satisfaction and motivation.
- Leadership substitute theory identifies aspects of the situation that make leadership behaviour redundant or irrelevant.
- The multiple linkage model uses a model of group performance with six intervening variables to explain leadership effectiveness.

Table 1. Comparison of six contingency theories (Sources: Yukl, 1994, p. 311).

Contingency Theory	Leader Traits	Leader Behavior	Situational Variables	Intervening Variables	Validation Results
Path-Goal Theory	None	Instrumental, Supportive, Participative, Achievement	Many aspects	Expectancies, Valences, Role Ambiguity	Many Studies, some support
Leadership substitutes Theory	None	Instrumental, Supportive	Many aspects	None	Few Studies, inconclusive
Multiple linkage model	None	Many aspects	Many aspects	Effort, ability organization, teamwork, resources, external coordination	Few Studies, inconclusive
LPC Contingency Model	LPC	None	Task Structure, L-M Relations	None	Many studies, some support
Cognitive Resource Theory	Intelligence, Experience	Directive	Stress Group Support	None	Few studies, some support
Normative Decision Theory	None	Decision Procedures	Many aspects	Decision Quality and acceptance	Many studies, moderate support

- Fieldler's LPC model deals with the moderating influence of three situational variables on the leadership between a leader trait (LPC) and subordinate performance.
- Cognitive resources theory examines the conditions under which cognitive resources such as intelligence, experience, and technical expertise are related to group performance.

Conclusion

Organizations constantly encounter forces driving them to change. Because change means doing something new and unknown, the natural reaction is to resist it. Extension programme managers must overcome this resistance and adopt innovative and efficient management techniques to remain high performers. They must improve their personal,team, and cultural management skills if they hope to adapt themselves to a changing world. Overwhelmingly, current management wisdom touts the goal of getting decisions made as low down in the organization as possible.

The basic idea is that since people closest to the work are likely to know the most about solving problems in their areas, they should be involved in the decisions concerning those areas. An added benefit is that they are more motivated if they have some control over their work and over their own destinies.

There is still considerable discussion as to whether management is an art or a science, a philosophy or a skill.No one sustainable model can holistically encompass all management situations and environments. Management can be defined as the rational assessment of a situation and the systematic selection of goals and purposes; the systematic development of strategies to achieve these goals; the marshalling of the required resources, the rational design, organization, direction, and control of the activities required to attain the selected procedures (McNeil & Clemmer, 1988). Managers typically engage in a large number of discrete activities each day, and the average number of activities appears to increase at lower levels of management. The activities, however, are usually very brief in duration (Mintzberg, 1973).

To carry out their responsibilities, managers need to obtain recent, relevant information that exists in books, journals, and people's heads who are widely scattered within and outside the organization. They have to make decisions based on information that is both overwhelming and incomplete. In addition, managers need to get cooperation from subordinates, peers, superiors, and people over whom they may have no formal authority. Factors that affect managers include level of management, size of the organizational unit, function of the unit, lateral interdependence, crisis conditions, and stage in the organizational life cycle.

Despite all these demands and constraints, managers do have some alternatives. They have a choice in what aspects of the job to emphasize and how to allocate their time. Generally managers are engaged in four types of activities: 1) building and maintaining relationships, (2) getting and giving information, (3) influencing people, and (4) decision making.

References

Bonoma, T., & Slevin, D. P. (1978). *Executive survival manual.* Boston: CBI Publishing Company.

Buford, J. A., Jr., & Bedian, A. G. (1988). *Management in extension.* Alabama: Auburn University Press, p. 263.

Chambers, R. (1993). *Challenging the profession—Frontiers for rural development.* London: Intermediate Technology Publications, p. 67.

Cherrington, D. J. (1989). *Organizational behavior: The management of individual and organizational performance.* Boston: Allyn and Bacon.

Claar, J. B., & Bentz, R. P. (1984). Organizational design and extension administration. In B. E. Swanson (Ed.), *Agricultural extension: A reference manual.* Rome: FAO.

Clark, R.C. (1973). *Basic concepts and theories of administration and supervision* (revised edition). Madison: Department of Agricultural and Extension Education, University of Wisconsin at Madison.

Covey, S. R. (1992). *The 7 habits of highly effective people.* Toronto: Simon and Schuster, p. 29.

Drucker, P. F. (1966). *The Effective executive.* New York: Harper & Row.

Drucker, P. F. (1977). *People and performance.* New York: Harper's College Press.

Elkins, A. (1980). *Management structures, functions, and practices.* Massachusetts: Addison-Wesley, p. 539.

Evans, J. P., & Dahl, D. T. (1984). Organizing for extension communication. In B. E. Swanson (Ed.), *Agricultural extension: A reference manual.* Rome: FAO.

Flemming, P. (1989). *Strategic planning: Fact sheet.* Ontario: Ministry of Agriculture and Food, Rural Organizations and Services Branch.

Gulick, L., & Urwick, L. (1959). *Papers on the science of administration in extension.* Madison: National Agricultural Extension Center for Advanced Study, University of Wisconsin at Madison.

Ivanevich, J. M., Donnelly, J. H., Jr., & Gibson, J. L. (1980). *Managing for performance.* Georgetown, Ontario: Irwin Dorsey.

Koontz, H., & O'Donnel, C. (1978). *Essentials of management.* New Delhi: Tata McGraw-Hill, p. 104-105.

Marshall, P. (1992). Introduction to the management process. In *Managing people at work.* Guelph: University of Guelph Press.

Massie, J. L. (1979). *Essentials of management.* New Delhi: Prentice-Hall.

McGregor, D. (1960). *The human side of enterprise.* New York: McGraw-Hill.

Mintzberg. H. (1973). *The nature of managerial work.* New York: Harper & Row.

Mintzberg, H. (1988). *Mintzberg on management.* New York: Free Press.

Moris, J. (1991). *Extension alternatives in tropical Africa.* London: Overseas Development Institute, p. 184.

McNeil, A., & Clemmer, J. (1988). *The V.I.P. strategy—leadership skills for exceptional performance.* Toronto: Key Porter Books Ltd.

Ottaway, R. N., & Terjeson, L. (1986). A proposed change evaluation scaling method. *Journal of Managerial Psychology, 1*(2), ii.

Senge, P. M. (1990). *The fifth discipline: The art and practice of the learning organization.* New York: Doubleday.

Stevenson, W. J. (1989). *Introduction to management science.* Boston: Irwin.

Swanson, B. E. (Ed.). (1984). *Agricultural extension: A reference manual.* Rome: FAO.

Van Den Ban, A. W., & Hawkins, H. S. (1988). *Agricultural extension.* New York: Longman Scientific and Technical, p. 268-271.

Vroom, V. H., & Yetton, P. W. (1973). *Leadership in decision-making.* Pittsburgh: University of Pittsburgh Press.

Waldron, M. W. (1994a). Models for the future. In M. Brooke & M. Waldron (Eds.), *University continuing education in Canada: Current challenges and future opportunities.* Toronto: Thompson Educational Publishing.

Waldron, M. W. (1994b). Management and supervision. In D. Blackburn (Ed.), *Extension handbook: Processes and practices.* Toronto: Thompson Educational Publishing.

Waldron, M. W. (1984). Management of adult and extension education. In D. Blackburn (Ed.), *Extension handbook.* Guelph: University of Guelph Press, p. 127-140.

Yukl, G. (1994). Effective leadership in open systems. *Leadership in organizations* (3rd ed.). Englewood Cliffs, NJ: Prentice-Hall.

Chapter 14
Managing Human Resources within Extension

K. Vijayaragavan and Y. P. Singh

One of the most significant developments in the field of organization in recent times is the increasing importance given to human resources. More and more attention is being paid to motivational aspects of human personality, particularly the need for self-esteem, group belonging, and self-actualization. This new awakening of humanism and humanization all over the world has in fact enlarged the scope of applying principles of human resource management in organizations. The development of people, their competencies, and the process development of the total organization are the main concerns of human resource management (Pareek & Rao, 1992).

Extension organizations in developing countries face the major problems of professional incompetence and lack of motivation among their employees. Further, many of the agricultural extension departments of these countries do not have a well-defined system of human resource management. Proper planning and management of human resources within extension organizations is essential to increase the capabilities, motivation, and overall effectiveness of extension personnel. Keeping this in view, this chapter discusses the various dimensions of human resource management as applicable to extension organizations: human resource planning for extension, job analysis, recruitment and training of extension personnel, performance appraisal, supervision, management of rewards and incentives, improvement of the quality of work life, and organizational development for extension.

Human Resource Planning for Extension

Human resource planning forecasts the future personnel needs of extension organizations. With the rapid changes in technology, needs of farmers, market situation, and competitive environment, planning for human resources has become an important, challenging task for extension. Human resource planning involves plans for future needs of personnel, their required skills, recruitment of employees, and development of personnel (Miller, Burack, & Albrecht, 1980). Human resource forecasting and human resource audit are the two most important components of this type of planning. Human resource forecasting refers to predicting an organization's future demand for number, type, and quality of various categories of employees. The assessment of future needs has to be based on analysis of present and future policies and growth trends. The techniques of forecasting include the formal expert survey, Delphi technique, statistical analysis, budget and planning analysis, and computer models. The human resource audit gives an account of the skills, abilities, and performance of all the employees of an organization (Werther & Davis, 1982).

Job Analysis

Job analysis traditionally was done for purposes connected with recruitment, pay, administration, and supervision. But the increasing complexity of work has made job analysis an important instrument for developing people in organizations. Job analysis requires a systematic collection, evaluation, and organization of information about the job. This information is collected through interviews, mailed questionnaires, observation, study of records, and similar methods. The collected information becomes a basis for preparing job descriptions and specifications. The job description, or job profile, is a written statement which includes detailed specifications of duties to be performed, responsibilities, and working conditions and indicates what is expected of a job holder. A job specification is a profile of the human characteristics needed for the job, such as education, training, skills, experience, and physical and mental abilities (Werther & Davis, 1982).

Extension organizations in developing countries do not have clearly defined job descriptions or job specifications for extension personnel. The training and visit system of extension considerably improved the preparation of job charts, work plans, and time-bound work for different categories of extension personnel. However, the actual utility of job descriptions in extension organizations is complicated by factors such as work overload, seasonality of extension, the range of cropping systems, and distribution of extension service over a large area (Hayward, 1990). Studies analysing the role of extension agents reveal that they face work-related problems such as role ambiguity and lack of job authority, expertise, and accountability (Vijayaragavan & Singh, 1989). This shows that job analysis is needed to improve the performance and effectiveness of extension

K. Vijayaragavan is a Senior Scientist in the Division of Agricultural Extension at the Indian Agricultural Research Institute, New Delhi, India. **Y. P. Singh** is a Professor in the Division of Agricultural Extension at the Indian Agricultural Research Institute, New Delhi, India.

employees. Job analysis can more effectively contribute towards the development of extension personnel by adopting the following procedures which involve identifying key performance areas (KPAs) and critical attributes.

Key Performance Areas for Various Categories of Extension Personnel

A job description consists of many details, but does not specify key areas which need attention. Further, it gives the details of what is expected from the current jobholder. On the other hand, key performance areas are specific and show the critical functions relevant at present and for the future to achieve the objectives (Pareek & Rao, 1992). The identification of key performance areas helps in role clarity as well as in delegation of functions. This in turn aids in performance appraisal and training. Generally, four or five key areas for a job are identified. The core extension personnel of developing countries consists of village extension workers, subject-matter specialists, and supervisory staff or extension officers. Examples of key performance areas of core extension personnel are given below.

Village Extension Workers. People in this category (1) make regular and systematic visits to villages and farms to develop rapport with the clientele and to understand their problems; (2) undertake educational activities in the form of meetings, campaigns, demonstrations, field days, training sessions, and exhibitions; and (3) provide advisory services to the farmers and solve their production problems.

Subject-Matter Specialists. Their role is to (1) keep abreast of current recommendations and findings related to farm production by maintaining continuous contact with agricultural research stations; (2) provide feedback to the research system about farmers' problems which need solutions; and (3) train and backstop village extension workers on the latest farm technology and help them in solving field problems.

Supervisory Staff or Extension Officers. People holding these positions (1) plan, organize, coordinate, and implement extension programmes and activities; (2) supervise and monitor the work of field staff, providing guidance, motivation, and evaluation of performance; and (3) coordinate the programme with inter- and intradepartmental agencies.

Critical Attributes for Extension Personnel

The key performance areas indicate the important roles and contributions of different categories of extension personnel. Once the roles are delineated, they can be analysed to indicate the attributes which can discriminate an effective from an ineffective role occupant. These critical attributes consist of qualities such as educational qualifications, skills, experience, physical characteristics, mental abilities, values, and attitudes needed for extension. The critical attributes needed for field-level and supervisory extension staff are necessary formal training in agriculture, practical skills and experience in farming, and knowledge of modern farm practices. Abilities in group dynamics, human relations, and communication are also important. Basic skills related to management and leadership are needed by extension supervisors. Values and attitudes such as faith in rural people, commitment to agricultural development, and concern for the whole community are important for all extension personnel (Gupta, 1963; Bhasin, 1976).

The importance of assessing personal and professional attributes for selecting productive extension personnel has been reported by several researchers (Gupta, 1963; Perumal, 1975). Assessment is essential because an unsatisfactory educational level of extension staff is one of the most serious problems of extension in countries like Bangladesh, Botswana, Kenya, Malaysia, Sudan, and Zambia (Blanckenburg, 1984). A worldwide analysis of the status of agricultural extension reveals the low level of formal education and training of field extension agents in developing countries (Swanson, Farner, & Bahal, 1990).

Recruitment and Training of Extension Personnel

Recruitment is important in selecting the right kind of extension personnel. Since the job of extension personnel calls for technical skills as well as commitment and willingness to educate rural people, an appropriate selection system is essential to ensure the right selection. The success of extension depends heavily upon selection of qualified and motivated personnel. Extension organizations in developing countries use two major sources of recruitment: from outside and from within. Entry-level positions such as village extension workers and agricultural extension officers are filled by outside recruitment, using the services of government placement agencies. Other channels of recruitment are advertisements, private placement agencies, professional search firms, and educational institutions. In some countries, farmers are recruited to help extension agents (Adams, 1982). In Israel, volunteers with practical experience in farming, usually a couple, were recruited as extension workers to help the immigrants. These agents were found to be enthusiastic; they lived with the farmers, set a personal example, and were effective instruments for making desired changes (Blum, 1987).

Most of the extension departments in developing countries have the policy of promoting or recruiting within for middle-level and top-level positions. For example, in India, positions like deputy director, joint director, and additional director of extension are filled through promotion (Vijayaragavan, 1994). The advantages of this policy are that it promotes loyalty and provides opportunities for existing extension staff to get high-level positions. However, its greatest disadvantage is that it prevents the lateral entry of talented extension personnel and promotes complacency because seniority ensures promotion.

Methods and Techniques for Selecting Extension Staff

The selection of extension staff starts with making the job opportunities known to all potential applicants through advertisement. The help of extension workers' training centres, agricultural colleges, rural institutions, and local government agencies may be sought to give wide publicity, as well as to inform candidates living in rural areas. This is followed by screening applicants to short-list suitable candidates and by evaluating potential candidates through various tests.

A typical selection process consists of the following steps: completed job application, initial screening, testing, indepth selection interview, physical examination, and job offer (French, 1982). In general, extension organizations in developing countries use a simple knowledge test and a brief interview to select extension personnel. By using the above method, it is impossible to discriminate an effective candidate from an ineffective candidate, because selecting extension personnel demands thorough, indepth testing of cognitive and noncognitive abilities.

Testing cognitive ability includes a knowledge test, a skill or ability test, and an aptitude test. A noncognitive test is a measure of behavioural dimensions which are important for field-level extension personnel, including concern for and commitment to rural people, empathy, problem-solving orientation, high motivation to influence and educate farmers, ability to work under unsupervised and difficult village conditions, patience and persistence, and team spirit. A good example of selecting village-level extension workers on the basis of behavioural characteristics is provided by the extension project of Allahabad Agricultural Institute (Bathgate, 1956). In response to an advertisement for 27 posts of village guides, 700 to 800 candidates had applied. The final selection procedure consisted of five days of testing skills and attitudes in actual village situations. The test included testing attitudes towards menial tasks like cleaning a cattle shed or digging a compost pit. The candidates' responses to emergency situations were also tested by dropping them into isolated villages.

The assessment centre approach, originally used during World War II, can be used to select extension staff. In this approach, an organization develops its internal resources for assessing new staff. The candidates to be recruited go through a number of simulation exercises, and an expert assesses their behaviour. The techniques used are a psychological test, role play, in-basket exercise, group discussion, projective test, knowledge test, and interviews.

Training and Development

The training of extension personnel contributes directly to the development of human resources within extension organizations. "Training programmes are directed towards maintaining and improving current job performance, while development programmes seek to develop skills for future jobs" (Stoner & Freeman, 1992, p. 388). Training has to start with the identification of training needs through job analysis, performance appraisal, and organizational analysis. Once the training needs of extension personnel have been identified, the next step is to organize training programmes. Methods such as games, role playing, simulation exercises, and case study can be used in extension organizations to create learning situations based on experience (Lynton & Pareek, 1990). Training based on actual field experience should be emphasized. Emerging new farm technologies such as integrated pest management and improved practices in horticulture call for actual field experience. Extension agents need training not only in the technological aspects but also in human relations, problem solving, sensitivity towards disadvantaged groups, and the basic concepts of management (Hayward, 1990).

Management Development Programmes

Management development programmes are meant to improve the managerial skills of senior-level extension officers and to prepare them for future roles. There is a great need for management development programmes in extension organizations because they face complex situations due to changing agricultural scenarios. Further, extension managers have to be exposed to modern management techniques and methods. Management development programmes have to be suited to the needs of top-level extension managers and should be based on needs analysis. Methods such as coaching, job rotation, training sessions, classroom instruction, and educational institute-sponsored development programmes are used to train managers. In India, a separate institute called MANAGE has been established to train senior extension managers in managerial skills and human relations.

Performance Appraisal

In the previous sections, we discussed how extension personnel are recruited and trained and become part of a work group. These are all vital activities. However, the ultimate measure of effective human resources within an extension organization is the performance of extension personnel. Thus performance appraisal is important for effective human resource management. Performance appraisal is a process of evaluating employee performance in order to guide and develop the employee's potential. In many extension organizations which are government departments, the performance appraisal is nothing more than a confidential judgement of work done and a character report used to facilitate disciplinary action or promotion. The employees do not get feedback about their performance. Extension organizations need to have an open appraisal system to provide feedback and opportunities for open discussion with employees on their performance, because they have immense potential to grow and develop. This system can create a healthy working climate and employee motivation.

The performance appraisal which aims at facilitating employee development has the following major purposes: (1) to provide feedback and guidance, (2) to set performance goals, (3) to identify training needs, and (4) to provide inputs for management of pay administration, rewards, and promotion. The steps involved in effective performance appraisal are (1) identification of key performance areas and setting yearly objectives under each KPA, (2) identification of critical attributes for effective performance, (3) periodic review of performance, (4) discussion of performance with employees, and (5) identification of training and developmental needs (Pareek & Rao, 1992).

Potential Appraisal

The potential appraisal is a future-oriented appraisal by which the potential of an employee to occupy higher positions and to assume higher responsibilities is evaluated. The potential appraisal can help the extension staff to know their strengths and weaknesses and can motivate them to further develop their skills. Thus the potential appraisal helps in planning overall career development of employees. Some of the techniques used for the appraisal are self-appraisals, peer rating, the management by objectives (MBO) approach, psychological test and simulated work exercises, case analyses, and leadership exercises.

Performance Review and Counselling

An important purpose of the performance appraisal is to counsel and guide employees towards greater job effectiveness. Thus a system of performance counselling is needed in extension organizations. Performance counselling is provided by the manager to the subordinates to help them in the analysis of job performance, identification of training needs, and finding solutions to the problems which hinder job effectiveness. Counselling is an art of communication involving two people—manager and employee. Counselling differs from training in that the former involves a dyadic relationship and establishes more mutuality and confidentiality. The success of performance counselling depends upon the employee's interest, a climate of openness and mutuality, and the counselling process. Extension managers can use directive, nondirective, and cooperative counselling (Werther & Davis, 1982).

Supervision

Two major functions of supervision are task orientation and concern for employees. Therefore, direction and organization of activities, motivation of employees, and management of work groups are the important functions of extension supervisors.

Direction and Organization

Extension supervisors have to plan the work and maintain a high standard of performance. The whole process of job analysis, identification of key performance areas, and performance appraisal will help in planning and organizing extension work. The training and visit system of extension has introduced mechanisms for defining goals, planning, and scheduling work at the field level with provisions for monitoring and evaluation. Some of the management techniques used by extension organizations in overall planning and management of programmes are the programme evaluation and review technique (PERT/critical path method (CPM) (Wiest & Levy, 1982), management by objectives (MBO) (McConkey, 1983), programme and performance budgeting system (PPBS), and time management techniques. These techniques have been practised by extension organizations in Asian and African countries with varying success. Personal computers offer good scope for extension managers to increase certain managerial skills.

A study of supervisory practices to improve field performance of agricultural extension in Kenya, Malawi, and the Philippines revealed the following effective supervisory practices (Honadle, 1982): (1) use of collaborative, realistic, and result-oriented target setting and a daily activity plan; (2) a needs-based participatory evaluation system; (3) involvement of farmers in decision making and a reachable service target under local constraints; and (4) effective communication and use of simple proforma and report procedures.

Studies of agricultural extension in Asia and Africa show that extension supervisors must be considerate as well as task oriented, involving subordinates in decision making and treating the employees with more interpersonal co.npetence (Leonard, 1977; Vijayaragavan & Singh, 1991). Up to 86 per cent of field agents in Southeast

Asia reported "friendship" as the most effective way for supervisors to ensure extension workers' reliable performance (Goodell, 1983). Thus if extension managers are to be effective, they have to give supportive evaluation by way of enhancing employee motivation and improving the functioning of work groups.

Motivating the Extension Personnel

The work motivation and morale of extension staff, as reported earlier, are very poor in many countries. The reasons are many. The bureaucratic structure of extension administration, lack of rewards and incentives, poor facilities, poor promotional avenues, and the low esteem given to extension are the major causes of poor motivation and morale. Extension supervisors should have the ability to motivate and lead the field extension workers so that the field agents perform more than routine jobs, and supervisors should be involved in attaining excellence in extension work. This calls for extension managers having an understanding of various theories of motivation as applicable to frontline extension agents. Therefore, a knowledge of major theories of motivation such as Maslow's hierarchy of needs theory, Herzberg's two factory theory, McClelland's need theory, theory X and theory Y, and expectancy theory of motivation is essential (Stoner & Freeman, 1992). Special training for developing motivation among fieldworkers has to be undertaken by supervisors.

Work-Group Management

Every organization has formal and informal groups. Formal groups are established by the management, while informal groups are spontaneous and developed to satisfy mutual interest of the members. Because work groups have a considerable influence on the work situation, supervisors should be sensitive to the needs of the group and develop skills to guide and achieve the group's goal, which will benefit the organization and the members. Effective extension supervision can use work groups in problem solving because they can provide many creative solutions. One way to improve supervisory effectiveness in extension work is to develop a leadership style which represents the extension workers' group interest at the higher level of organization. This will increase the confidence and morale of the work group. An understanding of group dynamics and their implications for increasing work-group performance is essential for extension supervisors. For example, in the "Hawthorne Effect," increased performance due to special treatment of the group can be effectively used in extension organizations (Honadle, 1982). Studies have pointed out that well-developed group dynamics result in increased extension performance (Leonard, 1977).

Management of Rewards and Incentives

An important aspect of human resource management which needs special attention in extension organizations is the development of a reward system which will attract, retain, and motivate extension personnel, as well as provide training and promotional opportunities. Extension organizations in Asian and African countries have a poor reward system (Vijayaragavan, 1994; Swanson, Farmer, & Bahal, 1990). The extension agents are not only poorly paid but are paid late and after reminders or visits to head-quarters (Wiggins, 1986). Most of the extension services are run by government agencies and operate under rules and regulations of public administration. These rules do not have provisions for rewarding superior performance or for a wage system based on merit. Promotion criteria are based on seniority and length of service. Thus the bureaucratic structure of extension services is a basic hindrance to designing a better reward system. Among many of the government departments, the agricultural department and extension service have a low public esteem and poor pay structure (Vijayaragavan & Singh, 1992).

The rewards and incentive system can be improved in several ways.

Rewarding Superior Performance. Extension organizations have to develop a reward system which encourages superior performance so that pay and wage administration will be an effective tool to promote performance, motivation, and satisfaction. A clear job description, performance standards, and performance appraisal will help in evaluating extension work and rewarding people for meritorious service. Ways and means have to be found within the existing framework of public administration for basing pay on performance. For example, extension workers on the basis of their performance can be sent for higher education. Nonmonetary rewards such as recognizing the good ideas of field workers or awarding honourable titles will also help in improving performance. Extension personnel may also be encouraged to form professional societies to develop and communicate high standards, as well as to recognize superior performance. A professional monthly journal or newsletter can help extension agents to communicate innovative ideas and reinforce superior performance.

Improved Working Conditions at the Field Level. The reward system must also be internally equitable. The relative importance of field-level extension functionaries has to be realized in terms of pay compensation and other amenities. Lower level extension workers often have to work under unpleasant and isolated conditions. A carefully planned system of field allowance will compensate this (Baxter, 1990). The living conditions of field extension workers must be improved by providing adequate facilities for housing, transport, and medical and educational allowances for children.

Career Planning and Development for Extension Personnel. A career refers to all of the jobs that people hold during their working lives. Career planning is the process by which employees plan their career goals and paths. Career development refers to all of the technical and managerial skills employees acquire to achieve their career plans. Career advancement, which gives a picture of future opportunities in terms of promotion, is a motivating factor for performance and development of skills. Unfortunately, no career structure exists for extension personnel in many organizations. In developing countries like India, there are many cases where one joins as a village extension worker and retires in the same position after serving thirty to thirty-five years. As part of improving the rewards and incentives system, extension organizations have to develop suitable career paths and advancement for different categories of extension personnel on a systematic basis.

As part of career development, extension personnel should be provided with opportunities to develop their technical and managerial skills to enable them to occupy higher positions. Extension personnel should have a salary structure as well as promotion opportunities comparable to other professions like health or engineering. In Kenya, the pay and career opportunities of extension workers are comparable to other government employees (Onyango, 1987). Recommendations have already been made to equate the status of agricultural extension with that of agricultural research by offering an equal salary structure, professional advancement, and incentives and rewards (FAO, 1985).

Improvement of the Quality of Work Life

The earlier approach to human resource development emphasized individual development through training and proper supervision. However, with the increasing complexity of organizations and society, it was soon realized that training individuals plays only a limited role in the development of organizations. The need for improving the quality of work life through making the job more satisfying and productive has been greatly felt. Factors such as the nature of the job or the role and involvement of employees in work decisions are important for improving the quality of work life. The methods used to do so are job enrichment, job design, and role interventions (Pareek, 1993). An understanding of these methods and their application in extension organizations are essential for extension managers to improve the performance of extension agents. Studies have shown that the work environment of extension organizations is poor and needs improvement (Jhamtani & Singh, 1989, 1992).

Job Enrichment and Job Design

Job enrichment refers to detailed analysis of the work to know the factors which make it a satisfying experience. Job enrichment uses the job as the medium of developing employees and changing organizational practices. Some of the factors which increase job satisfaction are a sense of achievement in the job, recognition for the job, the nature of the work itself, and opportunities to learn new things and grow. The principles of job enrichment, according to Herzberg (1966), are removing controls while retaining accountability, introducing new tasks, giving a complete unit of work, granting job freedom, and helping employees to become expert in their tasks. These principles can be practised by extension managers to increase the quality of work and job satisfaction among extension personnel.

Job enrichment programmes were successful in improving the quality of work and job satisfaction. However, it was found that job enrichment had a limited view of the job, and the need for greater emphasis on human values was realized. This led to the concept of job design, which refers to structuring a job to satisfy the technical, organizational, social, and human requirements of the person performing the work (Davis & Taylor, 1979). Based on the humanization of work, job design aims at increasing the quality of work life through treating the employees as human beings and emphasizing their development and involvement in work decisions. It emphasizes the use of extrinsic and intrinsic job factors, employee participation in management, autonomy, adaptability, and variety. The concept of job design can be used by extension managers to increase participation of extension personnel in the planning and management of extension programmes, which will improve the quality of their work life.

Role Interventions

The study of roles, which are the positions employees hold in an organization, as defined by the expectations of significant persons and the individuals occupying the positions, is a comparatively neglected aspect of organizations. Roles are an important dimension in increasing organizational effectiveness. Through their roles, people are linked with the organization. This linkage increases organizational effectiveness by integrating the individuals with the organization. Such integration increases mental well-being and personal effectiveness (Pareek, 1993). The purpose of role-based intervention is to increase the mutuality of roles in organizations. Role-based interventions are done through learning situations such as process laboratory, group discussion, and use of questionnaires and schedules. Role-based interventions in extension organizations will result in increased work commitment, motivation, creativity, and team spirit.

Organizational Development

An efficient extension organization needs to develop the capability of responding to changes in relation to its environment. Extension organizations have to cope with changes within and outside the organization, such as changes in farm technology, communication methods, needs of farmers, rural situations, export and import of farm produce, and market economy. Organizational development allows for planned changes in the organization's tasks, techniques, structure, and people. Attitudes, values, and practices of the organization are changed so that it can cope with changing situations. The employees also gain greater skills to deal with new problems.

Also focussing on team building and conflict management (Chattopadhyay & Pareek, 1982), organizational development is a planned effort and is done with the help of an external expert in the behavioural sciences. The process consists of diagnosis of the problem, data collection, feedback of data to the organization, introduction of specific interventions, evaluation, and follow-up. Techniques such as sensitivity training, transactional analysis, and team-building exercises are used to develop interpersonal relationships. Organizational development is an effective approach that can be used by extension organizations to bring about planned changes and to increase the interpersonal relationships among the employees.

Conclusion

The key factor in the success of extension organizations is improving their human resources. This chapter has discussed various dimensions of human resource management which will help extension managers improve their human resource system. The proper planning and implentation of the human resource system will result in overall development of extension personnel. This will also enable extension organizations to adapt to the rapid changes occurring in the extension environment of developing countries.

References

Adams, M. E. (1982). *Agricultural extension in developing countries.* Burnt Mill, Harlow, Essex: Longman

Bathgate, F. (Ed.). (1956). *Experiment in extension-The Gaon Sathi.* London: Oxford University Press.

Baxter, M. (1990). Agricultural extension investment. In D. C. Misra (Ed.), *Training and visit system of agricultural extension in India in action.* New Delhi: Ministry of Agriculture, Government of India.

Bhasin, K. (1976). *Participatory training for development.* FFHC/AD. Bangkok: FAO.

Blanckenburg, P. V. (1984). *Agricultural extension systems in some African and Asian countries.* FAO Economic and Social Development Paper 46. Rome: FAO.

Blum, A. (1987). The Israeli experience in agricultural extension and its application to developing countries. In W. M. Rivera & S. G. Schram (Eds.), *Agricultural extension worldwide—Issues, practices and emerging priorities.* London: Croom Helm.

Chattopadhyay, S., & Pareek, U. (1982). *Managing organizational change.* New Delhi: Oxford & IBH Publishing Company.

Davis. L. E. & Taylor, C. (Eds.). (1979). *Design of jobs.* Santa Monica, CA: Good Year.

FAO (1985). *Report of the expert consultation on linkages of agricultural extension with research and agricultural education.* Bangkok: RAPA/FAO.

French, W. L. (1982). *The personnel management process* (5th ed.). Boston: Houghton Miffin.

Goodell, G. E. (1983). Improving administrators' feedback concerning extension, training and research relevance at the local level: New approaches and findings from South-east Asia. *Agricultural Administration, 13,* 39-55.

Gupta, T. S. (1963). *Characteristics of effective village level workers.* Unpublished master's thesis, New Delhi, Division of Agricultural Extension, Indian Agricultural Research Institute.

Hayward, J. A. (1990). Agricultural extension: The World Bank's experiences and approaches. In B. E. Swanson (Ed.), *Report of the global consultation on agricultural extension* (p. 115-134). Rome: FAO.

Herzberg, F. (1966). *Work and nature of work.* Cleveland: World Publishing.

Honadle, G. (1982). Supervising agricultural extension: Practices and procedures for improving field performance. *Agricultural Administration, 9,* 29-45.

Jhamtani, A., & Singh, Y. P. (1989). Perceptual variations of individuals on organizational environment dimensions of a development department. *Indian Journal of Extension Education, 25,* 53-58.

Jhamtani, A., & Singh, Y. P. (1992). Inter-unit variation in organizational environment of a development department. *Manpower Journal, 28,* 17-30.

Leonard, D. K. (1977). *Reaching the peasant farmer: Organizational theory and practice in Kenya.* Chicago: University of Chicago Press.

Lynton, R., & Pareek, U. (1990). *Training for development* (2nd ed.). New Delhi: Sage.

McConkey, D. D. (1983). *How to manage by results* (4th ed.). New York: American Management Association.

Miller, E. L., Burack, E. H., & Albrecht, M. H. (1980). *Management of human resources*. Englewood Cliffs, NJ: Prentice-Hall.

Onyango, C. A. (1987). Making extension effective in Kenya: The district focus for rural development. In W. M. Rivera & S. G. Schram (Eds.). *Agricultural extension worldwide* (p. 149-162). London: Croom Helm.

Pareek, U., & Rao, T. V. (1992). *Designing and managing human resource systems*. New Delhi: Oxford & IBH Publishing Company.

Pareek, U. (1993). *Making organizational roles effective*. New Delhi: Tata McGraw-Hill.

Perumal, G. (1975). *A study of job performance, psychological characteristics, communication behaviour and training aspects of the agricultural extension officers of Tamil Nadu*. Unpublished doctoral thesis, New Delhi, Division of Agricultural Extension, Indian Agricultural Research Institute.

Stoner, J. A. F., & Freeman, R. E. (1992). *Management* (5th ed.). New Delhi: Prentice-Hall of India.

Swanson, B. E., Farmer, B. J., & Bahal, R. (1990). The current status of agricultural extension worldwide. In B. E. Swanson (Ed.), *Report of the global consultation on agricultural extension*. Rome: FAO.

Vijayaragavan, K. (1994). *Agricultural administration in India*. New Delhi: Concept Publishing Company.

Vijayaragavan, K., & Singh, Y. P. (1989). Job design and unit structural characteristics of department of agriculture. *Indian Journal of Extension Education, 25*, 1-12.

Vijayaragavan, K., & Singh, Y. P. (1991). Supervisory behaviour in agricultural departments. *Indian Journal of Extension Education, 27*, 16-22.

Vijayaragavan, K., & Singh, Y. P. (1992). Pay administration in agricultural departments. *Indian Journal of Extension Education, 28*, 60-64.

Werther, W. B., Jr., & Davis, K. (1982). *Personnel management and human resources*. Tokyo: McGraw-Hill International Book Company.

Wiest, J. D., & Levy, F. K. (1982). *A management guide to PERT/CPM*. New Delhi: Prentice-Hall of India.

Wiggins, S. (1986). Agricultural policy and agricultural extension: The African experience. In G. E. Jones (Ed.), *Investing in rural extension: Strategies and goals* (p. 99-106). London: Elsevier Applied Service Publishers.

Chapter 15
Training and Professional Development

Abdul Halim and Md. Mozahar Ali

Training is the process of acquiring specific skills to perform a job better (Jucious, 1963). It helps people to become qualified and proficient in doing some jobs (Dahama, 1979). Usually an organization facilitates the employees' learning through training so that their modified behaviour contributes to the attainment of the organization's goals and objectives. Van Dersal (1962) defined training as the process of teaching, informing, or educating people so that (1) they may become as well qualified as possible to do their job, and (2) they become qualified to perform in positions of greater difficulty and responsibility.

Flippo (1961) differentiated between education and training, locating these at the two ends of a continuum of personnel development ranging from a general education to specific training. While training is concerned with those activities which are designed to improve human performance on the job that employees are at present doing or are being hired to do, education is concerned with increasing general knowledge and understanding of the total environment. Education is the development of the human mind, and it increases the powers of observation, analysis, integration, understanding, decision making, and adjustment to new situations.

Learning Theories and Training

Learning theories are the basic materials which are usually applied in all educational and training activities. The more one understands learning theories, the better he or she will be able to make decisions and apply them to achieving the objectives. The behaviourists, the cognitivists, and the humanists emphasize different aspects of the teaching-learning process in their approaches. While the behaviourists stress external conditions (environment) resulting in observations and measurable changes in behaviour, the cognitivists are more concerned with how the mind works (mental processes such as coding, categorizing, and representing information in memory). The humanists, on the other hand, emphasize the affective aspects (e.g., emotions, attitudes) of human behaviour that influence learning (IRRI, 1990). In extension systems, effective training must be able to take care of all the theories of learning in order to change the *action*, *belief*, and *knowledge* components of a trainee simultaneously. *Andragogy* (a theory of adult learning) is usually used rather than *pedagogy* (a theory of child learning) in extension training.

Training Approach

There are three approaches to training: (1) the traditional approach, (2) the experiential approach, and (3) the performance-based approach (Rama, Etling, & Bowen, 1993). In the traditional approach, the training staff designs the objectives, contents, teaching techniques, assignments, lesson plans, motivation, tests, and evaluation. The focus in this model is intervention by the training staff. In the experiential approach, the trainer incorporates experiences where in the learner becomes active and influences the training process. Unlike the academic approach inherent in the traditional model, experiential training emphasizes real or simulated situations in which the trainees will eventually operate. In this model, the objectives and other elements of training are jointly determined by the trainers and trainees. Trainers primarily serve as facilitators, catalysts, or resource persons. In the performance-based approach to training, goals are measured through attainment of a given level of proficiency instead of passing grades of the trainees. Emphasis is given to acquiring specific observable skills for a task. This performance-based teacher education (PBTE) model, developed by Elam (1971), is mostly task or skill centred and is also applicable to nonformal educational organizations such as extension.

Extension Personnel around the World in Need of Training

Worldwide, there are currently more than 600,000 extension workers comprised of administrative staff, subject-matter specialists (SMS), fieldworkers, and some multipurpose unidentified people; the Asian and Pacific countries have absorbed more than 70 per cent of them (Bahal, Swanson, & Farner, 1992). The percentage of extension personnel by position, as reported by Swanson, Farner, and Bahal (1990), was 7 per cent administrative, 14 per cent SMS, and 79 per cent field staff, with regional differences. Almost 13 per cent of extension workers are women, with significant regional differences (Bahal et al., 1992). The ratio of SMS to field staff is also low in Asia, Africa, the Near East, and Latin American countries, varying from about 1:11 to 1:14. The ratio for countries of Europe and North America varies from 1:1.5 to 1:1.6. The worldwide ratio of SMS to field staff is 1:11.5 (Swanson et al., 1990).

Abdul Halim is a Professor in the Department of Agricultural Extension Education and Director of the Extension Centre, Bangladesh Agricultural University, Mymensingh, Bangladesh. **Md. Mozahar Ali** is an Assistant Professor (Agricultural Extension) in the Graduate Training Institute, Bangladesh Agricultural University, Mymensingh, Bangladesh.

Deficiencies in knowledge, skills, and ability among extension personnel, particularly those of Asia, Africa, and Latin America, are remarkable. About 39 per cent of the extension personnel worldwide have a secondary-level and 33 per cent an intermediate-level education (Bahal et al., 1992). Moreover, within each region, there is a lot of variation in basic academic qualifications of the frontline extension workers, SMS, and administrators. Differences in training received are also wide. In Africa, most frontline extension workers still have only a secondary school diploma (Bahal et al., 1992). The poor educational background of extension personnel necessitates regular training.

Types of Training

Training may broadly be categorized into two types: preservice training and inservice training. Preservice training is more academic in nature and is offered by formal institutions following definite curricula and syllabuses for a certain duration to offer a formal degree or diploma. Inservice training, on the other hand, is offered by the organization from time to time for the development of skills and knowledge of the incumbents.

Preservice Training

Preservice training is a process through which individuals are made ready to enter a certain kind of professional job such as agriculture, medicine, or engineering. They have to attend regular classes in a formal institution and need to complete a definite curriculum and courses successfully to receive a formal degree or diploma. They are not entitled to get a professional job unless they can earn a certificate, diploma, or degree from the appropriate institution. Preservice training contents emphasize mostly technical subject matter such as crops, animal husbandry, and fisheries as well as pedagogical skills to prepare the students to work in agriculture.

In general two types of preservice training are available for agricultural staff. These are (1) degree level (at least a bachelor's degree in agriculture or related field), which is usually offered for four years by a university or agricultural college; and (2) diploma level, which is mostly offered by the schools of agriculture for a period of two to three years. The entry point for the former is normally twelve years of schooling and for the latter ten years of schooling.

Inservice Training and Staff Development

Inservice training is a process of staff development for the purpose of improving the performance of an incumbent holding a position with assigned job responsibilities. It promotes the professional growth of individuals. "It is a program designed to strengthen the competencies of extension workers while they are on the job" (Malone, 1984, p. 209). Inservice training is a problem-centred, learner-oriented, and time-bound series of activities which provide the opportunity to develop a sense of purpose, broaden perception of the clientele, and increase capacity to gain knowledge and mastery of techniques.

Inservice training may broadly be categorized into five different types: (1) induction or orientation training, (2) foundation training, (3) on-the-job training, (4) refresher or maintenance training, and (5) career development training. All of these types of training are needed for the proper development of extension staff throughout their service life.

Induction or Orientation Training. Induction training is given immediately after employment to introduce the new extension staff members to their positions. It begins on the first day the new employee is on the job (Rogers & Olmsted, 1957). This type of training is aimed at acquainting the new employee with the organization and its personnel. Induction training for all new personnel should develop an attitude of personal dedication to the service of people and the organization. This kind of training supplements whatever preservice training the new personnel might have had (Halim and Ali, 1988). Concerning the characteristics of a new employee, Van Dersal (1962) said that when people start to work in an organization for the first time, they are eager to know what sort of outfit they are getting into, what they are supposed to do, and whom they will work with. They are likely to be more attentive and open-minded than experienced employees. In fact, the most favourable time for gaining employees' attention and for moulding good habits among them is when they are new to the job.

Foundation Training. Foundation training is inservice training which is also appropriate for newly recruited personnel. Besides technical competence and routine instruction about the organization, every staff member needs some professional knowledge about various rules and regulations of the government, financial transactions, administrative capability, communication skills, leadership ability, coordination and cooperation among institutions and their linkage mechanism, report writing, and so on. Foundation training is made available to employees to strengthen the foundation of their service career. This training is usually provided at an early stage of service life.

Maintenance or Refresher Training. This training is offered to update and maintain the specialized subject-matter knowledge of the incumbents. Refresher training keeps the specialists, administrators, subject-matter officers, extension supervisors, and frontline workers updated and enables them to add to the knowledge and skills they have already. Maintenance or refresher training usually deals with new information and new methods, as well as review of older materials. This type of training is needed both to keep employees at the peak of their possible production and to prevent them from getting into a rut (Van Dersal, 1962).

On-the-Job Training. This is ad hoc or regularly scheduled training, such as fortnightly training under the training and visit (T&V) system of extension, and is provided by the superior officer or the subject-matter specialists to the subordinate field staff. This training is generally problem or technology oriented and may include formal presentations, informal discussion, and opportunities to try out new skills and knowledge in the field. The superior officer, administrator, or subject-matter specialist of each extension department must play a role in providing on-the-job training to the staff while conducting day-to-day normal activities.

Career or Development Training. This type of in-service training is designed to upgrade the knowledge, skills, and ability of employees to help them assume greater responsibility in higher positions. The training is arranged departmentally for successful extension workers, at all levels, for their own continuing education and professional development. Malone (1984) opined that extension services that provide the opportunity for all staff to prepare a plan for career training will receive the benefits of having longer tenured and more satisfied employees, which increases both the effectiveness and efficiency of an extension service. Malone stated that "career development is the act of acquiring information and resources that enables one to plan a program of life-long learning related to his or her worklife" (p. 216). Although extension workers are responsible for designing their own career development education, the extension organization sometimes sets some criteria and provides opportunities for the staff by offering options.

Phases of Training

Training is a circular process that begins with needs identification and after a number of steps ends with evaluation of the training activity. A change or deficiency in any step of the training process affects the whole system, and therefore it is important for a trainer to have a clear understanding about all phases and steps of the training process. In the broadest view, there are three phases of a training process: planning, implementation, and evaluation.

Planning Phase

The planning phase encompasses several activities, two of which—training needs identification and curriculum development—are very important.

Training Needs Identification. Training need is a condition where there is a gap between "what is" and "what should be" in terms of incumbents' knowledge, skills, attitudes, and behaviour for a particular situation at one point in time. This gap is called "a problem," which usually occurs when a difference exists between "desired performance" and "actual performance." The needs identification process assists trainers in making sure that they have matched a training programme to a training problem. For example, agricultural extension officers (AEOs) have been giving training to village extension workers (VEWs), but performance of the VEWs is not improving. The reasons may be:

1. The AEOs lack subject-matter knowledge.
2. The AEOs do not conduct training well.
3. The training centre lacks training facilities.
4. The VEWs are organized not to work properly until their demands are satisfied by the government.

The first two problems are related to knowledge and skills and can be solved effectively by a training programme, but the third and fourth problems need government attention to solve.

Training needs identification is possible through different analytical procedures. The major procedures used in determining training needs are the following:

Organizational analysis determines where training emphasis should be placed within the organization and is based on the objectives of an organization. Concerning what one should do in analysing an organization, McGhee and Thayer (1961) suggest four steps:

1. Stating the goals and objectives of an organization
2. Analysing the human resources
3. Analysing efficiency indices
4. Analysing the organizational climate

The results of these analyses are then compared with the objectives of the organization. These comparisons point to specific areas in which training is needed.

Individual analysis aims at identifying specific training needs for an individual or group of employees so that training can be tailored to their needs. This analysis centres on individuals and their specific needs concerning the skills, knowledge, or attitudes they must develop to perform their assigned tasks. The possible methods or techniques for individual analysis include performance appraisal, interviews, questionnaires, tests, analysis of behaviour, informal talks, checklist, counseling, critical incidents, recording, surveys, and observations.

Group analysis includes a number of techniques in which a group of well-informed employees discuss different aspects of the organization, the employees, and the tasks to identify the major discrepancies in achieving predetermined targets for each of them with a view to assessing training needs as distinguished from other necessary changes for removing these discrepancies. The major techniques which are used in this approach are brainstorming, buzzing, cardsorts, advisory committee, conferences, problem clinic, role playing, simulation, task forces, workshops, and so forth.

Many problems exist in an organization, but some problems cannot be solved by training. After a preliminary needs analysis, which gives probable causes and solutions, the results should be verified with the concerned personnel of the organization to determine whether training is an appropriate action to solve that problem.

Curriculum Development. This is the most important part in a training programme after a need for training has been identified. The curriculum specifies what will be taught and how it will be taught. It provides the framework and foundation of training. The first phase of curriculum development determines what will be taught, that is, the training content.

Once training needs have been identified and training activities have been decided as part of the solution, a *needs analysis* should be done to determine knowledge, skills, and attitude requirements and performance deficiencies. The needs analysis procedure involves breaking down the "training problem" into its basic parts in different successive phases to identify and understand the important components in each phase. Ultimately it leads to identifying and understanding the training content. The training needs analysis process can be divided into three distinct analytical phases: job analysis, task analysis, and knowledge and skiil-gap analysis.

A. *Job analysis.* Job analysis is a method of determining major areas of tasks where training may be needed (see JA Worksheet). It involves the dissecting of a job into its component events or parts. This analysis allows a trainer to better understand what an employee does in an organization. Job analysis involves the "task identification" of a particular job (Wentling, 1992). The techniques used in task identification include job questionnaire, interview, participant observation, work sampling, job audit, and small-group discussion. The following steps may.provide a guide for completion of job analysis:

1. Identify the job that is to be the subject of the analysis. This involves defining the focal point for the job analysis. It may include the entire job of a group of employees or only a specific segment of their job.
2. Prepare a list of tasks which can be done following different approaches and methods.Four approaches can be used to identify job tasks: (1) expertsidentify and list critical tasks, (2) observations and interviews are conducted with employees, (3) meetings are held with group representatives, and (4) a tentative list of task is reviewed by employees and their supervisors.
3. Verify the tasks. The draft list of tasks should be verified by experts, workers, and supervisors in the analysis process. This can be done through expert review, small-group discussions, and inter views. When the tasks are verified, a final list of job tasks is prepared.
4. Determine the frequency. The workers and super visors can fill in a form indicating how frequently each task in a job is performed. Different scales such as "seldom," "occasionally," "weekly to monthly," "daily to weekly," and "daily" can be used to quantify the intensity of a task accomplished.

Job Analysis Worksheet

Job: Agriculture Extension Officer

Tasks:	Frequency performed[a]	Importance[b]	Learning difficulty[c]	Total score	Focus
1. Supervision	4	3	1	8	...
2. Conducting training	4	3	3	10	yes
3. Planning programmes	2	3	2	7	...
4. Research trial	2	2	1	5	...
....	...	...	...	...	...
....	...	...	...	...	...
....	...	...	...	...	...

[a] 1 = Seldom
2 = Occasionally
3 = Weekly to monthly
4 = Daily to weekly
5 = Daily

[b] 1 = Marginally important
2 = Moderately important
3 = Extremely important
4 = Extremely difficult

[c] 1 = Easy
2 = Moderately difficult
3 = Very difficult

5. Determine the importance. Not all tasks are equally important to a job. An occasionally performed task may be very important. Therefore, a relative importance rating is useful along with frequency rating. A scale such as "marginally important," "moderately important," and "extremely important" may be used to determine the relative importance of the job tasks.
6. Estimate the learning difficulty. An estimate of learning difficulty is another dimension of the job-task analysis. It shows the trainer the employees' perception of difficulty, which may be differ ent from the trainer's own perception. A scale such as "easy," "moderately difficult," "very difficult," and "extremely difficult" may be used to determine the difficulty indices of job tasks.
7. Calculate the total score. This can be done by simply adding the scores for frequency, importance, and learning difficulty for each task. The column for total score in a worksheet indicates the priority tasks for training if these are training problems.
8. Review the findings. The results of the job-task analysis should be discussed with significant people in the training system, including government leaders, programme directors, and others interested in related training.

B. Task analysis. The output of the job analysis is a list of broad job tasks, based on importance, learning difficulty, and frequency of doing the task. Each task is a complex set of procedures in itself, and therefore it needs further analysis to find out which specific segment of the of the task is critical in designing a training programme (see Task Analysis Worksheet). To do this, it is necessary to follow a method called task analysis, which is similar to job analysis.

Task analysis procedures include preparing a blank task analysis worksheet, writing down the name of the job at the top of each sheet, and then making copies. Each of these forms will be used for breaking down and analysing each of the most important job tasks. Therefore, it is necessary to write one important task identified for training on each of the task analysis worksheets and to list all component parts of each task on its respective task analysis worksheet. This is followed by the steps used for job analysis to find out the frequency, importance, and learning difficulty for each step of the tasks. Then the score for each component part is put in the "total score" column, and the results are discussed with concerned personnel in the organization. The job analysis and task analysis processes are similar to each other, so the model for both worksheets is the same.

The important difference between these two steps of analysis is that "the job analysis helps us identify major blocks of content to include in training; the task analysis helps us understand what comprises an individual block" (Wentling, 1992). Both are very important to the curriculum development process. What needs to be taught and what steps are involved in the process are completed by these analyses and comprise the major steps in curriculum development.

Task Analysis Worksheet

Job: Agriculture Extension Officer

Task: Conducting training

Components/steps	Frequency Performed[a]	Importance[b] Difficulty[c]	Learning Score	Total	Focus
Establishing rapport	5	3	1	9	-
Introducing the topic	-	-	-	-	-
Presenting the subject	5	3	1	9	-
Maintaining sequence	-	-	-	-	-
Maintaining eye contact	-	-	-	-	-
Using A/V aids in time	5	3	4	12	yes
________	-	-	•	-	-
________	-	-	-	-	-
Summarizing the lecture	5	3	3	11	yes

[a] 1 = Seldom
2 = Moderately important
3 = Weekly to monthly
4 = Daily to weekly
5 = Daily

[b] 1 = Marginally important
2 = Moderately important
3 = Extremely important
4 = Extremely difficult

[c] 1 = Easy
2 = Moderately difficult
3 = Very difficult

Skill-Gap Analysis Worksheet

Job: Agriculture Extension Officer
Task: Delivering lecture in VEWs training

Steps-components	Level of proficiency	Is proficiency a problem?	Can problem be solved by training?
Establishing rapport	1 2 3 (4) 5	[ ]	[ ]
Introducing the topic	1 2 (3) 4 5	[Y]	[Y]
Presenting the subject	1 2 3 (4) 5	[ ]	[ ]
Maintaining sequence	1 2 (3) 4 5	[ ]	[ ]
Maintaining eye contact	1 (2) 3 4 5	[Y]	[Y]
Using A/V aids in time	1 (2) 3 4 5	[ ]	[ ]
Supplying handouts	1 2 (3) 4 5	[Y]	[N]
....	1 2 3 4 5	[ ]	[ ]
Summarizing the lecture	(1) 2 3 4 5	[Y]	[Y]

1 = Cannot do at all
2 = Can do less than half of the task
3 = Can do more than half but less than total
4 = Can do total but cannot maintain time schedule
5 = Can do within time schedule

C. *Knowledge and skill-gap analysis.* The knowledge or skill-gap analysis is a process of determining the training needs of individual employees in relation to the important tasks-steps or components of tasks identified for training (see Skill-Gap Analysis Worksheet). The skill-gap analysis determines how skilled or proficient individual employees are on these tasks-steps or components, how much individuals differ from desired performance, and whether or not they need training. It would be a waste of resources and frustrating to the trainer and trainees to design and deliver training on topics and skills where the trainees are already able and proficient. A priority list of the tasks identified for training according to the total score in the job analysis is made. Then, the steps or components that were identified on each task analysis worksheet are listed on the skill-gap analysis worksheet. This is followed by rating each step-component in terms of the trainee's current proficiency on a scale of 1 to 5, as shown in the legend of the worksheet. Identifying the steps-components that appear to have low proficiency is required because there is a gap between what is desired and the current situation. After this, a review is done to ponder whether the gap can be decreased or removed through training or whether training is the most appropriate method. There may be some steps-components for which measures other than training are more appropriate. At this stage, key personnel such as subject-matter specialists, supervisors, and extension-training experts should discuss the findings before finalizing the curriculum. This helps to identify different perspectives and to avoid unnoticed mistakes or biases in curriculum development.

The training needs analyses provide many things to a trainer. The analyses determine the training contents and how deficient the trainees are in these contents, and the sequence of tasks provides the sequence of training activity.

Selecting a Training Method

A training programme has a better chance of success when its training methods are carefully selected. A training method is a strategy or tactic that a trainer uses to deliver the content so that the trainees achieve the objective (Wentling, 1992). Selecting an appropriate training method is perhaps the most important step in training activity once the training contents are identified. There are many training methods, but not all of these are equally suitable for all topics and in all situations. To achieve the training objective, a trainer should select the most appropriate training method for the content to involve the trainees in the learning process. Four major factors are considered when selecting a training method: the learning objective, the content, the trainees, and the practical requirements (Wentling, 1992). According to Bass and Vaughan (1966), training methods should be selected on the basis of the degree to which they do the following:

1. Allow active participation of the learners.
2. Help the learners transfer learning experiences from training to the job situation.
3. Provide the learners with knowledge of results about their attempts to improve.

4. Provide some means for the learners to be reinforced for the appropriate behaviour.
5. Provide the learners with an opportunity to practise and to repeat when needed.
6. Motivate the learners to improve their own performance.
7. Help learners increase their willingness to change.

These criteria indicate that a single training method will not satisfy the objectives of a training programme.

A variety of training methods are available to a trainer. The most commonly used methods include:

1. *Instructor presentation.* The trainer orally presents new information to the trainees, usually through lecture. Instructor presentation may include classroom lecture, seminar, workshop, and the like.
2. *Group discussion.* The trainer leads the group of trainees in discussing a topic.
3. *Demonstration.* The trainer shows the correct steps for completing a task, or shows an example of a correctly completed task.
4. *Assigned reading.* The trainer gives the trainees reading assignments that provide new information.
5. *Exercise.* The trainer assigns problems to be solved either on paper or in real situations related to the topic of the training activity.
6. *Case study.* The trainer gives the trainees information about a situation and directs them to come to a decision or solve a problem concerning the situation.
7. *Role play.* Trainees act out a real-life situation in an instructional setting.
8. *Field visit and study tour.* Trainees are given the opportunity to observe and interact with the problem being solved or skill being learned.

Implementation Phase

Once the planning phase of a training programme is complete, then it is time to implement the course. Implementation is the point where a trainer activates the training plan, or it is the process of putting a training programme into operation.

The first step towards implementing a training programme is publicity. Most of the well-established training centres develop training brochures which contain course descriptions, prepare an annual calendar of training opportunities, and inform concerned organizations, agencies, or departments well ahead of time about their training plans. Once the training centre and concerned organizations agree to implement training, the next step is to arrange available resources such as sufficient funds for the course and facilities for food, lodging, transportation, and recreation. All these resources need to be well managed and coordinated to run the programme smoothly.

Evaluation Phase

Evaluation is a process to determine the relevance, effectiveness, and impact of activities in light of their objectives. In evaluating an extension training programme, one needs to consider that most training activities exist in a larger context of projects, programmes, and plans. Thus Raab et al. (1987, p. 5) define training evaluation as "a systematic process of collecting information for and about a training activity which can then be used for guiding decision making and for assessing the relevance and effectiveness of various training components."

Kirkpatrick (1976) suggested four criteria to evaluate training programmes: (1) reaction, (2) learning, (3) behaviour, and (4) results. Each criterion is used to measure the different aspects of a training programme. *Reaction* measures how the trainees liked the programme in terms of content, methods, duration, trainers, facilities, and management. *Learning* measures the trainees' skills and knowledge which they were able to absorb at the time of training. *Behaviour* is concerned with the extent to which the trainees were able to apply their knowledge to real field situations. *Results* are concerned with the tangible impact of the training programme on individuals, their job environment, or the organization as a whole.

Types of Evaluation

On the basis of the time dimension, evaluation may be classified as (1) formative evaluation and (2) summative evaluation. *Formative evaluation* involves the collection of relevant and useful data while the training programme is being conducted. This information can identify the drawbacks and unintended outcomes and is helpful in revising the plan and structure of training programmes to suit the needs of the situation. *Summative evaluation* is done at the end of the programme and makes an overall assessment of its effectiveness in relation to achieving the objectives and goals.

Raab et al. (1987), however, classified evaluation into four major types: (1) evaluation for planning, (2) process evaluation, (3) terminal evaluation, and (4) impact evaluation.

Evaluation for planning provides information with which planning decisions are made. Training contents and procedures (methods and materials) are usually planned at this stage in order to choose or guide the development of instructional aids and strategies. *Process evaluation* is conducted to detect or predict defects in the procedural design of a training activity during the implementation phase (Raab et al., 1987). Through this process the key elements of the training activities are systematically monitored, problems are identified, and attempts are made to rectify the mistakes before they become serious. Process evaluation is periodically conducted throughout the entire period of the programme.

Terminal evaluation is conducted to find out the effectiveness of a training programme after it is completed. The objectives of terminal evaluation are to determine the degree to which desired benefits and goals have been achieved, along with the causes of failure, if any. *Impact evaluation* assesses changes in on-the-job behaviour as a result of training efforts. It provides feedback from the trainees and supervisors about the outcomes of training. It measures how appropriate the training was in changing the behaviour of participants in real-life situations.

References

Bahal, R., Swanson, B. E., & Farner, B. J. (1992). Human resources in agricultural extension: A worldwide analysis. *Indian Journal of Extension Education, 28* (3, 4), 1-9.

Bass, B. M., & Vaughan, J. A. (1966). *Training in industry: The management of learning.* Belmont, CA: Wadsworth Publishing.

Dahama, O. P. (1979). *Extension and rural welfare.* New Delhi: Ram Parsad and Sons.

Elam, S. (1971). *Performance based teacher education: What is the state of the art.* Washington, DC: AACTE.

Flippo, E. B. (1961). *Principles of personnel management.* New York: McGraw Hill.

Halim, A., & Ali, M. M. (1988). Administration and management of training programmes. *Bangladesh Journal of Training and Development, 1* (2), 1-19.

IRRI. (1990). *Training and technology transfer course performance objectives manual.* Manila: International Rice Research Institute.

Jucious, M. J. (1963). *Personnel management* (5th ed.). Homewood, IL: Richard D. Irwin.

Kirkpatrick, D. (1976). Evaluation of training. In R. L. Craig (Ed.), *Training and development handbook.* New York: McGraw Hill.

Lynton, R. P., & Pareek, U. (1990). *Training for development.* West Hartford, CT: Kumarian Press.

Malone, V. M. (1984). Inservice training and staff development. In B. E. Swanson (Ed.), *Agricultural extension: A reference manual.* Rome: FAO.

McGhee, W., & Thayer, P. W. (1961). *Training in business and industry.* New York: John Wiley and Sons.

Raab, R. T., Swanson, B. E., Wentling, T. L., & Clark, C. D. (Eds.). (1987). *A trainer's guide to evaluation.* Rome: FAO.

Rama, B. R., Etling, A. W. W., & Bowen, B. E. (1993). Training of farmers and extension personnel. In R. K. Samanta (Ed.), *Extension strategy for agricultural development in 21st century.* New Delhi: Mittal Publications.

Rogers, F. E., & Olmsted, A. G. (1957). *Supervision in the cooperative extension service.* Madison, WI: National Agricultural Extension Center for Advanced Study.

Swanson, B. E., Farner, B. J., & Bahal, R. (1990). The current status of agricultural extension worldwide. In B. E. Swanson (Ed.), *Report of the Global Consultation on Agricultural Extension.* Rome: FAO.

Van Dersal, W. R. (1962). *The successful supervisor.* New York: Harper and Row.

Wentling, T. L. (1992). *Planning for effective training: A guide to curriculum development.* Rome: FAO.

Chapter 16
Acquiring and Managing Financial Resources

Robert P.Bentz[1]

Just as there are several organizational models for delivering extension services to the public, there are a number of ways to finance those services and to keep track of the money. Sound financial management may be fundamental to success. Poor financial management, on the other hand, often accompanies and contributes to failure.

This chapter focuses its attention on principles related to money matters. It is hoped that readers will apply these concepts as appropriate to situations they face in their own organizations and in accordance with the rules and customs of their own countries. Because people are more readily inclined to apply ideas that deal with practical concerns, this discussion will be organized around practical problems. The chapter also will address some of the major methods found to be most successful in resolving those problems.

The major financial concerns of an extension organization can be described rather simply:

- Extension organizations have too few financial resources, given the nature and size of their missions.
- Managers and other staff members do not know and apparently cannot find out how many resources are available for use in dealing with a particular problem or program activity.
- They do not know the costs of carrying out various activities or whether a particular approach is a cost-effective way to deliver programming.
- Money may be committed almost totally to salaries, leaving little to cover operating costs.
- Financial allocation decisions are made by people who are not in a position to know best how the money should be spent.
- Little is known about whether work is being carried out efficiently or what has been the impact of organization programmes.
- Financial resources are wasted.
- Resources are misappropriated.

Most of what can go wrong in an extension organization when it comes to money is covered by one or more of the above problems. Managers may be inclined to say that their problems would be solved if they just had more money to work with. And having more money to use is certainly better than having too little. But more money may not always result in greater impact if the money is not well managed.

There may be little connection between quality of programme delivery and an organization's system for managing money. A specialist's message getting through to a farmer does not totally depend on how much money is spent on the process. However, organizations that are effective tend to know how their money is being spent. A case can easily be made that sound financial management is the first step toward quality programming.

Obtaining Financial Resources

There is an almost universal reaction of managers when it comes to money, namely, they do not have enough of it. They need and want more.

> Concept No. 1. Leaders are responsible for acquiring and maintaining resources for their organization. These efforts tend to be more successful when requests deal with issues of high national priority and when an explanation is provided as to how the additional resources will be used.

Extension organizations are supported financially from a variety of sources. Publically funded organizations receive support from:

- Public appropriations
- Contracts with other agencies, donors, or both
- Loans
- Gifts
- Income from sales
- User fees

Private extension organizations are financed by:

- Endowment income
- Private investment
- Income-producing activities
- The same sources as public extension organizations

Depending on custom and regulations, extension organizations may receive support from any combination of the above. Currently, there is interest in some countries in "privatizing" some functions—that is, shifting certain responsibilities (and the costs associated with them) to privately funded enterprises (NGOs) or directly to users.

Robert P. Bentz is Senior INTERPAKS Adviser in the Office of International Agriculture, University of Illinois at Urbana-Champaign, Urbana, Illinois.

But whether the extension effort is private or public, it is not uncommon for financial support to be provided from more than one source. Obviously, the more sources of funding, the more complicated administration is, because most funding sources have their own expectations of outcomes, reporting requirements, and timetables.

Despite administrative complexity, managers often find that having more than one source of funding is advantageous. Several sources provide a hedge against adverse treatment by a single funding partner or source. Special funding sources often help the organization to focus on emerging, high priority concerns or to test new approaches. A practical goal might be to have a resource commitment from the principal funding partner (the most dependable source of continuing funding) that will maintain the "core" or basic structure of the organization (e.g., selected staff salaries and other expenses). Support for whatever is beyond the core might then come from the so-called softer resources or some combination of funding sources as described previously. Temporary resources permit experimentation and the carrying out of special projects.

No single ideal ratio of core to total support can be applied everywhere. This is partly because organizations will define what is part of the core and what is left for special or temporary support differently. As a general rule, managers might reasonably seek something like an 80-20 split (core to soft resources).

Keeping Track of Financial Resources

The function of management is to plan, organize, staff, lead, and control. Every one of these functions is influenced to a great degree by how much money there is. Managers and programme staff simply cannot carry out their assigned responsibilities effectively without understanding their financial constraints:

> Concept No. 2. Leaders of extension organizations are accountable for the financial resources assigned. It is incumbent upon them to establish workable systems that will enable staff members to know how many resources they have to carry out their work.

Managers need to have some means for knowing what is happening with respect to their financial resources if they are to make informed management decisions. The notion that leaders of extension organizations are accountable to funding partners is one of the reasons managers need to keep track of how money is spent. The organization will be expected to report how much money there was, how much was spent, what it was spent for, and how much is left.

This responsibility is carried out by installing and managing a financial accounting system. That system may well be automated at some point, but a manual system will serve most needs at the outset. But regardless of how reports are produced and records maintained, they should be accurate and produced in a timely fashion so that staff can base their decisions on good information. A number of acceptable computer software programmes are available to meet this accounting need. It is generally unnecessary for most organizations to spend time and resources designing and implementing a unique system.

Predicting Organizational Costs

Resources to be committed to any activity are always finite. An extension organization's need for money and the benefits that accrue from that expenditure compete in theory for resources against all other potential uses.

> Concept No. 3. Budgeting the use of resources to particular purposes and for specific accounting periods is an essential part of the planning and managing processes. This process combined with cost accounting makes it possible to make decisions based on expected costs and returns.

Governments are expected to provide essential services and to use resources for the betterment of citizens. Other funding sources seek to maximize benefits in accordance with their particular goals. And no matter how well endowed the funding sources are, there are always limits to how much they can provide. Budgeting is the technique used by organizations to help determine how many resources will be needed and how they will be used. The budget also is used to help manage the rate of expenditures as the year progresses.

Budgets are typically subdivided into categories that have meaning in the local situation and that help managers and potential funders to understand how money is to be spent. Budget information when compared with actual expenses enables managers and programme staff to stay within assigned allocation levels as they accomplish programme goals.

One way to subdivide budget categories is by planned activities, such as:

On-Farm Trials	xxx
Field Demonstrations	xxx
Workshops	xxx
Mass Media	xxx
Evaluation	xxx
Total	xxx

A second way is to separate planned expenditures according to some standard accounting system nomenclature:

Salaries and Wages	xxx
Supplies	xxx
Materials	xxx
Transportation	xxx
Printing Services	xxx
Communications	xxx
Total	xxx

The two approaches may be combined:

On-Farm Trials

Salaries and Wages	xxx
Supplies	xxx
Materials	xxx
Etc.	xxx
Total	xxx

This latter approach makes it possible to obtain information on the costs of various activities and to consider different combinations of expenditure that might achieve the same result. And of course, all of the above approaches may be further subdivided by time periods (e.g., by month). The budget then becomes a tool for managers to control rates of expenditure.

The budgeting cycle perhaps can best be presented pictorially. The process may take one year or longer, depending on local practice and regulation, but it should include all of the following elements (See figure 1):

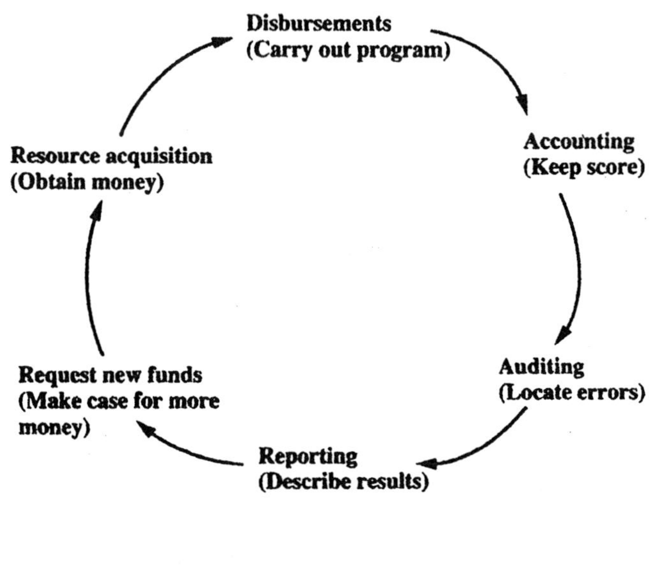

Figure 1. A typical budgeting cycle.

Maintaining a Balance in How Resources Are to Be Used

Some extension organizations simply do not have enough resources to cover normal operating costs. This may be a result of previous decisions that had the effect of protecting staff salaries and staff size at the expense of allocations for operating costs. Or it may be that salary support is provided by a central authority and is not even a part of the organization's budget. In this situation, when budgets are cut (with little or no change in staff size), it merely means that staff members have fewer resources to use to carry out their work. Or it may be that funds have been provided, but that increases have not kept pace with inflation. The result is that too few resources are available to carry out the work of the organization. In some organizations, almost no resources exist to cover travel, publications, or any of a wide variety of other possible operating costs.

> Concept No. 4. Managers should seek to align resources in ways that make it possible to accomplish organizational objectives. This usually suggests the need to maintain an appropriate balance between salaries and operating expenses.

It does little good to have a large staff if resources are too limited to provide that staff with the means to get something accomplished. Managers might logically address this problem by reducing the number of positions and increasing expense allocations, thus changing the ratio between salaries and operating expense items.

However, such a move might make it necessary to change the way programmes are delivered. For example, staff may be too few in number to permit visits to individual farmers. If farmers are to be reached in this situation, it may be necessary to bring them together in groups. Or the result may be that only a small number of individual farmers can be reached. In some countries, national regulations do not make it easy to exchange personnel resources for money to use for operating costs, or vice versa, or even to avoid unfunded programme responsibilities. In those situations especially, managers might seek help from alternative funding sources. However, if the problem can be resolved simply by reallocation, that is often the most straight forward approach. Establishing user fees and charging for publications are other ways to obtain additional resources when reallocations between salaries and expenses are not possible. The problem with the latter approaches is that targeted audiences may not have resources to pay user fees or to buy publications. Consequently, these approaches can have negative effects on programme impact.

There is probably no single "best" ratio to maintain between salaries and expense resources. The percentage of total resources spent on salaries might vary with the mixture of staff members (e.g., the number of village workers or paraprofessionals and the number of subject-matter specialists). The amount of operating resources required to carry out the mission will be affected by geography, size of country, and method of programme delivery. However, as a general guide, if salary costs require more than 75 to 80 per cent of total resources, this is a fairly good indication that operating funds may be beginning to limit success.

Decentralizing the Decision-Making Process

The purpose of an extension organization is to deliver practical problem-solving educational programmes and to carry out various other designated activities. These activities tend to be very location specific. Yet the tendency is to structure extension organizations according to a "command" or top-down decision-making model. However, this approach has been found wanting time and again. In fact, the opposite condition appears to yield the best results when it comes to actually carrying out programmes. It appears that decentralized decision making is almost essential if an organization is to be responsive to the people being served. In short, central staff should limit their decision making to determining financial commitments to programme areas and leave decisions regarding how money is spent in support of programmes to others closer to the scene.

> Concept No. 5. The determination of how money is to be spent may best be made by those who are directly carrying out activities that serve intended clientele.

Who better than the people delivering programmes know what is needed? And who would know better how to spend an additional resource (were it to be available) to achieve greater benefit? Research suggests that the answer to these questions is that teams of programme leaders, specialists, and others in direct contact with clientele are in better positions than central staff are to make informed programme decisions and to spend money wisely.

If this kind of decentralized decision making is to be accomplished, organizational resources will have to be subdivided among programme units. And programme leaders and others will need to be given authority and responsibility to make decisions about how money is to be spent. If this is simply not possible under existing rules, then such decisions should be made by staff members who are as close to the programme level as is possible.

Decisions regarding centrally managed financial reserves should be made so that needs with the highest priority can be addressed. And of course, it is a wise practice to maintain a reserve fund centrally to deal with emerging and unforeseen events.

Using Information to Improve Efficiency

There has been much discussion in the literature about whether extension organizations are appropriate vehicles to use to bring about agricultural development, to avoid rapid urbanization, and even what kinds of activities to include under the extension rubric. A lot of research has been conducted that is designed to determine the "outcomes" from extension education. A considerable body of research deals with these issues. But managers cannot get by leaving those issues entirely to research institutions. They are expected by stakeholders to efficiently manage resources assigned to them.

> Concept No. 6. Efficiency is a relative measure. Efficiency can be implied by sets of physical data compared over time or among categories (e.g., number of pages produced per editor, this time period compared with an earlier time period).

Efficiency can also measured in monetary terms (e.g., cost of delivering a particular message per 100 users compared with another delivery method or another time period).

Organizations typically create some kind of feedback mechanism to collect needed data from staff and others when they want to take measurements of efficiency. Many organizations assign staff and other resources to coordinate the task of collecting the necessary information. Often, data collection and analysis are a full-time responsibility of a programme monitoring unit or a management information system (MIS) unit.

The monitoring unit will need to obtain appropriate data from the organization's financial information system to determine costs of various undertakings and to compute input-output ratios. These cost estimates or ratios, when compared over time and between different approaches, help staff to know which methods or approaches are more efficient.

It is possible to collect almost unlimited amounts of data about programme activities. Management's responsibility is to select from a large potential array what is important enough to collect (as well as how often) and how best to collect and analyse it. Organizations should avoid collecting more data than they can effectively use, because this needlessly wastes staff time and resources.

Using Information to Increase Effectiveness

While it is important for organizations to efficiently carry out their missions, they also need to be concerned about effectiveness. An organization is said to be effective if it accomplishes its stated objectives. Ideally, an organization should be both effective and cost efficient.

> Concept No. 7. Effectiveness is a measure of programme impact as compared with the intended goals. To measure effectiveness, organizaitons must collect programme impact data. An organization that is both effective and cost efficient is achieving its goals, and the benefits obtained are greater than the costs involved.

To determine whether an organization is effective, it is necessary to look at programme impact.

To obtain programme impact measurements, organizations either use the services of external specialists (to avoid the appearance or the reality of bias), or they create a capability within their organization to measure programme impacts. But whether external or internal resources are used, the effort needs to be conducted professionally and in an unbiased fashion.

The programme impact unit or external service will need to measure programme impact in context—that is to say, a low-budget undertaking with a very limited goal can be said to have been effectively carried out if the stated goal was achieved.

Any number of programme impact studies can be conducted. But they cost money and staff time. The management responsibility is to determine which outcomes are important enough to measure at a particular point in time, to ensure that efforts to measure those outputs are properly carried out and analysed, and then to use the information in decision making. The first such use should be directed at making changes that will cause future programme efforts to be more successful.

Those who manage or who provide resources to extension organizations are interested in two kinds of measurements—those that indicate the organization is productive and efficient and those that help to determine if the organization is effective. The former tend to be measurements of inputs and outputs, and the latter tend to be project impacts.

Wasting Resources

Waste can be found in all kinds of organizations: public and private, profit and nonprofit. A commonly held view is that public organizations tend to "waste" more resources than other kinds do. This view holds that private organizations are motivated by profits, and consequently they avoid waste to a greater extent than "nonprofits." Alternatively, public organizations often are required to adhere to a higher standard of accountability. Consequently, the profit-nonprofit question may not be the major determining factor. But whether public or private, profit or nonprofit, waste increases costs and decreases benefits.

> Concept No. 8. Resources can be wasted if decisions are poor or are not made in a timely fashion. They are wasted if they are not used to best advantage (e.g., if more is bought than is needed or if the purchase price was not competitive).

Organizations can avoid waste by creating appropriate policies. Those policies then need to be translated into procedures designed to achieve policy goals. Finally, managers need to administer the procedures fairly and consistently, and staff members affected by the procedures need to cooperate in the effort.

Organizations need something more than high-sounding platitudes along the line of "waste not, want not." Policies and operating procedures need to be couched in terms of real-world situations faced by the organization. For example, organizations need to establish policies and procedures that determine how purchases are to be made (including when and by whom) and how sales, if any, are to be handled (including what happens to the money). It should be clear to staff members what their authorities and responsibilities are with respect to spending money. And those staff members who are not allowed to spend the organization's money need to know who they should turn to for a decision. To the extent there is confusion, there is likely to be waste.

Misappropriating Resources

Organizational resources sometimes are used by staff members or others for personal gain. Resources may simply be stolen. Misappropriation of resources negatively affects an organization's reputation, and its ability to carry out its mission. Costs are increased, and thus its ability to accomplish its mission and to attract additional resources is jeopardized. To protect against these eventualities, organizations should institute and administer policies and procedures designed to discourage and prevent these behaviours. There should be appropriate supervision and a periodic auditing of financial records and procedures.

> Concept No. 9. Organizational resources should be used to further organizational goals. Staff members should not be permitted to use their positions for personal gain.

The problem of misappropriation of organizational resources appears to be widespread. It is found in both developed and developing countries. It may entail pilfering supplies or taking money. It may involve payments to individual staff members who require such payments before they will render services or for special treatment.

To avoid waste, loss of productivity, negative impacts on staff morale, and loss of confidence by clients being served, organizations need to take steps to avoid losses from misappropriation. They should administer procedures fairly, and then act in ways appropriate to regulation and law.

Summary

A simple listing of the principles discussed above will serve to summarize the concepts addressed in this chapter. The principles are as follows:

1. Organization leaders are generally held responsible for obtaining new resources from funders and stakeholders and for maintaining current levels of financial support. It is necessary to support a request for funding with an explanation of how the additional resources will be used. Requests that are consistent with high national priorities are more likely to be funded than requests dealing with less important matters.
2. Leaders of extension organizations are accountable for financial resources assigned to their organizations. It is incumbent upon them to establish workable systems that will enable staff to keep track of those resources.
3. Budgeting the use of resources to particular purposes and for specific accounting periods is an essential part of the planning and control processes.
4. If resources are too limited to provide support for staff salaries and the various direct and indirect costs of carrying out the work of the organization, then resources should be reallocated. It does little good to have a large staff without sufficient resources to carry out their work. Reallocation can be based on changing the way programmes are delivered (e.g., group meetings versus individual contacts) or by focusing on only the highest priority concerns (narrowing the mission of the organization).
5. The determination of how money is to be spent often is best made by staff members who are directly involved in programme delivery. This suggests that decision making be decentralized, with resources being subdivided among the various units of the organization.
6. Little is known about what is being accomplished by the organization or whether work is being carried out efficiently. Programme monitoring efforts combined with information from the accounting system will provide information about efficiency. Programme impact studies will provide answers to questions about results.
7. Some ways of carrying out work are more effective than others. For example, clientele may learn faster and they may more readily adopt proposed technology when certain teaching methods are used as opposed to others. Society benefits little from the continued use of ineffective or less effective methods. An ongoing research effort is needed to help staff to know which techniques to use in which situations. The measurement of effectiveness ultimately turns on what happens as a result of a particular programme effort.
8. Resources can be wasted if decisions are not made in a timely fashion or if the wrong decision is made. Often waste occurs because inappropriate methods are used. Organizations can protect against waste if they institute practices designed to protect against it. Staff training can help to avoid waste. Proper administrative procedures and practices help to reduce waste.
9. Organizational resources sometimes are used by staff members or others for personal gain. Resources may simply be stolen. Misappropriation of resources negatively affects an organization's ability to carry out its mission. Organizational reputation is adversely affected, and thus its ability to attract additional resources. To protect against these eventualities, organizations should conduct appropriate audits of financial records and procedures and take steps to correct problems as they arise.

References

Bruneau, Hilarion. (1994) *Managing Financial Resources-Principles and Recommendations for Agricultural Research Managers*. The Hague: International Service for National Agricultural Research (ISNAR).

Buford, J., Jr. & Bedeian, A. G. (1988). *Management in Extension* (2nd ed.). Alabama: Alabama Cooperative Extension Service, Auburn University.

CGIAR Secretariat. (1988). *Financial Guidelines series No. 1—Financial management*. Washington, D.C.

CGIAR Secretariat. (1993). *Financial Guidelines series No. 2—Accounting policies & reporting practices manual*. Washington, D.C.

Denhardt, K. G. (1989). *The ethics of public service: resolving moral dilemmas in public organizations*. Greenwood Press.

Drucker, P. F. (1992). *Managing the non-profit organization: practices and principles*. New York: Harper.

Osborne, D. & Gaebler, T. (1993) *Reinventing government—How the entrepreneurial spirit is transforming the public sector.* Reading, Mas: Addison-Wesley P. C. Inc.

United Nations. (1991). *Government financial management in least developed countries.* New York: United Nations Department of Technical Cooperation for Development.

World Bank. (1994). *Agricultural extension: lessons from completed projects, Operations Evaluation Department Report No. 13000.* Washington, D.C.: World Bank.

World Bank (1994 unpublished). *Fiscal Systems for Agricultural Extension.*

Chapter 17
Monitoring Extension Programmes and Resources

D.C.Misra

The Global Consultation on Agricultural Extension observed that monitoring and evaluation are important yet frequently neglected functions in most organizations (FAO, 1990, p. 27). In the worldwide survey of national extension systems, it was found that only about one half of all national extension systems have some type of monitoring and evaluation (M & E) capacity. The consultation noted that in many cases the M & E units are weak and are limited to ad hoc studies. Frequently, these M & E units are abandoned when project funding terminates. In many organizations, monitoring and evaluation have a negative image because these units may concentrate on problems, exposing weaknesses and failures. Instead, monitoring and evaluation should be used in a positive manner to improve extension's performance and increase its efficiency. Therefore, attitudes about and uses of M & E must be changed if this capacity is to be used to advantage in strengthening extension's performance and impact (p. 27-28).

The consultation recommended that national extension systems should be strongly encouraged to establish and use monitoring procedures and evaluation studies both to improve extension performance and to communicate the results of extension programmes to policy makers and clientele being served (p. 29).

Extension systems need to carefully consider monitoring, management information needs, sources of information, and a management information system. The word "monitor" is derived from the Latin word meaning to warn, and "evaluate" stems from the word value (Hortan, Peterson, & Ballantyne, 1993, p. 5).

Monitoring is an integral and important part of a management information system (see Chapter 17). Managers require information to keep track of an extension programme and to guide its course of action.

Management information needs include six kinds of information: (1) diagnostic information (why a situation is as it is), (2) implementation information (physical and financial or input information), (3) utilization information, (4) impact information (Murphy, 1993, p. 5-6), (5) situation information, and (6) information for review.

Sources of information cover a spectrum of sources which supply information to management, ranging from informal (say, an unscheduled encounter) to formal sources (say, a sample survey). Management should be ever alert to receive feedback on extension programmes —whatever the source. Monitoring, it is useful to realize, is only one of management's sources of information.

A management information system is a scheme by which the "right" information is obtained in the right amount at the right time and is made available to the right person or persons (Bloom, 1980, p. 28). An information system is usually created in modern organizations to cater to the information needs of management. A monitoring system is a subsystem of a management information system and has several distinguishable components.

A Conceptual Framework for Monitoring

A conceptual framework for monitoring consists of four principal components (Figure 1): an extension organization, a monitoring and evaluation (M & E) unit, an information needs matrix, and a monitoring and evaluation cycle. Top management receives information from the monitoring unit and from other formal and informal sources. This influences programme implementation, leads to better programme planning, and ensures sustainability of extension programmes. Ultimately this leads to institutional development, which has been defined as "the process of improving the ability of institutions to make effective use of available human and financial resources" (Israel, 1987, p. 3).

Approaches to Extension Monitoring

Several approaches to extension monitoring are available. They arose, notably during the 1980s, as a response to the need to improve extension performance. Basically similar, these approaches, however, differ significantly in their emphasis. All of them advocate simplicity and timeliness, essential requirements of good monitoring. Half a dozen approaches are described briefly below.

Traditional (Administrative) Approach

Based on routine administrative reporting, this approach is concerned with physical and financial achievements in a programme. Its primary weaknesses include multiplicity of reports by programme personnel and absence or neglect of beneficiary contact. It is being increasingly replaced by other approaches.

D. C. Misra is Chief Secretary and Secretary (Agriculture), Government of Goa, Panaji, Goa, India.

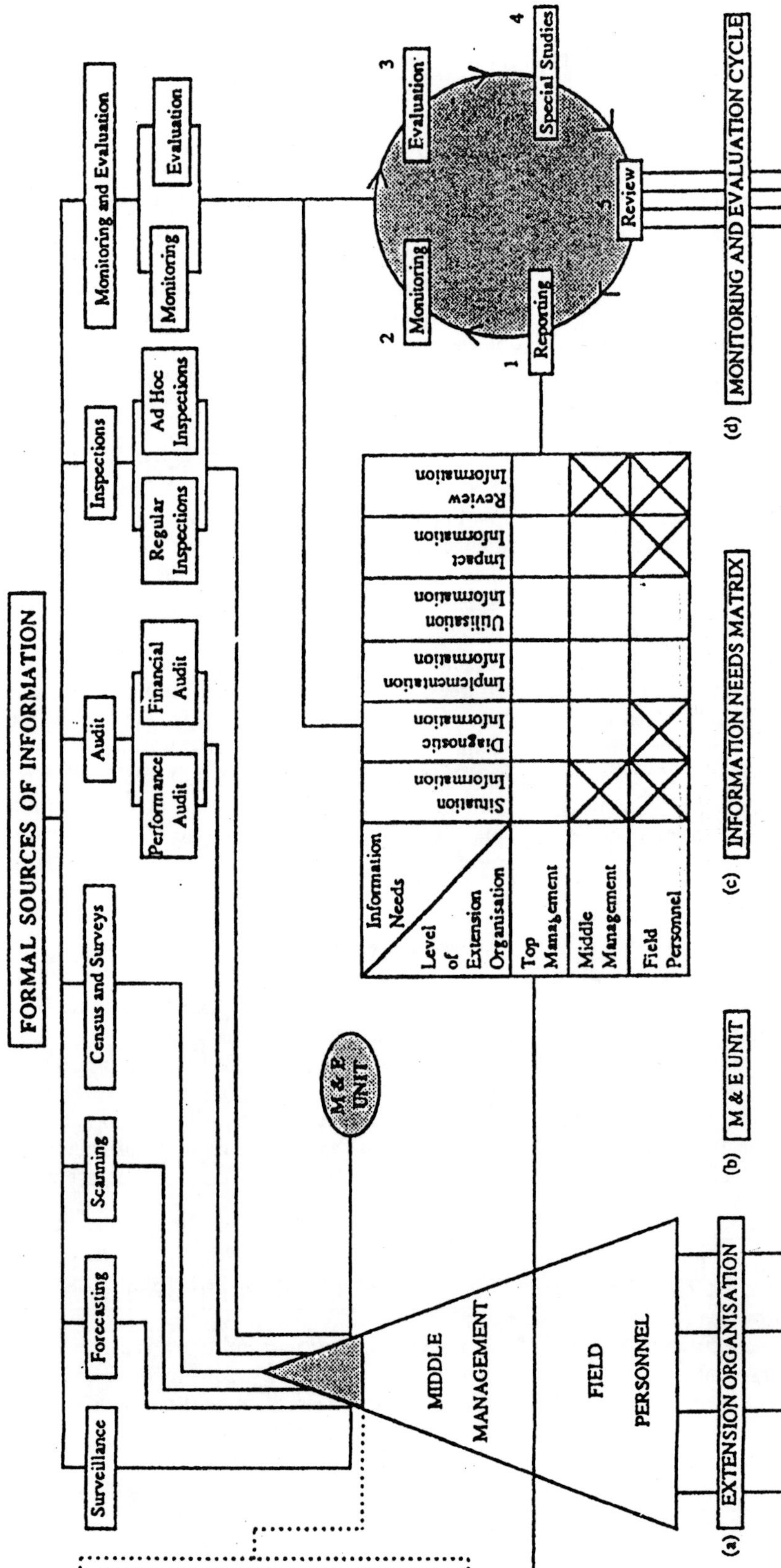

Figure 1 : Schematic Arrangement of Monitoring and Evaluation for Agricultural Extension Programmes

Zones-of-Concentration Approach

The Cernea and Tepping approach, introduced in 1977, concentrates on three zones: (1) visits, as the final outcome of extension efforts; (2) recommendations, as the content of the visit and means towards the end yields; and (3) yields, as the eventual consequence of the development effort (Cernea & Tepping, 1977, p. 19-20).

Methodological Approach

The Slade and Feder approach, introduced in 1981, builds upon the zones-of-concentration approach and suggests a monitoring survey early in each cropping season, a monitoring-cum-evaluation survey in each cropping season at the time of harvest, specific indicators, and reporting. Working manuals are a characteristic feature of this approach (Slade & Feder, 1985).

Expanded Monitoring Approach

Casley and Kumar suggested in 1987 an expansion of the monitoring function to cover not only physical and financial information, but also beneficiary contact information and project diagnostic studies (p. 5). Under this approach, there is greater emphasis on monitoring and less on evaluation. Project diagnostic studies are a novel feature of this approach.

Adoption Rates Approach

Murphy and Marchant suggested in 1988 an approach which concentrates on adoption rates as key indicators. This approach moves away from trying to monitor agricultural results and concentrates on directly monitoring the provision of and response to project services (p. 11). Under this approach, extension service is treated as the "medium" and the recommendations as the "message."

Marketing Approach

Lee (1990) has suggested an approach which is based on market segmentation, a standard technique in marketing. Under this approach, the need and likely demand for new technology are first assessed, and then target market segments are predicted. For example, 10 per cent of the targeted farmers are likely to readily adopt a new technology; a further 15 per cent are likely to adopt the technology when returns for wheat reach x dollars per ton; 40 per cent of the targeted farmers are likely to adopt the technology when they perceive that their peers have adopted it; and 35 per cent of the target market is not likely to adopt the technology at all in the next five years (p. 26) (enumeration supplied).

Among these approaches, Lee's approach has much to commend it, because it is based on careful assessment of the need for new technology, an exercise seldom under-taken by extension organizations.

Operationalizing the Definition of Monitoring

Monitoring is a specialized, dynamic, semiautonomous, and institutionalized management resource. Preferably it is computerized. Monitoring helps to ensure the implementation of extension programmes in accordance with with their design and takes into account the interests of various stakeholders.

The definition of monitoring can be operationalized by establishing principles to follow; setting up a data collection system; establishing relationships among the monitoring unit, management, the extension staff, and extension's clients (farmers); and making appropriate use of nongovernmental organizations (NGOs) (Figure 2).

Principles of Monitoring

In light of past experience in monitoring and its designated role, it is possible to lay down the following ten principles of monitoring:

1. *Monitoring must be simple.* A complex or complicated monitoring system is self-defeating. The basic task of monitoring is to simplify the field-level complexity, sifting the more important concerns from the less important.
2. *Monitoring must be timely.* Timeliness is of the essence in monitoring. Management requires input from the monitoring system so that timely action may be taken. Also, timeliness is closely related to the credibility of monitoring itself.
3. *Monitoring must be relevant.* It must concern itself only with parameters which are relevant to programme objectives. This also ensures that monitoring does not generate information that is not used or is not usable by management.
4. *Information provided through monitoring should be dependable.* Management will rely on monitoring findings only if the information is believed to be reasonably accurate.
5. *Monitoring efforts should be participatory.* Effort should be made to ensure participation by all concerned with extension, be they field-level personnel, subject-matter specialists, or extension's clients (the farmers).
6. *Monitoring must be flexible.* It is iterative in nature. It also gets routinized with the passage of time. These two features should not, however, lead to rigidity.
7. *Monitoring should be action oriented.* Monitoring often leads to action. Consequently, it should follow

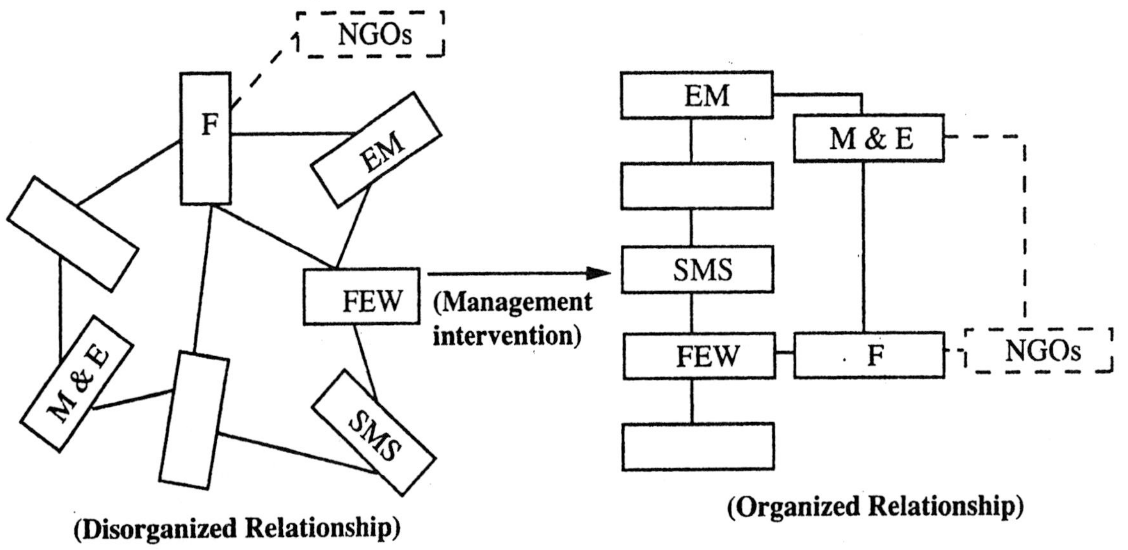

Key: EM = Extension Management; M&E = Monitoring and Evaluation Unit; F = Farmers; FEW = Field Extension Workers; SMS = Subject Matter Specialists;
▭ = Other Levels of Extension Personnel; NGOs = Non Government Organizations

Figure 2. Organized relationship in monitoring and evaluation due to management intervention.

pragmatic approaches, keeping the requirements of extension's clients uppermost in view. Generating information for which there is no intended use should be assiduously avoided.

8. *Monitoring must be cost-effective.* Monitoring efforts cost money and time. It is therefore essential to make it cost-effective. While principles of simplicity, time-liness, relevance, and accuracy will lead to cost-effectiveness, computerization also can help to make monitoring more cost-effective by reducing staff hours in data processing.
9. *Monitoring efforts should be top management oriented.* Monitoring units should keep in mind the requirements of top management when designing and operating a monitoring system. Yet at the same time, monitoring must take into account the fact that those who provide information to the system also must benefit or the quality of the information provided will decline.
10. *Monitoring units represent specialized undertak ings.* Monitoring is not merely concerned with the collection and analysis of data, but with diagnosing problems and suggesting alternative practical solutions.

Monitoring and Its Main Stakeholders

There are many stakeholders in monitoring efforts. The funding agency, usually the ministry of finance or planning, wants to know that money has been spent properly. The ministry of agriculture wants to be able to say it has been efficient and effective. The target beneficiaries, that is, small and marginal farmers, want help from the organization and want to improve the quality of their lives. And nongovernmental organizations (NGOs) need information generated by monitoring efforts and also want to contribute information. Thus there are four main stakeholders in monitoring—funding agencies, the implementing agency, beneficiaries, and NGOs—and their interests need to be taken into consideration when designing a system of monitoring.

Frequency of Monitoring

Monitoring is an ongoing, continual exercise. Data collection involving crop production should be undertaken twice in a cropping season: at the time of sowing to obtain benchmark information and again at the time of harvest. The period of recall for collection of data should not be more than a month.

Monitoring Unit

The monitoring unit should be staffed by technical personnel having specialized skills. The staff often consists of extension specialists, economists, sociologist or anthropologists, statisticians, computer programmers, and supporting staff. The head of the monitoring unit may come from any of these disciplines. The staff should be no larger than necessary to accomplish the work of the unit. The leader of the monitoring unit should report to one of the top managers in the organization.

Monitoring Indicators

Indicators, as the term suggests, are variables that help to measure changes in a given situation (ACC, 1984, p. 37). They are tools for monitoring and evaluating the effects of an activity. Indeed, indicators are the principal means by which a monitoring unit keeps track of extension's capability, effectiveness, and efficiency. Any monitoring system will therefore incorporate the use of appropriate indicators in these four aspects.

There are two approaches to indicator development: the inductive and the deductive. In the inductive approach, a system of social, economic, and demographic statistics is created and a wide range of indicators is developed on the basis of the statistics available. This is the approach followed in the case of United Nations Social Indicators (FAO, 1988, p. 5). In the deductive approach, the areas of interest are first identified, and then requisite indicators are developed. Socioeconomic indicators for the World Conference on Agrarian Reform and Social Development (WCARD) follow this approach (FAO, 1988, p. 5).

In extension monitoring, both the inductive and the deductive approaches are followed; that is, indicators are developed on the basis of available statistics, and some (for example, performance indicators) are developed as a result of specially collected data.

There are different types of indicators, for example, development indicators, socioeconomic indicators, agricultural development indicators, and extension indicators. They range from general to specific concerns.

Indicators can be categorized into direct and indirect or proxy indicators (Clayton, 1983; ACC, 1984); single and unitary or composite indicators; quantitative and qualitative indicators; primary, core, and supplementary indicators (FAO, 1988); input and output indicators; and monitoring and evaluation indicators.

The criteria for selecting indicators depend upon the purpose, resources, and time available. The following criteria are usually suggested:

- Simplicity: The indicator should be simple enough to be understood by nonspecialists (FAO, 1988, p. 8).
- Unambiguous definition: It should be clearly defined (Casley & Lury, 1982, p. 31; Casley & Kumar, 1987, p. 59).
- Ready determination: The data can be obtained without undue difficulty (WHO, 1989, p. 11). This is also referred to as "timely" (ACC, 1984, p. 38) and "feasible" (FAO, 1988, p. 8; Gha, Hopkins, & McGranahan, 1988, p. 11).
- *Accurate measurement:* The indicator should be measured accurately, which is often difficult when dealing with farming communities (Casley & Lury, 1982, p. 32).
- *Validity:* The indicator should actually measure what it is supposed to measure (ACC, 1984, p. 38; FAO, 1988, p. 7; see also Gha et al., 1988, p. 13).
- *Relevance:* It should be geared to the specific needs of decision makers (Pétry, 1983, p. 38) and be relevant to project objectives (ACC, 1984, p. 38).
- *Specifity:* It should reflect changes only in the situation concerned (WHO, 1989, p. 19) and should measure specific conditions that the project aims to change (Casley & Kumar, 1987, p. 59).
- *Consistency:* The value of indicators should stay constant so long as they are collected in identical conditions, no matter who does the collecting (Casley & Kumar, 1987, p. 69). Indicators should be objective and verifiable (FAO, 1988, p. 8).
- *Sensitivity:* Indicators should be sensitive to changes in the situation being observed (ACC, 1984, p. 38). They should be sensitive enough to reflect changes in the phenomenon (FAO, 1988, p. 8).
- *Prioritization:* Indicators should be prioritized and a minimum feasible list prepared (Gha et al., 1988, p. 11).

Capability, Effectiveness, Efficiency, and Impact

These four concepts are basic to monitoring and evaluation. They correspond, respectively, to operational investment (e.g., investment in agricultural extension per farm family), operational efficiency (e.g., the number of visits, meetings, demonstrations, and trials, per extension worker), technical efficiency (e.g., the number of adopters, hectorage, output, and value added), and extension-induced changes (e.g., production, productivity, income, and income distribution) (Ruthenburg, 1985, p. 120).

Capability, effectiveness, and efficiency fall in the monitoring domain. Impact falls in the evaluation domain.

Capability is the command that extension has over physical, financial, and human resources, enabling it to serve its clients (the farmers). It is reflected by extension's outreach, intensity, technical competence, and physical and financial resources. Extension performance depends directly upon its capability.

Effectiveness is defined by a handbook on productivity management as "the degree to which goals are attained" (Prokopenko, 1987, p. 9). Agricultural extension has many goals such as social goals (e.g., farmer welfare) and economic goals (e.g., increased income).

Among these, operational goals (e.g., physical and financial targets) are of special significance because their attainment makes realization of other goals possible.

Efficiency in extension is usually measured by the rates at which farmers adopt recommended practices. Adoption rates of varying degrees of complexity can be conceived (Casley & Lury, 1982, p. 37).

Impact in extension can be measured by a simple indicator, like yield of a crop per hectare, or by constructing simple productivity indices. Such indicators provide ultimate tests for the success of extension programmes.

Extension Monitoring Indicators

Extension monitoring indicators can be grouped into two categories: (1) extension capability indicators, and (2) extension performance indicators. Both should be generated by the monitoring unit.

Extension Capability Indicators

Extension capability indicators must be monitored regularly not only to know the status of extension's capability at a certain point in time, but also to determine changes in it over time. These indicators should be calculated annually. They involve only desk work because they are based on in-house data.

Extension Performance Indicators

Extension performance indicators reflect extension's operational and technical efficiency. They can be grouped into two categories.

Extension Effectiveness Indicators. These can again be grouped into two subcategories: (1) single indicators and (2) unitary or composite indicators. By definition, a single indicator will reflect an aspect of extension performance, while a unitary or composite indicator will reflect two or more aspects of extension performance. It may be useful to construct a unitary or composite indicator to provide a consolidated view of extension effectiveness to management, because management is often interested in having an overall view of extension effectiveness. These indicators reflect extension's cooperational efficiency (Box 1).

Extension Efficiency Indicators. These indicators are based on adoption rates of recommended practices and reflect extension's technical efficiency (Box 2).

Extension performance indicators, as suggested here, should be calculated separately for contact farmers, other (noncontact) farmers, and all (contact and noncontact) farmers.

Extension Evaluation Indicators

Yield and productivity indices occupy the central position in these groups of indicators. They are calculated on the basis of crop-cutting experiments which are conductebby provincial revenue or economics and statistics departments (Box 1).

Action Plan and Chain of Events

Monitoring requires collection of data and their analysis. For this, an action plan in a monthly, quarterly, and annual timeframe is essential. Data collections will require choosing appropriate methods, which will depend upon the monitoring unit's time and physical and financial resources such as computerization and trained technical personnel for collecting and analysing data.

Monitoring methodologies can be perceived as a chain of distinguishable and sequential events comprising (1) planning and design of the study, (2) desk research, (3) selection of methods, (4) data collection and analysis, (5) report writing, (6) report presentation, and (7) follow-up action. A detailed study programme should be prepared within the prescribed time limit (Figure 3).

An array of methodologies is now available, ranging from casual, informal interactions to highly structured sample surveys, including emerging methodologies of rapid appraisals. These methodologies can be grouped into two categories: informal and formal.

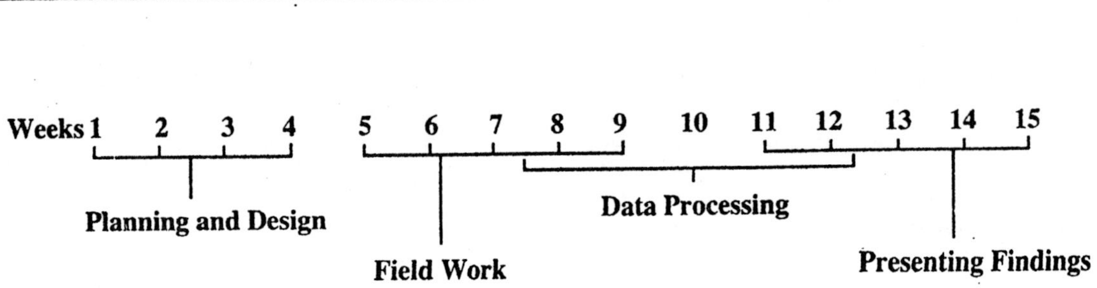

Figure 3. A typical summary programme for the study. Source: Nichols (1991, p. 22).

Box 1. Extension Performance Indicators.

	(I) Extension Effectiveness Indicators
	(a) Single Indicators
1. Awareness	Number of farmers aware of Village Extension Worker (percentage)
2. Visit	Number of visits by Village Extension Worker to farmers a) twice a month, b) once a month, and c) no visit (average)
3. Field Meetings	Number of meetings of Village Extension Worker with farmers in their fields (percentage)
4. Regularity	Number of meetings of Village Extension Worker with farmers on the fixed day (percentage)
5. Field Day	Number of field days organized by Village Extension Worker a) in preceding month, b) quarterly, and c) annually (average)
6. Demonstration	Number of a) method demonstrations, b) result demonstrations, and c) method-cum-result demonstrations organized by Village Extension Worker (i) in preceding month, (ii) quarterly, and (iii) annually
7. Supervision	Number of supervisory visits from Agricultural Extension Officers to Village Extension Worker in the field per month (average)
8. Research-Extension Linkage	Number of research-extension linkage workshops organized per month (average)
9. Farmer Training	Number of farmers trained in farmers' training centres (institutionalized training courses) per year (average)
	(b) Unitary or Composite Indicators
10. Extension Effectiveness	Arithmetic average of selected extension effectiveness indicators, say, Awareness Indicator (know the Village Extension Worker), Visit Indicator (number of visits twice a month), Field Indicator (meeting place at farmers' fields), and Regularity Indicator (visit on the same day) (Misra, 1994)
	(ii) Extension Efficiency Indicators
11. Performance Index	Actual number of farmers reached out of the target number which should be reached (Casley and Lury, 1982, p.7) (percentage)
12. Penetration Index	Number of farmers adopting the recommended practice out of the actual number reached (ibid. p.37) (percentage)
13. Achievement Index	Number adopting the recommended practice out of the target number of farmers (ibid. p.37) (percentage). Note that (13) =(12) x (11)
	Extension Productivity Indicators
1. Yield	Yield per hectare for main crop(s) (average)
2. Productivity Index	Increase in yield over base year compared with base year (percentage)

Box 2. Extension Capability Indicators.

1.	Coverage	Area under cultivation per Extension Worker
2.	Intensity	Number of Farm Families per Extension Worker
3.	Competence	Number of Graduate Extension Workers out of total number of Extension Workers (percentage)
4.	Subject-Matter Specialist	Number of Subject-Matter Specialists per hundred Extension Workers
5.	Research-Extension Ratio	Number of Agricultural Scientists per hundred Extension Workers
6.	Monitoring	Number of Monitoring Unit Personnel per thousand Extension Workers
7.	Gender Ratio	Number of Female Extension Personnel out of total number of Extension Personnel (percentage)
8.	Equity	Number of Small and Marginal Farmers out of total number of Contact Farmers (percentage)
9.	Mass Contact	Number of group meetings held per month per Extension Worker in a year (average)
10.	Computerization	Number of personal computers in Extension Organization per thousand Extension Personnel
11.	Print Media	Number of leaflets/pamphlets distributed per month per Extension Worker in a year (average)
12.	Audio-Visual Media	Number of audio-visual (cinema/television) shows organized per month per Extension Worker in a year (average)
13.	Training	Number of Extension Personnel out of total number of Extension Personnel trained in specialized training courses in a year (percentage)
14.	Finance	Budgetary expenditure on Agricultural Extension out of total budgetary expenditure on agriculture per year (percentage)
15.	Investment	Expenditure on Agricultural Extension as percentage of Agricultural Gross Domestic Product per year
16.	Transport	Number of (i) bicycles, (ii) motorcycles, and (iii) 4-wheel vehicles, per thousand Extension Workers

Notes: 1. Extension Personnel means total personnel engaged in extension, including Subject-Matter Specialists, whereas Extension Worker means Village Extension Worker.

2. The data source for all the indicators will be the offices of District Agriculture Officers and Provincial Directors of Agriculture.

3. At the national level, the indicators are required to be aggregated by the Central Monitoring Unit in the Ministry of Agriculture.

Informal Methods

Informal methods include participant observation, case studies, key informants, individual interviews or discussions, group interviews or discussions, oral testimonoral and life histories, longitudinal studies, cross-sectional studies, interdisciplinary terms, sondeo or reconnaissance survey, diagnostic studies, rapid rural appraisal, and participatory rural appraisal (Casley & Lury, 1987; Nichols, 1991; Pratt & Loizos, 1992; Hildebrand, 1981; FAO, 1992; Beebe, 1987; Kumar, 1993). (For details on rapid rural appraisal, see chapter 6.).

Formal Methods

Formal methods include using the population and agricultural census; sample surveys such as random sampling, including simple, systematic, stratified, cluster, and multistage; nonrandom sampling, including purposive, quota, and accidental (Shaner, Phillipp, & Schmehl, 1982); and special studies.

Among these, the sample survey occupies a unique place. A sample, as a representative part of the population, enables us to draw reasonable inferences about the characteristics of the population. As such, the sample survey will constitute the "handover" of monitoring methodologies. Other methods, notably the emerging rapid rural appraisal, can be used to advantage for exploratory, diagnostic, complementary, and supplementary purposes for decision making.

Conclusion

Regular and systematic monitoring provides timely information to management for corrective action, justification for investment in agricultural extension, inducement to extension personnel to perform their designated tasks, a sense of participation to farmers in extension when approached for responses to questionnaires, and direction for development of future extension policies and programmes. Experience of the last two decades in extension monitoring, however, indicates that monitoring has become a routine, mechanical exercise confined to attainment of programme objectives to the neglect of the wider objectives of meeting the needs of extension's clients.

Monitoring efforts should be broadly conceptualized so that they incorporate not only farm production and productivity, but also natural resource conservation, beneficiary participation, and institution-building. This will require monitoring units to upgrade their technical competence. There is need for systematic and regular training of personnel, introduction of new electronic technology, particularly computers (see, for example, Wentling & Wentling, 1993), and appropriate use of methodologies like rapid rural appraisal that yield quick results. Investment in monitoring should then yield rich dividends through improved extension performance.

References

ACC (Administrative Committee on Coordination), United Nations (1984). *Guiding principles for the design and of monitoring and evaluation in rural development projects and programmes*. Rome: International Fund for Agricultural Development (IFAD).

Beebe, J. (1987). Rapid appraisal: The evolution of the concept and the definition of issues (p. 47-68). In *Proceedings of the 1985 international conference on rapid rural appraisal*. Khon Kaen, Thailand: Khon Kaen University.

Bloom, C. (1980). The managerial role (p. 21-46). In L. J. Goodman & R. N. Love (Eds.), *Project planning and management: An integrated approach*. New York: Pergamon Press.

Casley, D. J., & Kumar, K. (1987). *Project monitoring and evaluation in agriculture*. Baltimore and London: Johns Hopkins University Press (published for the World Bank).

Casley, D. J., & Lury, D. A. (1982). *Monitoring and evaluation of agriculture and rural development projects*. Baltimore and London: Johns Hopkins University Press (published for the World Bank).

Casley, D. J., & Lury, D. A. (1987). *Monitoring and evaluation of agriculture and rural development projects*. Baltimore and London: Johns Hopkins University Press (published for the World Bank).

Cernea, M. M., & Tepping, B. J. (1977). *A system for monitoring and evaluating agricultural extension projects*. World Bank Staff Working Paper No. 272. Washington, DC: World Bank.

Clayton, E. (1983). Role, characteristics, and operational features of agricultural monitoring systems (p. 1-9). In E. Clayton & F. Pétry (Eds.), *Monitoring systems for agricultural and rural development projects* (vol. 1). Rome: FAO.

FAO (1988). *Socio-economic indicators for monitoring and evaluating agrarian reform and rural development*. WCARRD: Ten years of follow-up (1979-1989). Rome: Author.

FAO (1990). *Report of the global consultation on agriculture extension*. Rome: Author.

FAO (1992). *Sociological analysis in agricultural investment project design*. FAO Investment Centre Technical Paper 9. Rome: Author.

Gha, D., Hopkins, M., & McGranahan, D. (1988). *Some reflections on human and social indicators for development*. Discussion Paper No. 6. Geneva: United Nations Research Institute for Social Development (UNRISD).

Hildebrand, P. E. (1981). Combining disciplines in rapid appraisal: The sondeo approach. *Agricultural Administration, 8*, 423-432.

Hortan, D., Peterson, W., & Ballantyne, P. (1993). M & E principles and concepts (p. 5-19). In D. Hortan et al. (Eds.), *Monitoring and evaluating agricultural research: A sourcebook*. Oxon, UK: CAB International.

Israel, A. (1987). *Institutional development: Incentives to performance*. Baltimore and London: Johns Hopkins University Press (published for the World Bank).

Kumar, K. (Ed.). (1993). An overview of rapid appraisal methods (p. 1-22). In K. Kumar (Ed.), *Rapid appraisal methods*. Washington, DC: World Bank.

Lee, J. W. (1990). *More than accountability: Evaluating agricultural extension programmes*. TRDC Publication No. 172. Amidale, Australia: University of New England, Rural Development Centre.

Misra, D. C. (1994). *Agricultural extension effectiveness in India (mimeo)*. University of Oxford, Queen Elizabeth House, International Development Centre.

Murphy, J. (1993). Good enough, soon enough. *The Rural Extension Bulletin, 1*, 27-33.

Murphy, J., & Marchant, T. J. (1988). *Monitoring and evaluation in extension agencies*. World Bank Technical Paper No. 79. Washington, DC: World Bank.

Nichols, P. (1991). *Social survey methods: A field guide for development workers*. Development Guidelines No. 6. Oxford: Oxfam.

Pétry, F. (1983). Designing a monitoring and evaluation system (p. 20-49). In E. Clayton & F. Pétry (Eds.), *Monitoring systems for agricultural and rural development projects* (vol. 1). Rome: FAO.

Pratt, B., & Loizos, P. (1992). *Choosing research methods: Data collection for development workers*. Oxford: Oxfam.

Prokopenko, J. (1987). *Productivity management: A practical handbook*. Geneva: International Labour Office (ILO).

Ruthenberg, I. (1985). *Innovation policy for small farmers in the tropics: The economics of technical innovations for agricultural development*. Oxford: Oxford University Press.

Shaner, W. W., Phillipp, P. F., & Schmehl, W. R. (1982). *Farming systems research and development: Guidelines for developing countries*. Boulder, CO: Westview Press.

Slade, R. H., & Feder, G. (1985). *Training and visit extension: A manual of instruction (mimeo)*. Washington, DC: World Bank.

WHO (World Health Organization) (1989). *Evaluation of programmes to ensure food safety*. Geneva: Author.

Chapter 18
Establishing a Management Information System

A. Ramesh Babu, Y. P. Singh, and R.K.Sachdeva

Information is a *critical resource* in the operation and management of organizations. Timely availability of relevant information is vital for effective performance of managerial functions such as planning, organizing, leading, and control. An information system in an organization is like the nervous system in the human body: it is the link that connects all the organization's components together and provides for better operation and survival in a competitive environment. Indeed, today's organizations run on information.

The term *information system* usually refers to a computer-based system, one that is designed to support the operations, management, and decision functions of an organization. Information systems in organizations thus provide information support for decision makers. Information systems encompass transaction processing systems, management information systems, decision support systems, and strategic information systems.

Information consists of data that have been processed and are meaningful to a user. A system is a set of components that operate together to achieve a common purpose. Thus a management information system collects, transmits, processes, and stores data on an organization's resources, programmes, and accomplishments. The system makes possible the conversion of these data into management information for use by decision makers within the organization. A management information system, therefore, produces information that supports the management functions of an organization (Davis & Olson, 1985; Lucas, 1990; McLeod, 1995).

Basic Concepts

Data versus Information

Data refers to raw, unevaluated facts, figures, symbols, objects, events, etc. Data may be a collection of facts lying in storage, like a telephone directory or census records.

Information is data that have been put into a meaningful and useful context and communicated to a recipient who uses it to make decisions. Information involves the communication and reception of intelligence or knowledge. It appraises and notifies, surprises and stimulates, reduces uncertainty, reveals additional alternatives or helps eliminate irrelevant or poor ones, and influences individuals and stimulates them to action. An element of data may constitute information in a specific context; for example, when you want to contact your friend, his or her telephone number is a piece of information; otherwise, it is just one element of data in the telephone directory.

Computers have made the processing function much easier. Large quantities of data can be processed quickly through computers aiding in the conversion of data to information. Raw data enter the system and are transformed into the system's output, that is, information to support managers in their decision making.

Characteristics of Information

The characteristics of good information are relevance, timeliness, accuracy, cost-effectiveness, reliability, usability, exhaustiveness, and aggregation level. Information is relevant if it leads to improved decision making. It might also be relevant if it reaffirms a previous decision. If it does not have anything to do with your problem, it is irrelevant. For example, information about the weather conditions in Paris in January is relevant if you are considering a visit to Paris in January. Otherwise, the information is not relevant.

Timeliness refers to the currency of the information presented to the users. Currency of data or information is the time gap between the occurrence of an event in the field until its presentation to the user (decision maker). When this amount of time is very short, we describe the information system as a *real-time* system.

Accuracy is measured by comparing the data to *actual* events. The importance of accurate data varies with the type of decisions that need to be made. Payroll information must be exact. Approximations simply will not suffice. However, a general estimate of how much staff time was devoted to a particular activity may be all that is needed.

Value of Information

Information has a great impact on decision making, and hence its *value* is closely tied to the decisions that result from its use. Information does not have an absolute universal value. Its value is related to those

---

A. Ramesh Babu is Senior Scientist (Agriculture Extension); **Y. P. Singh** is Professor, Division of Agricultral Extension, Indian Agricultural Research Institute, New Delhi; and **R. K. Sachdeva** is Professor of Management and Head, Computer Centre, Indian Insitute of Public Administration, New Delhi, India.

who use it, when it is used, and in what situation it is used. In this sense, information is similar to other commodities. For example, the value of a glass of water is different for someone who has lost his way in Arctic glaciers than it is to a wanderer in the Sahara Desert.

Economists distinguish value from *cost* or *price* of a commodity incurred to produce or procure the commodity. Obviously, the value of a product must be higher than its cost or price for it to be *cost-effective*.

The concept of *normative value* of information has been developed by economists and statisticians and is derived from decision theory. The basic premise of the theory is that we always have some preliminary knowledge about the occurrence of events that are relevant to our decisions. Additional information might modify our view of the occurrence probabilities and consequently change our decision and the expected payoff from the decision. The value of additional information is, hence, the difference in expected payoff obtained by reduced uncertainty about the future event.

Information supports decisions, decisions trigger actions, and actions affect the achievements or performance of the organization. If we can measure the differences in performance, we can trace the impact of information, provided that the measurements are carefully performed, the relationships among variables are well defined, and possible effects of irrelevant factors are isolated. The measured difference in performance due to informational factors is called the *realistic value* or *revealed value* of information.

For most information systems, particularly those supporting middle and top management, the resulting decisions often relate to events that are not strictly defined and involve probabilities that cannot be quantified. The decision-making process often is obscure and the outcomes are scaled by multiple and incomparable dimensions. In such cases, we may either attempt to perform a multiattribute analysis or derive an overall *subjective value*. The *subjective value* reflects people's comprehensive impression of information and the amount they are willing to pay for specific information (Ahituv, Neumann, & Riley, 1994).

Information as an Aid to Decision Making

Simon (1977) describes the process of decision making as comprising four steps: intelligence, design, choice, and review. The *intelligence* stage encompasses collection, classification, processing, and presentation of data relating to the organization and its environment. This is necessary to identify situations calling for decision. During the *decision* stage, the decision maker outlines alternative solutions, each of which involves a set of actions to be taken. The data gathered during the intelligence stage are now used by statistical and other models to forecast possible outcomes for each alternative. Each alternative can also be examined for technological, behavioural, and economic feasibility. In the *choice* stage, the decision maker must select one of the alternatives that will best contribute to the goals of the organization. Past choices can be subjected to *review* during implementation and monitoring to enable the manager to learn from mistakes. Information plays an important role in all four stages of the decision process. Figure 1 indicates the information requirement at each stage, along with the functions performed at each stage and the feedback loops between stages.

Classification of Management Information Systems

There are various types of management information systems. Mason and Swanson (1981) describe four categories of management information systems: (1) databank information system, (2) predictive information system, (3) decision-making information system, and (4) decision-taking information system. The classification is based on the level of support that the information system provides in the process of decision making. Sachdeva (1990) comprehensively presents these four types of systems:

Databank Information System. The responsibility of this information system is to observe, classify, and store any item of data which might be potentially useful to the decision maker. Examples of the kind of data that might be recorded in such a database for a given village, region, or area are as follows:

- Number of farms
- Number of units of arable land (hectares, fedans, acres)
- Average farm size
- Amounts of selected farm inputs applied annually
- Production per year on a unit of land for selected crops

A second example of data that might be recorded in a database (this time involving data internal to the organization) is as follows:

- Number of extension staff by category and assigned to a particular village, region, or area
- Number of work hours devoted by staff to selected concerns for a particular village, region, or area
- Total extension salary costs and other expenses by village, region, or area
- Number of demonstrations conducted for selected farm technologies by village, region, or area
- Number of on-farm trials conducted for selected farm technologies by region or area
- Number of radio, TV, and print media releases regarding selected farm technologies by time period and region or area

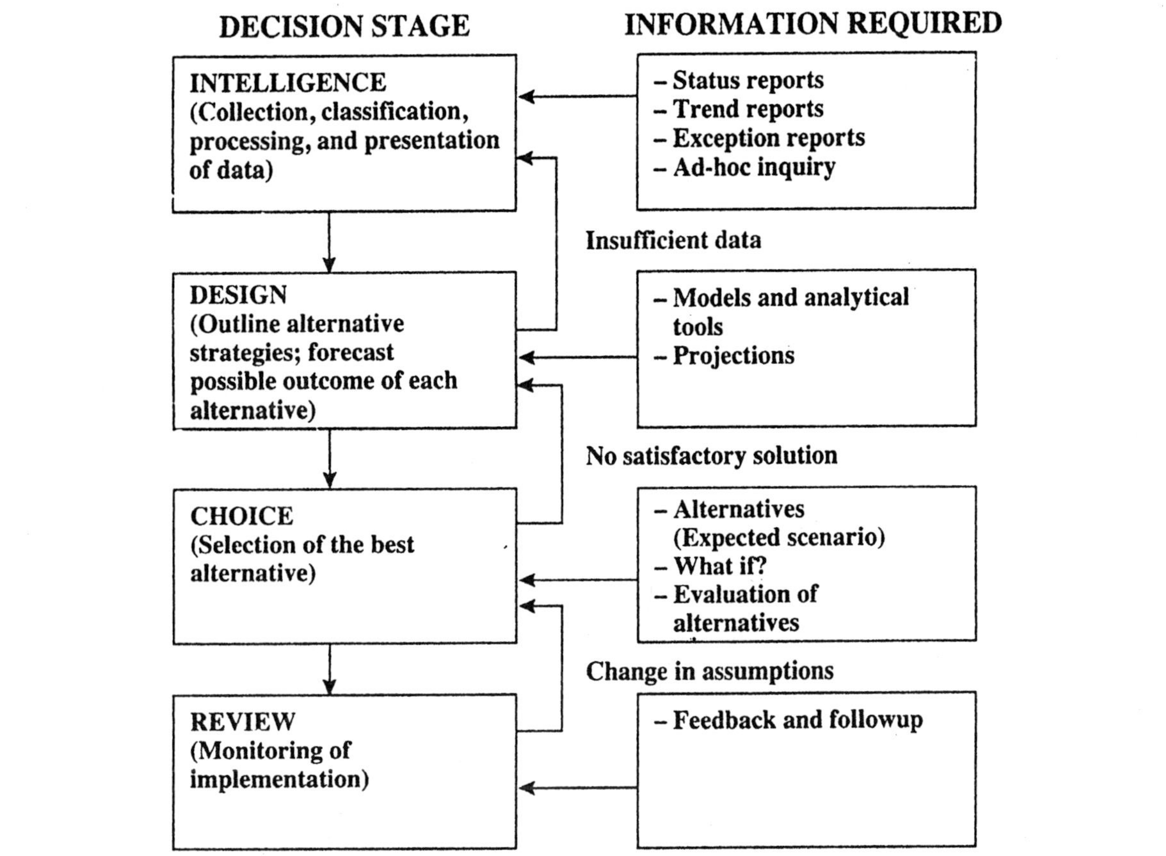

Figure 1. Role of information in the decision process.

Each of these databases can be summarized and converted to single tabular presentations of information of interest to management. When information from two or more time periods is compared, trends can be observed.

Predictive Information System. This system moves beyond pure data collection and the determination of trends over time. Predictive information systems provide for the drawing of inferences and predictions that are relevant to decision making. If data from the above examples were to be used in this way, it is possible to obtain information useful for making predictions or for drawing inferences. For example, tables containing the following information for a given village, region, or area might be produced:

- The ratio between the number of farms and the various categories of extension staff members
- The ratio between the amount of farmland and the various categories of extension staff members
- Amount of extension financial operating resources allocated per year to selected farmer problems or concerns
- Amount of extension financial resources, both salary and operating expenses, allocated per year to selected extension approaches to solving different farmer problems or concerns

Information obtained from these kinds of analyses is normally summarized in a two-way tabular format. And likewise, the information often is compared over time. Managers can then use such information to make predictions, for example to forecast costs of particular undertakings for budgeting purposes or as a basis for predicting results if a given change is made, such as change in the number of demonstrations with a given change in staffing.

Decision-Making Information System. This system goes one step further in the process of decision making and incorporates the value system of the organization or its criteria for choosing among alternatives. An extension organization's values are many and varied. They include concerns for resolving farmer problems, increasing and providing for stability of farmer incomes, and improving the quality of farm life. But they also including

Table 1. Information Groups in India's Agricultural Extension System.

Levels	Groups	Types of Information Needed
Central	Extension commissioner, joint commissioners, directors, joint directors, etc. of the directorate of extension, ministry of agriculture	(1) Information on human resources, plans, and budgets for various extension services (2) Statewide monitoring and evaluation of activities completed
State	Director of agriculture, additional director, joint directors, etc. of the state department of agriculture	(1) Districtwide information on extension programmes,activities, expenditures, etc (2) Research-extension linkages and coordination with other allied departments such as animal husbandry and horticulture
District	District agricultural officers (DAOs)	(1) Information on extension resources and constraints at subdivision and block levels (2)Training requirements of staff at subdivision and block levels
Subdivision	Subdivisional agricultural officers	(1) Field demonstration programmes, activities planned and implemented by subject-matter spcialists (SMSs) (zone) at the block level (2) Technical programme and constraints identified at the block level
Block (county)	Agricultural extension officers	(1) Performance of VEWs in terms of achievements in extension activities (2) Field-level problem of assessment of beneficiaries' response to various extension programmes

and providing for stability of farmer incomes, and improving the quality of farm life. But they also include an intent to provide well for staff members (training, adequate salaries, etc.) and to aid in the process of bringing about rural economic development.

Information regarding these various attributes helps managers to make more enlightened decisions. Examples of ways that an extension organization uses information from a decision-making information system are as follows:

- Change in specific farm outputs (yields, practices) following selected extension activities
- Change in staff productivity following selected interventions (in-service training, better transport, etc)
- Comparison of relative costs and relative effectiveness of alternative extension delivery methods
- Analysis of economic returns to farmers who adopt recommended practices as compared to those who do not

Decision-Taking Information System. Examples of decision-taking information systems are not usually found in an extension organization. This is a decision system in which the information system and the decision maker are one and the same. Management is so confident in the assumptions incorporated in the system that it basically relegates its power to initiate action to the system itself. Airplanes carry automatic pilot systems, which are an example of a decision-taking system. Once activated, the system itself keeps the plane on course and at the proper speed and altitude (according to parameters determined by the pilot). Another example of decision-taking information systems is found in modern factory production. In automobile production, continuous inventories of parts are maintained by computer as cars move down an assembly line. Orders are placed automatically by the computer when additional parts are needed. This is done without the intervention of a manager.

The choice of an appropriate management information system (MIS) category primarily depends on the nature of the decisions it supports. While unstructured decisions may use MIS-category (I), the highly structured ones, such as production schedules in an industry, may use MIS-category (iv). Further, Banerjee and Sachdeva (1995) observe that "as the deep structure of the decision problem becomes more and more understood, we may move to higher level of MIS i.e., from MIS-category (I) to MIS-category (ii); and MIS-category (ii) to MIS-category (iii); and so on."

Role of MIS in the Management of Agricultural Extension Programmes

National agricultural extension systems, especially in developing countries, tend to be very large. For example, in India, the national agricultural extension system employs about 125,000 people. Extension managers at various levels need relevant information in order to make effective decisions. In the absence of such information, they act only on the basis of their intuition and past experience. Data that have been processed, stored, and presented properly will aid them in analysing situations and to make effective decisions.

As suggested above, at every phase of the management process, managers need information in order to make effective decisions. This we call *management information*. It does not include purely functional information or technical information, such as packages of practices for rice or wheat cultivation. Management information is the information required by managers as they make their decisions, such as the number of extension personnel employed by category, their training requirements, career development plans, job descriptions, budgets, forecasts, benchmark surveys, reports on socioeconomic conditions of people served, and existing facilities (Ramesh Babu & Singh, 1987).

The main purpose of management information systems is to provide management information to decision makers at various levels in the organization. Specifically, in an agricultural extension organization, MIS is needed:

1. To plan the most effective allocation of resources, for example, the allocation of extension personnel under a T & V extension system, the need for communications and training equipment and facilities, mobility, the amounts of required operational resources
2. To choose between alternative courses of action, whether to conduct a study on the impact of the T & V system with the resources on hand or hire an expert to investigate
3. To control day-to-day operations, for example, comparing the actual results achieved and those planned under the T & V system.

Design of a MIS in an Agricultural Extension Organization

The following are suggested steps to follow when designing a MIS for a national agricultural extension system.

Step One: Assessing Information Needs for Planning, Monitoring, and Evaluation

An investigation needs to be conducted into the types of decisions that extension managers have to make. For example, village extension workers (VEWs) seek solutions to their problems from their supervisors. In turn, supervisors need to be in a position to resolve these problems and to document how problems were solved for future reference.

State-level managers also need information to resolve problems. They are concerned with implementing extension programmes district by district. They need information on staffing, transport, research-extension linkages, staff training activities, and successes (or lack of them) in solving technical problems. Feedback is needed from field staff and farmers on farmer problems and on which recommended practices are helpful. State-level managers need to know something about the amounts, kinds, and combinations of media support (i.e., print, radio, television) that have been used for various efforts. They need to know if external factors have limited the success of particular efforts such as supply of credit or farm inputs and they need some assessment of farmers' responses to extension programmes (Raheja & Jai Krishna, 1991, p. 84).

Step Two: Deciding the Levels of Information Groups, Information Frequency, and Content

The number of information groups within an agricultural extension organization has to be decided because each group potentially will require a different type of information. As an example, in India, the reorganized national agricultural extension system can be grouped as shown in Table 1.

Data processing consists of identifying each item of data and systematically placing it within a scheme that categorizes data items on the basis of some common characteristic or feature. Data not organized into a meaningful pattern can serve almost no useful purpose to those who must use them to make decisions. A computer can help in processing the data effectively. Rao (1985) suggested the use of computers in agricultural extension in India. He proposed that computer programmes be focused on district and subdivisional levels. In that way, information collected can be viewed in terms of the crops that are likely to be grown, agroclimatic conditions, soil types, irrigation facilities, resources of the farmers, and availability of various farm inputs.

Documentation (storage and retrieval) involves storing items of information in an orderly manner. Storing information means recording it on storage media from which it can be made available when needed.

Storage media are materials such as ordinary office paper, magnetic tapes, magnetic disks, microfilms, film strips, and a few other devices. Once the information is recorded on these storage media, the system can generate, on demand, information required for making decisions, solving problems, or performing analyses and computations. Information retrieval refers to the ability to take different types of data in the storage media and to array information in some desired and meaningful format. A properly designed storage and retrieval system matches the related variables efficiently and accurately. In some cases, it even suggests alternative courses of action for management to take.

Presentation of information should be in a form and format suitable to the needs of extension managers. Generally, information is presented in reports, statistical summaries, analyses, and so forth in the form of text, figures, charts, tables, and graphs. The presentation of information should be precise, clear, and appealing.

Step Three: Ensuring System Flexibility and Adaptability

Flexibility means the ability to retrieve information from a system in whatever form it may be needed by decision makers. Therefore, data need to be collected in some detail so that they can be rearranged or summarized according to the needs of managers. But system design should not be too complex because it must first serve the needs of the lowest levels of management (i.e., subdistrict) that are likely to be instrumental in collecting important components of the original data. In addition, the system also must serve the needs of the district, regional, state or provincial, and national levels. Therefore, considerable care must be taken in assessing what types of information are required by management at the different levels. At the same time, effort must be made to ensure that the information collected meets acceptable standards of accuracy, timeliness, and coverage for each level.

Need for Automation

An automated MIS system contains data just as a manual system does. It receives input, processes input, and delivers the processed input as output. Some input devices allow direct human-machine communication, while others require data to be recorded on an input medium such as a magnetizable material (specially coated plastic flexible or *floppy* disks and magnetic tapes). The keyboard of a workstation connected directly to a computer is an example of a direct input device. Use of automation makes it possible to store immense quantities of information, to avoid many of the errors that find their way into manual records, and to make calculations and comparisons that would be practically impossible in a manual system.

Organization of a Database

Data are usually generated at the field level through transaction-processing systems, but once the data are captured, any echelon along the organizational hierarchy may use them, provided that information requirements have been well defined, appropriate programmes have been implemented, and a means has been arranged for the sharing of the data. This would imply that the same data can be used by different sets of programmes; hence we distinguish between the database (a set of data) and the applications (a set of programmes). In a decision support system (DSS), this set of programmes is the *model base* (Keen & Morton, 1978).

The term *database* may refer to any collection of data that might serve an organizational unit. A database on a given subject is a collection of data on that subject that observes three criteria: comprehensiveness (completeness), nonredundancy, and appropriate structure. Comprehensiveness means that all the data about the subject are actually present in the database. Nonredundancy means that each individual piece of data exists only once in the database. Appropriate structure means that the data are stored in such a way as to minimize the cost of expected processing and storage (Awad & Gotterer, 1992).

The idea of a large corporate database that can be flexibly shared by several applications or model bases has been realized by means of software packages specially devised to perform such tasks. These packages, called *database management systems* (DBMSs), are available in the market under different trade names such as ORACLE, SYBASE, INGRES, FOXBASE, and dBASE.

Networking and Interactive Processing

The two principal blocks that facilitate development and use of MIS are DBMS and *telecommunications*. The former makes data integration possible, while the latter brings information closer to the end users, who constitute nodes in a telecommunication network. The notion of telecommunications implies that some geographical distance exists between the computer site and the users' locations and that data are electronically transmitted between them. Remote applications may be executed between two floors in the same building, two offices in the same city, two offices on the same continent, or two places on opposite sides of the globe (Martin, 1990).

System Alternatives and Evaluation: Centralization versus Decentralization

A completely centralized information system handles all processing at a single computer site, maintains a single central database, has centralized development of applications, provides central technical services, sets development priorities centrally, and allocates computer resources centrally. The system's remote users are served by transporting input and output data physically or electronically.

A completely decentralized system may have no central control of system development, no communication links among autonomous computing units, and stand-alone processors and databases at various sites. Each unit funds its own information-processing activities and is totally responsible for all development and operation.

An advantage of centralized information systems is that they provide for standardization in the collection of data and the release of information.There also are some *economies of scale*. A centralized system reduces the need for multiple hardware, software, space, personnel, and databases. It may be possible to recruit more qualified personnel in a central facility.

Observations indicate that user motivation and satisfaction are increased under a decentralized environment. This is attained because users feel more involved and more responsible, systems are better customized to their specific needs, and they usually get better response time in routine operations as well as in requests for changes.

It is likely that for national agricultural extension systems, neither a completely centralized nor a completely decentralized system is desirable. While it may be useful to decentralize hardware and software resources at different locations, the development of applications and provision of technical services may better be centralized.

End-User Computing

The widespread use of personal computers and computer-based workstations has brought with it the age of end-user computing. End-user computing is a generic term for any information-processing activity performed by direct end users who actually use terminals or microcomputers to access data and programmes. The manager as end user may be provided with powerful

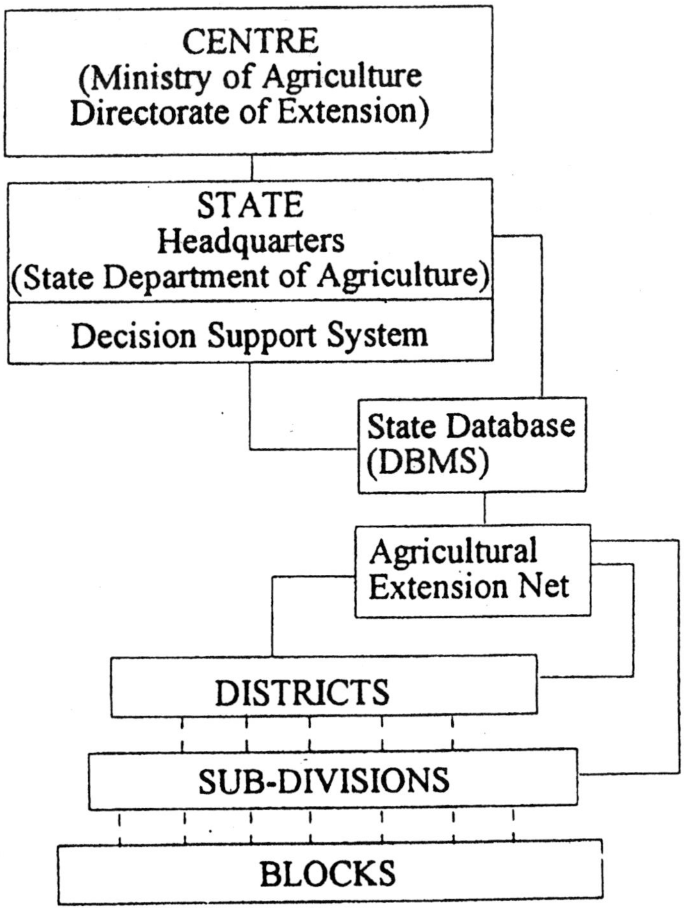

Figure 2. A typical MIS for a national extension system.

software (like DBMS) for accessing data, developing models, and performing information processing directly. This has brought computing directly under the control of the end users and eliminates their dependence on the information systems specialist and the rigidities of predesigned procedures. They may now make ad hoc queries of information and analyse it in various ways. They may write programmes, or may often use ready-made programmes stored in the computer, using the computing power of a local PC or the mainframe to which it is connected.

Illustrative Computer-Based MIS

A national agricultural extension system is a nationwide system managed by the national government. In India, agriculture is a state subject under the division of powers between the national and the state levels. Nevertheless, the national government supplements the financial resources of the states and provides coordination at the national level. The state's administrative machinery is divided into districts, districts into subdivisions, subdivisions into blocks. A block is a group of villages and the basic unit for the administration of an agricultural extension programme. Data collected at the block level need to be integrated at higher administrative levels to provide an integrated view at the district and state levels to support planning, monitoring, and decision making.

Keeping in view the requirements of the extension system and the budget constraints of the states, a typical design of the computer-based MIS is shown in Figure 2.

However, the actual design may vary with the size of the state and other considerations. An integrated database for the entire state may be supported by a mainframe/minicomputer at the state headquarters. Suitable programmes for the analysis of data may be designed to provide an interactive decision support system at the state level. Each district and subdivision may be provided with a mini/micro computer, depending on the volume of data to be handled. The computers in the districts and subdivisions may be networked with the state computer. The local data may be stored and processed in the district/subdivision, and the shared data with appropriate level of aggregation may be transmitted to the state headquarters to update the integrated database. The districts and subdivisions would have direct access to the integrated database with proper authorizations assigned to them through their passwords. The blocks may have only the input-output terminals connected to the subdivision computer to feed data to the subdivision and make on-line inquiries as and when necessary.

Summary

In this chapter, we have defined and described the basic concepts of a management information system. The characteristics of good information, namely, relevance, timeliness, accuracy, cost-effectiveness, reliability, usability, exhaustiveness, and aggregation level, have been described. The role of information systems in the process of decision making and the value of information have been explained. Four types of MIS, namely, databank information system, predictive information system, decision-making information system, and decisiontaking information system, have been presented. The role of MIS in management of agricultural extension programmes and the conceptual design of a MIS in an agricultural extension organization have been described.

Basic computer concepts have been explained. The advantages and disadvantages of centralized versus decentralized systems have been examined. The need for organizing databases and their integration and the need for programmes for decision analysis to evolve a decision support system have been explained. An assessment of hardware, software, and networking requirements for a typical computer-based MIS for a national agricultural extension system have been illustrated.

References

Ahituv, N., Neumann, S., & Riley, H. N. (1994). *Principles of information systems for management* (4th ed.). Dubuque, IA: Wm. C. Brown Communications.

Awad, E. M., & Gotterer, M. H. (1992). *Database management.* Danvers, MA: Boyd & Fraser.

Banerjee, U. K., & Sachdeva, R. K. (1995). *Management information system: A new frame work.* New Delhi: Vikas Publishing House.

Davis, G. B., & Olson, M. H. (1985). *Management information systems: Conceptual foundations, structure, and development.* New York: McGraw-Hill.

Imboden, N. (1980). *Managing information for rural development projects.* Paris: Organization for Economic Co-operation and Development.

Keen, P. G. W., & Morton, M. S. S. (1978). *Decision support systems.* Reading, MA: Addison-Wesley.

Lucas, H. C., Jr. (1990). *nformation systems concepts for management.* New York: McGraw-Hill.

Martin, J. (1990). *Telecommunications ar.d the computer* (3rd ed.). Englewood Cliffs, NJ: Prentice-Hall.

Mason, R. D., & Swanson, B. E. (1981). *Measurements for management decision.* Reading, MA: Addison-Wesley.

McLeod, R., Jr. (1995). *Management information systems: A study of computer-based information systems* (6th ed.). New Delhi: Prentice Hall of India.

Raheja, S. K., & Jai Krishna (1991). *Manual for monitoring and evaluation of T & V agricultural extension system.* New Delhi: Centre for Agricultural and Rural Development Studies.

Ramesh Babu, A., & Singh, Y. P. (1987). Management information system in an agricultural extension organization. In *Proceedings of the national seminar on management of information system in management of agricultural extension* (p. 1-15). Hyderabad: NIRD.

Ramesh Babu, A., & Singh, Y. P. (1990). Agricultural administration at block level: A case study. *Indian Journal of Extension Education, 26* (1 & 2), 88-90.

Rao, C. S. S. (1985). Agricultural extension management system in India: Past, present and modalities in future. *Indian Journal of Extension Education, 21* (1 & 2), 32-35.

Russell, H. M. (1979). A review of management information systems for agriculture. In H. M. Russell (Ed.), *Information for agriculture: Proceedings of the national workshop on agricultural information* (p. 41-51). Melbourne: Department of Agriculture, Victoria.

Sachdeva, R. K. (1990). *Management handbook of computer usage.* Oxford: NCC Blackwell.

Sanders, D. H. (1988). *Computers today* (3rd ed.). New York: McGraw-Hill.

Simon, H. A. (1977). The new science of management decision. New Jersey: Prentice-Hall.

Singh, Y. P., & Ramesh Babu, A. (1985). Basic management issues in extension. *Indian Journal of Extension Education, 21* (1 & 2), 20-31.

Wentling, T. L., & Wentling, R. M. (1993). *Introduction to microcomputer technologies. Rome: FAO.*

Chapter 19
Strengthening Research-Extension-Farmer Linkages

Burton E. Swanson

The lack of a close working relationship between national agricultural research and extension organizations, and with different categories of farmers and farm organizations, is one of the most difficult institutional problems confronting ministries of agriculture in many developing nations. Research and extension organizations generally compete over the same scarce government resources and, frequently, leaders of these institutions do not see themselves as part of a broader system: the agricultural technology system (ATS). Instead, they try to increase the flow of resources coming to their respective institutions and to solve day-to-day management problems, rather than ensuring that their respective organizations contribute to the broader goal of getting improved agricultural technology to all major categories of farmers. In addition, the leadership and staff of many research and extension organizations do not appreciate the important roles that farmers and farmer organizations can play, both in disseminating technology and, through effective feedback mechanisms, in helping set priorities and improving programme relevance.

The objective of this chapter is to outline an approach for identifying research-extension-farmer linkage problems, and then to describe different mechanisms that might be used to solve these problems. Three basic assumptions underlie this chapter. First, agricultural technology is a complex blend of materials, processes, and knowledge. Second, because of the complexity of agricultural technology, different institutional arrangements are needed to transfer different types of technology to technology users. And, third, most small-scale farmers in developing nations operate relatively complex farming systems in each agroecological zone (AEZ) of the country; consequently, farmers in different AEZs need access to a wide variety of locally validated technologies if they are to increase their productivity.

Agricultural Technology: Some Basic Concepts

At the most theoretical level, technology is the *application of knowledge for practical purposes*. Generally, technology is used to improve the human condition, the natural environment, or to carry out other socioeconomic activities. Technology can be classified into two major categories: (1) *material technology*, where knowledge is *embodied* into a technological product such as tools, equipment, agrochemicals, improved plant varieties or hybrids, improved breeds of animals (e.g., semen from progeny-tested sires used for artificial insemination), and vaccines; and (2) *knowledge-based technology* such as the technical knowledge, management skills, and other processes that farmers need to successfully grow a crop or produce animal products.

The transfer of *material* technology to farmers generally involves the production, distribution, and sale of seeds, implements, agrochemicals, and other production inputs. Therefore, the transfer process for material technology is generally simpler than training and disseminating technical knowledge and management skills to large numbers of poorly educated farmers who operate in different agroecological zones (i.e., the extension function). Also, the delivery systems needed for these different types of technologies are generally different. In most cases, the private sector is best suited to produce and distribute material technology. On the other hand, most *knowledge-based* technologies such as improved crop or livestock management practices, integrated pest management (IPM), and soil and water management practices are generally taught through vocational training programs for rural young people or disseminated through a publicly funded extension system for adult farmers.

At the same time, most material technology requires technical knowledge so that these products or tools can be used effectively. For example, to properly use an agrochemical in pest management, farmers need to know the proper application rates, the time and conditions for application, safety procedures, and so forth. In addition, if farmers use a sprayer (another type of material technology) to apply agrochemicals, then they need to know how to operate, adjust, calibrate, and clean the equipment to achieve the best results. Therefore, material and know-ledge-based technologies are generally closely intertwined. Private sector firms in developing countries have very limited technical capacity to train farmers in these product-related skills and knowledge; therefore, the transfer of most knowledge-based technologies is, by design or by default, left to the national or provincial extension system.

Burton E. Swanson is Professor and Director of the International Program for Agricultural Knowledge Systems (INTERPAKS), College of Agricultural, Consumer, and Environmental Sciences, University of Illinois at Urbana-Champaign.

Using Systems Analysis to Identify Linkage Problems

Systems analysis is an effective procedure to use in identifying linkage problems, since it is a *problem-solving methodology*. An outgrowth of operations research during World War II, systems analysis has been successfully applied in numerous fields, especially as a management tool to analyse, design, and implement complex technical processes. In short, systems analysis systematically examines a problem and makes each step of the analysis explicit. For example, in agricultural research, each disciplinary or commodity researcher (e.g., a plant breeder or agronomist) has relatively routine research methods that are commonly used in the process of developing specific technologies for a particular commodity (e.g., determining plant population, fertility, and other technical recommendations for a new cereal variety that has just been released for a particular AEZ). The problem is how to integrate these individual research efforts and relate them to broader system objectives such as increasing the productivity of different groups of farmers throughout the country. To achieve these objectives, the ATS will need to develop and transfer a package of recommendations for all of the economically important commodities being produced within different farming systems in each AEZ. By using a systems approach, it is possible to examine each system component and linkage at different levels within an ATS. In the process, specific system constraints and weaknesses can be easily identified, and then the most appropriate intervention strategy can be determined.

To understand what is meant by systems analysis, it might be helpful to start with some basic concepts. First, *a system is an organized set of functions and linkages that can be managed to achieve a specific goal or set of objectives*. Therefore, systems analysis helps make key functions and linkages within an ATS or research-extension system explicit or transparent. This procedure allows these different functions and linkages to be assessed and brought under management control, particularly in light of their overall contribution to ATS objectives. In short, technology development and transfer activities in agriculture involve very complex, interdisciplinary processes and interinstitutional relationships. Only by using systems analysis can managers begin to systematically analyse, manage, and monitor the various functions and linkages within their part of an ATS.

Second, what are the institutional components that typically make up an ATS? Figure 1 portrays the major institutional components that are found within most

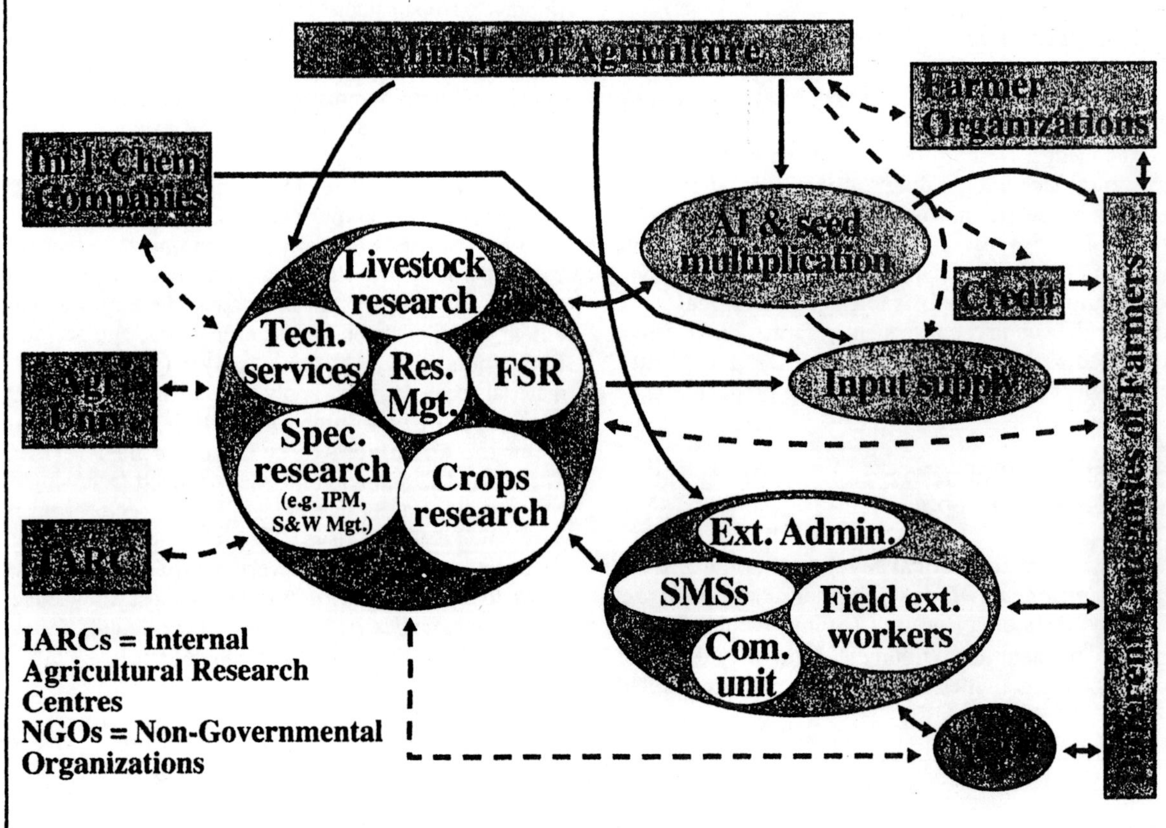

Figure 1. Institutional components of an agricultural technology system.

national ATSs, including a representation of some linkages used to help integrate this macrotechnology system. In this diagram, research and extension are divided into their respective subunits and operate within the policy direction and support of the ministry of agriculture. In addition, the production and distribution of genetic technology, agrochemicals, and other types of material technology are regulated by the ministry of agriculture. Also, as shown in Figure 1, farmers and their organizations provide feedback to research and extension and to input suppliers.

Third, one dimension of systems analysis is the use of key indicators to measure the inputs, activities, and outputs from each subsystem and component within an overall ATS. For example, human and financial resources allocated to various components or programmes within the research and extension system can be easily measured. Key input indicators include the percentage of the budget that is allocated to operational costs, the proportion and quality (educational qualifications and experience) of subject-matter specialists (SMSs) in relation to field staff, and so forth. For example, through research it has been determined that most research and extension organization allocate most of their financial resources to salaries, leaving too few resources for programme and operational costs (a target would be 35 per cent for operational costs). Also, most extension systems do not have sufficient numbers of competent SMSs to collaborate and link with their research counterparts (a target would be 20 per cent SMSs within the extension staff). Key performance indicators for crop management technology include, among others, the number of on-farm trials and demonstrations carried out in an AEZ. For more information on this aspect of analysing ATSs, see Swanson and Peterson (1989) and Swanson, Sands, and Peterson (1990).

Fourth, flow diagrams are used within systems analysis to map or illustrate the relationship of key functions and linkages for different types of technology within a national ATS. Therefore, through consultation with research and extension managers about key indicators and to develop a flow diagram for specific categories of technology, weak linkages or total gaps in the flow of technology can be easily identified and made transparent. Once problems are identified, this same procedure can be used to simulate the effect of alternative mechanisms that might be used to bridge these gaps and to strengthen weak relationships. Figures 2 and 3 illustrate both a *poorly* and a *well-linked* research-extension system for developing and disseminating crop management technology.

Figure 2 depicts a poorly linked research and extension (R&E) system, where the commodity research team does not collaborate directly with extension's subject-matter specialists (SMS) in planning, conducting, and interpreting data from agronomic trials. Also, these researchers do not interact with farmers or their organizations in the process of identifying production problems,

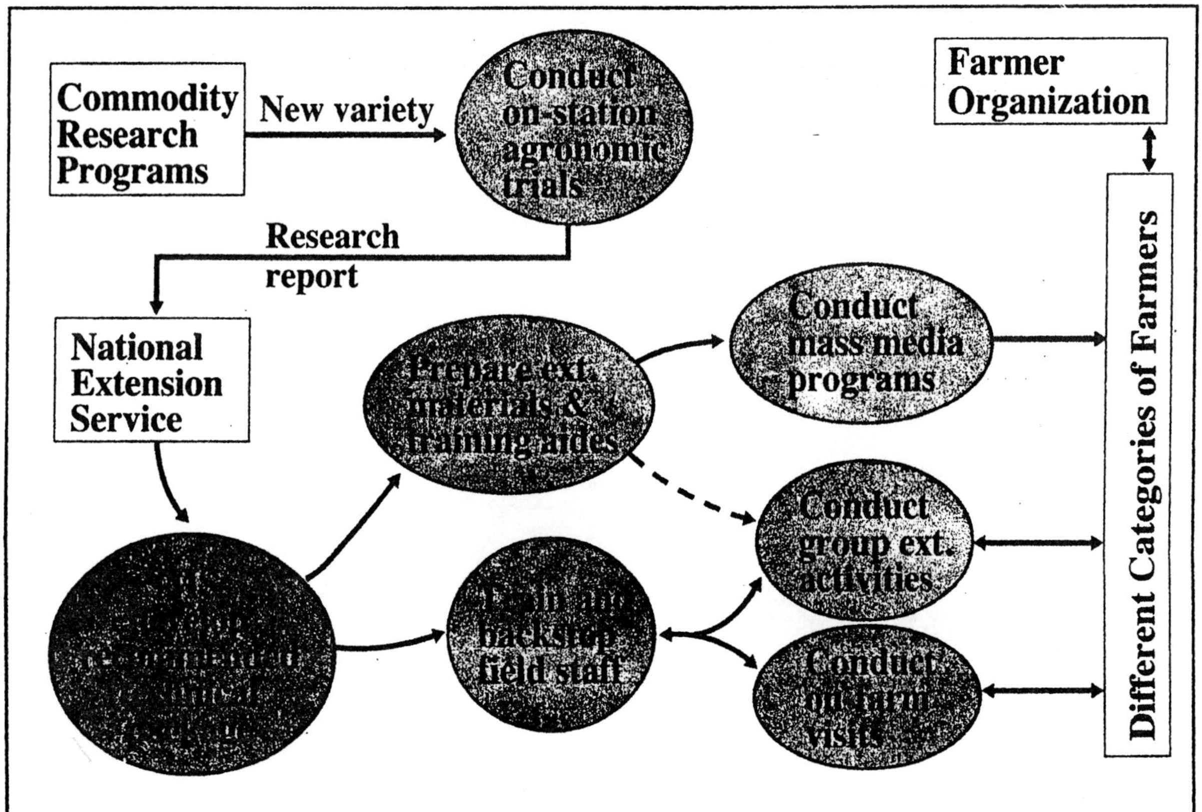

Figure 2. Poorly linked research and extension system.

setting research priorities, or in carrying out these trials. Furthermore, in this case, researchers do not discuss research findings with subject-matter specialists in the process of developing farmer recommendations; rather, they send a research report to extension that summarizes their findings. At this point, it is up to the SMS to interpret these findings and formulate a package of recommendations for dissemination to farmers. However, most SMSs lack sufficient analytical skills, including the ability to use economic criteria, in formulating technical recommendations. Recommended practices may thus reflect the government's goal of maximizing production, rather than the goals of farmers, which may be to maximize income. As a result, the package of recommendations may not be suitable for most farmers, especially resource poor farmers.

Figure 3 reflects a more fully integrated research and extension system for developing crop management technology for a specific commodity. The process begins with a joint assessment of farmer resources, constraints, and priorities. This needs assessment process (see chapter 5) could be carried out through a rapid or participatory rural appraisal (see chapter 6). Then, at each step in the process of developing technical recommendations, there is a close working relationship between researchers, SMSs, and farmers, until research findings are finally tested and demonstrated on farmers' fields. Technical recommendations that emerge from this collaborative process have the full confidence of researchers, extension personnel, and farmers, and have a much higher probability of being adopted once the extension field staff begins the widespread dissemination of this technology.

The advantage of using flow diagrams to map and analyse key functions and linkages within an agricultural technology system (ATS) is that each part or level of the system can be broken down into its constituent subsystems, components, functions, and corresponding linkages. Therefore, the research or extension manager can provide (management) oversight and coordination for each set of functions or linkages—for each system level—all the way down to individual researchers and SMSs who are carrying out their own programme of work in developing, testing, and disseminating a package of recommendations for a particular commodity or farming system. Contrast this functional depiction of various aspects of a technology system with information that is conveyed by examining organograms for research and extension organizations. These organizational charts primarily focus attention on the command and control structure of the organization, rather than on the technical or functional relationships that are necessary to ensure the flow of technology to farmers, plus farmer feedback throughout this process.

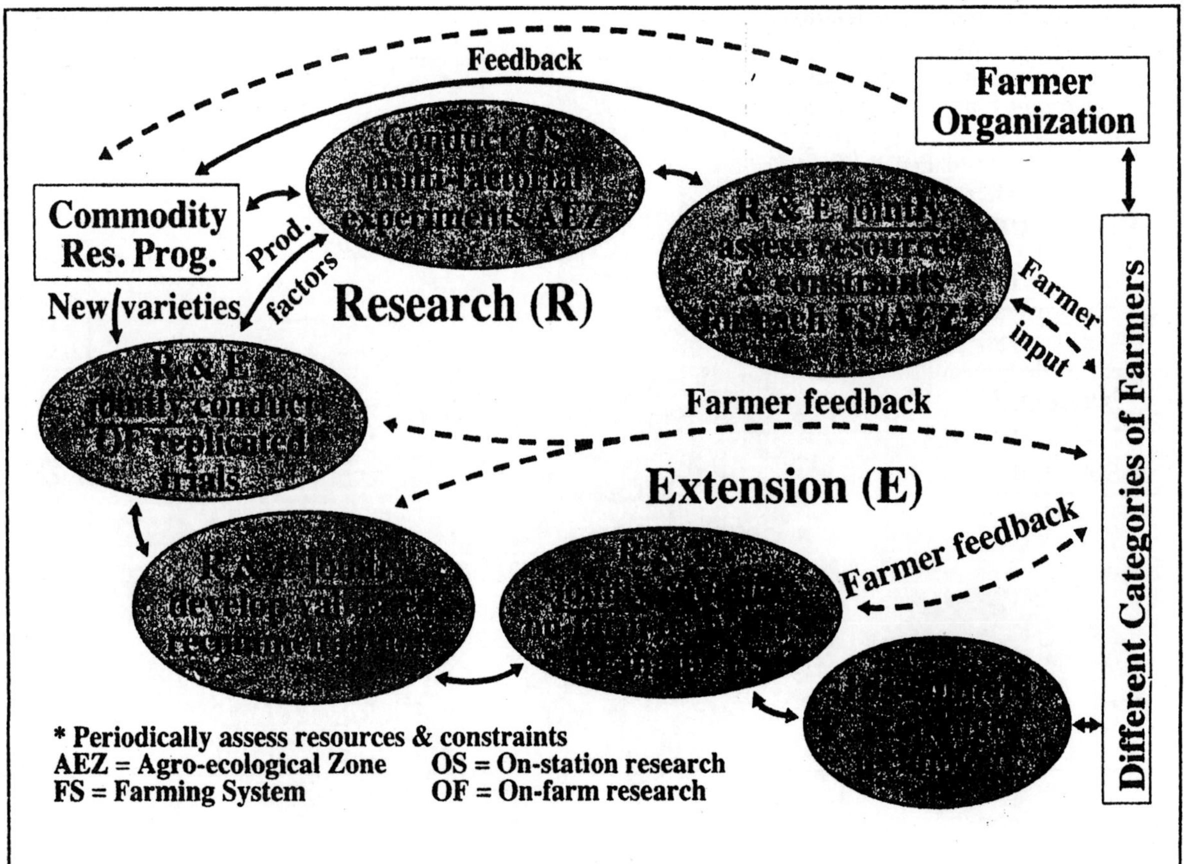

Figure 3. Linked research-extension-farmer system for crop management technology.

In management, there is a saying that *structure follows function.* In other words, the structure of an organization should reflect and support the work functions and processes that need to be carried out at each system level as the organization seeks to carry out its respective goals and objectives, as well as those of the larger (technology) system. In practice, most managers think about structure first, rather than trying to modify or build the structure (and organizational resources) around the key functions to be accomplished. It is not unlike the process of building a house; the process begins with the priorities of the homeowner, including how different functions will be arranged, the level of resources to be allocated for each function, and the key relationships that must exist between the different activity centres within the house. If an architect does not consider function first, then he or she might end up with a building that does not meet the needs of the homeowner.

Categories of Technology

To understand the different roles that research, extension, input supply dealers, and other system actors should play in the technology development and transfer process, it is essential to differentiate between the different categories of technology. Although there are some common elements, each category of technology has its own unique functional relationship within a national ATS. Flow diagrams can be used as a diagnostic tool to identify linkage problems in any ATS by tracking different types of technology through the research-extension system. Each category of technology follows a different channel as new technology is developed and transferred to farmers. In each case, it is possible to develop a functional map of the existing system to determine if serious linkage problems exist. Because of space limitations, it will not be possible to illustrate flow diagrams for different categories of technology (see Swanson, 1993), but the types that should be examined include the following:

Crops Technology
Genetic (improved varieties or hybrids)
Crop management practices
Plant protection (e.g., IPM)
Cropping systems

Livestock Technology
Genetic (breed improvement)
Livestock management practices
Animal health practices
Forage or range management

Other Types of Technology
Farming systems
Soil and water management practices
Farm mechanization
Agroforestry
Post-harvest technology
Farm management

It is important to note that different types of crop or livestock technologies have both *hardware* and *software* components. For example, a new crop variety, as a type of material technology, cannot be fully exploited without having a complementary set of agronomic or crop management practices, including pest management. Likewise, improved breeds of livestock generally require higher levels of management, including improved nutrition, housing, and preventive health practices. Therefore, the functional relationship or linkages, both within and across different categories of technology, must be carefully examined in identifying constraints that limit the flow of technology within an ATS.

Using Different Mechanisms to Solve Linkage Problems

Linkage mechanisms are used to channel information between groups and to coordinate required tasks in the process of getting relevant technologies to farmers. In the process, these linkage activities help to improve resource use by avoiding the duplication of effort and ensuring that critical tasks do not fall through the institutional cracks. In an effective ATS, numerous groups depend on one another to get improved technology to farmers. These groups have to work in concert and, as suggested for crop management technology in Figure 3, the failure of one link in the chain diminishes the overall performance of an ATS (Merrill-Sands, 1992).

There are two basic types of linkage mechanisms: organizational and managerial. Organizational mechanisms involve the structural modification of the research and/or extension organization or other organizations that are involved in an ATS. These modifications may range from the *formal merger* of research and extension at the broader system level, the merger of specific units within research or extension, or it could involve the creation of new positions, units, or permanent committees. In general, when reorganizing, be sure to combine those groups that are either dependent or that need to communicate with each other to get a job done. Also, put people together whose work might overlap to avoid duplication of activities (Merrill-Sands, 1992). On the other hand, in some cases it might be appropriate to create a *coordination position*, such as a research-extension liaison officer or to explicitly assign coordination functions to a specific position or unit (e.g., SMSs).

Alternatively, *permanent committees*, with representation from two or more units that need to collaborate, can be created and charged with the responsibility of coordinating and facilitating the flow of technology and feedback through the ATS. In creating a permanent committee, there must be agreement among all participating organizations or units that (1) such a committee is essential and that it should meet on a regular basis, (2) its members are senior enough to implement decisions and recommendations, (3) the committee's mandate is limited to programme coordination, and (4) the committee's work is visible and supported by senior management (Merrill-Sands, 1992).

The other major type of linkage mechanisms involves a range of managerial interventions. For example, research and extension may agree to collaborate on joint planning and review activities, such as conducting RRA or PRA in different agroecological zones to assess the farming systems and technology needs of different groups of farmers. These activities would likely result in joint priority-setting and in planning joint programme activities (e.g., on-farm trials and demonstrations). A second category of managerial linkages is when researchers and SMSs actually carry out *collaborative* programme activities together, such as on-farm trials and demonstrations, plus joint decision making on technical recommendations. In the process of carrying out these different programme activities, researchers and their extension counterparts share the responsibility for different tasks and regularly consult with each other on an informal basis. In short, through collaborative programme activities, research and extension personnel develop a positive professional relationship that is important, if not essential, in facilitating the flow of technology and feedback information within an ATS.

Other types of related managerial mechanisms include resource allocation procedures such as allocating time and financial resources for specific linkage activities. For example, an adaptive research or farming systems research team might be housed in a district or regional extension office so that they would have a closer working relationship with SMSs. Alternatively, SMSs might be assigned to a regional experiment station to work directly with one or more commodity research teams. Finally, various training and communications devices can be used to improve the flow of information and technology within an ATS. For example, a commodity research team and extension SMSs might collaborate on a joint farmer training activity; electronic mail is now being introduced in some countries to improve technical communications between research and extension personnel.

To illustrate both organizational and managerial linkages, two simplified organograms are presented in Figures 4 and 5. Figure 4 illustrates a variety of different managerial linkages, both within and between research and extension organizations. The actual types of linkage mechanisms have not been specified in the figure, but horizontal linkages between research and extension generally involve planning, review, and collaborative activities.

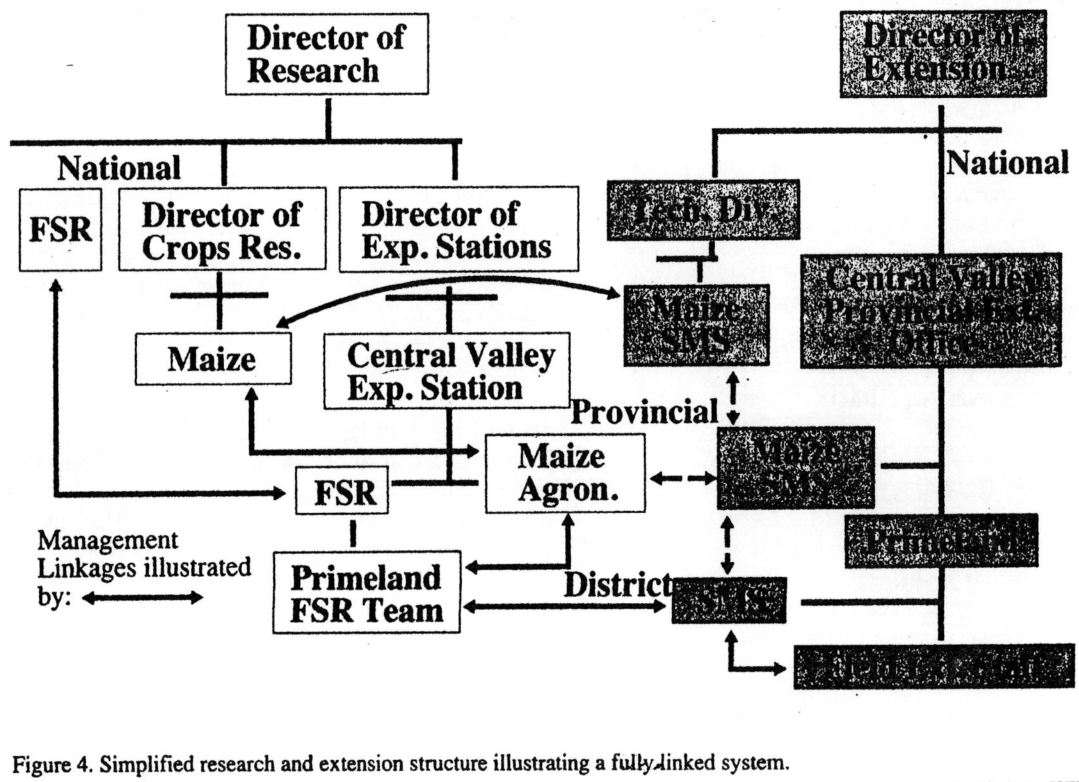

Figure 4. Simplified research and extension structure illustrating a fully-linked system.

Vertical linkages with each organization tend to involve resource allocation, training, and communications linkage activities. For example, assigning SMSs and FSR/commodity research personnel to carry out joint (linkage) activities and providing the travel and operational costs to successfully carry out these activities would demonstrate the commitment of both research and extension managers to these activities. In addition, there could be a permanent committee, at the national level, as an example of an organizational mechanism for each of the major crops and livestock species grown in the country.

Linkages with Farmers and Their Organizations

Figure 5 depicts organizational mechanisms within an extension system to improve planning and feedback linkages with farmers and their organizations (both general farm organizations and commodity groups). In this example, farmers are formally represented on *standing* (permanent) extension advisory committees at the district, provincial, and national levels. In general, members on these advisory committees would be elected by farmers through their respective organizations, from the bottom up. For example, different farmer groups or associations within a district would be allocated one or more seats on the extension advisory committee; therefore, they would elect a representative to that committee for a two-or three-year term (depending on the committee's by-laws). This district advisory committee, in turn, would elect a chairperson and/or a representative who would serve on the provincial extension advisory committee (along with chairpersons or representatives from the other district advisory committees in the province), and so forth up to the national level. If advisory committee membership is selected by the research or extension director, then the value of these advisory committees will be severely compromised; members of these committees must represent farmer interests and concerns.

In the case of research, each provincial and/or national commodity research programme should have an advisory committee composed of farmer-elected representatives who give each commodity research director specific direction and feedback on programme problems and priorities specific to each commodity. In addition, provincial or regional research stations should have an advisory committee that reflects the geographic interests, problems, and concerns of farmers within the province or region. Finally, there should be crop, livestock, and/or a general research advisory panel organized at the provincial and/or national level to provide the opportunity for stakeholder input into research policies, priorities, and other concerns that cut across the

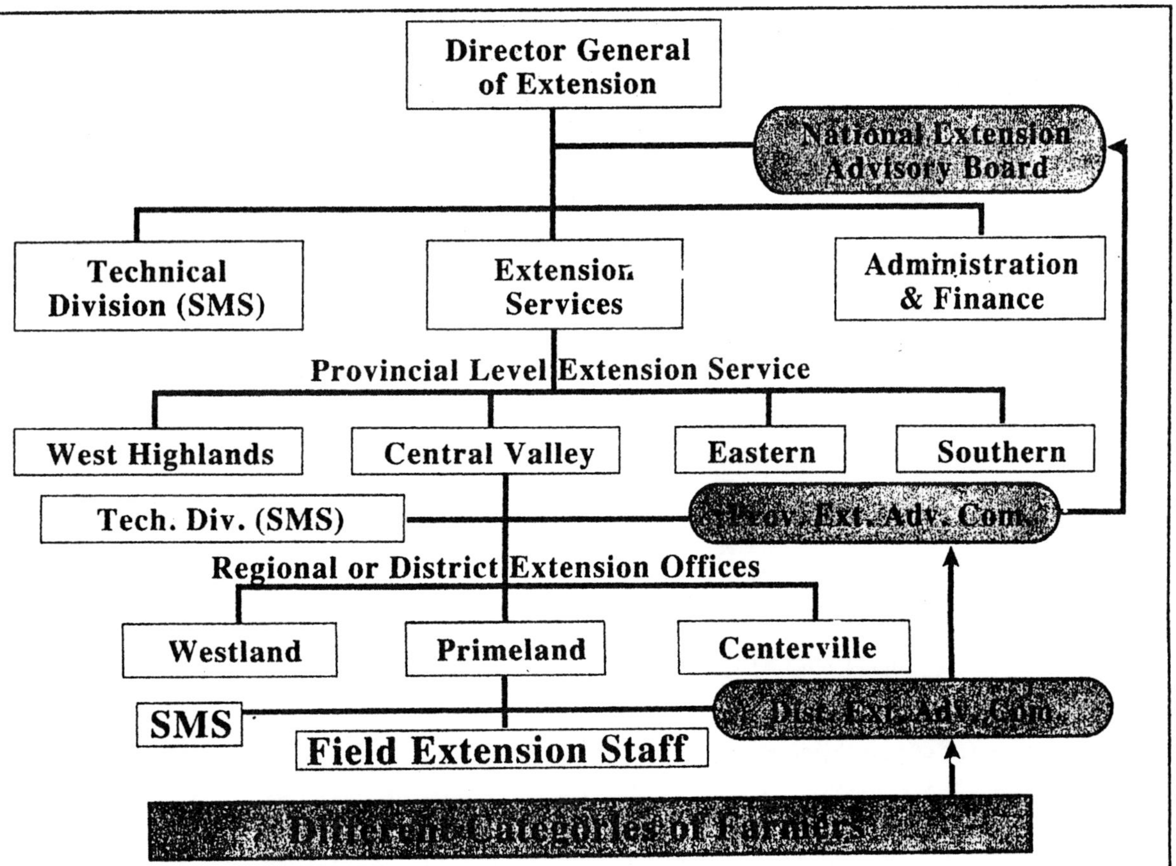

Figure 5. Simplified agricultural extension system showing farmer linkages through advisory committees.

research system. At these higher system levels, agribusiness representatives (input suppliers and processors) may join farmers in being represented on these research and extension advisory committees.

In addition to these formal linkage mechanisms, both research and extension personnel would be expected to have regular, informal contacts with different groups of farmers in their respective service area. As depicted in Figure 3, these linkages would occur through farmer participation in RRA or PRA activities (periodic needs assessment); also, they would occur while carrying out joint on-farm trials and demonstrations and during meetings and field days where farmers would have the opportunity to articulate different problems and concerns. The value of both formal and informal farmer feedback systems depends, in large part, on whether research and extension personnel, including senior management, are listening to what farmers and their representatives are saying. Too often, research and extension have become top-down, bureaucratic organizations that are not receptive or responsive to the needs of farmers. However, to become demand driven, research and extension organizations, directors, specialists and other research and extension personnel must be listening to what farmers are communicating through both informal and formal linkage mechanisms.

References

Merrill-Sands, D. (1992). *Managing links with technology users: A training module*. The Hague: International Service for National Agricultural Research.

Merrill-Sands, D., & Kaimowitz, D. (1989). *The technology triangle*. The Hague: International Service for National Agricultural Research.

Swanson, B. E. (1993). *Identifying linkage problems using systems analysis: A training module*. Urbana, IL: International Program for Agricultural Knowledge Systems.

Swanson, B. E., & Peterson, W. E. (1989). *A field manual for analyzing agricultural technology development and transfer systems*. Urbana, IL: International Program for Agricultural Knowledge Systems.

Swanson, B. E., Sands, C. M., & Peterson, W. E. (1990). Analyzing agricultural technology systems: Some methodological tools. In R. G. Echeverria (Ed.), *Methods for diagnosing research system constraints and assessing the impact of agricultural research (Vol. I)*. The Hague: International Service for National Agricultural Research.

Part IV

Current Trends and Development

Chapter 20
Extension's Role in Sustainable Agricultural Development

Niels Röling and Jules N. Pretty

Emerging Challenges for Sustainable Agriculture

During the past fifty years, agricultural development policies have been remarkably successful at emphasizing external inputs as the means to increase food production. This has led to growth in global consumption of pesticides, inorganic fertilizer, animal feedstuffs, and tractors and other machinery.

These external inputs have, however, substituted for natural processes and resources, rendering them less powerful. Pesticides have replaced biological, cultural, and mechanical methods for controlling pests, weeds, and diseases; inorganic fertilizers have substituted for livestock manures, composts, and nitrogen-fixing crops; information for management decisions comes from input suppliers, researchers, and extensionists rather than from local sources; and fossil fuels have substituted for locally generated energy sources. The basic challenge for sustainable agriculture is to make better use of these internal resources. This can be done by minimizing the external inputs used, by regenerating internal resources more effectively, or by combinations of both.

Evidence is now emerging that regenerative and resource-conserving technologies and practices can bring both environmental and economic benefits for farmers, communities, and nations. The best evidence comes from countries of Africa, Asia, and Latin America, where the concern is to increase food production in the areas where farming has been largely untouched by the modern packages of externally supplied technologies. In these complex and remote lands, some farmers and communities adopting regenerative technologies have substantially improved agricultural yields, often using only few or no external inputs (Bunch, 1991; GTZ, 1992; UNDP, 1992; Lobo & Kochendörfer-Lucius, 1992; Krishna, 1993; Shah, 1994; SWCB, 1994; Pretty, 1995).

But these are not the only sites for successful sustainable agriculture. In the high-input and generally irrigated lands, farmers adopting regenerative technologies have maintained yields whilst substantially reducing their use of inputs (Kamp, Gregory, & Chowhan, 1993; UNDP, 1992; Kenmore, 1991; van der Werf & de Jager, 1992; Bagadion & Korten, 1991). And in the very high-input lands of the industrialized countries, farmers have been able to maintain profitability even though input use has been cut dramatically, such as in Europe (Vereijken, 1992; Vereijken, Wijnands, Stol, & Visser, 1994; Van Weeperen, Röling, Van Bon, & Mur, 1995; Pretty & Howes, 1993; Jordan, Hutcheon, & Glen, 1993; El Titi & Landes, 1990) and in the United States (Liebhart et al., 1989; NRC, 1989; Hanson, Johnson, Peters, & Janke, 1990; Dobbs, Becker, & Taylor, 1991; Faeth, 1993).

All of these successes have three elements in common. They have made use of resource-conserving technologies such as integrated pest management, soil and water conservation, nutrient recycling, multiple cropping, water harvesting, and waste recycling. In all, there has been action by groups and communities at the local level, with farmers becoming experts at managing farms as ecosystems and at collectively managing the watersheds or other resource units of which their farms form a part. And there have also been supportive and enabling external government and nongovernment institutions, which have reoriented their activities to focus on local needs and capabilities.

Most successes, though, are still localized. They are simply islands of success. This is because an overarching element, a favourable policy environment, is missing. Most policies still actively encourage farming that is dependent on external inputs and technologies. It is these policy frameworks that are one of the principal barriers to a more sustainable agriculture (Pretty, 1994a). Figure 1 illustrates this chapter's area of discourse and its focus on the interfaces between natural resources, local stakeholders, supportive institutions, and the policy context.

Sustainability and Levels of Action

A necessary condition for sustainable agriculture is that large numbers of farming households must be motivated to use coordinated resource management. This could be for pest and predator management, nutrient management, controlling the contamination of aquifers and surface water courses, coordinated livestock management, conserving soil and water resources, and seed stock management. The problem is that, in most places, platforms for collective decision making have not been established to manage such resources (Röling, 1994a, 1994b). The success of sustainable agriculture therefore depends not just on the motivations, skills, and knowledge

---

Niels Röling is Extra-ordinary Professor of agricultural knowledge systems, Department of Communication and Innovation Studies, Wageningen Agricultural University, Wageningen, Netherlands. **Jules N. Pretty** is the Director of Sustainable Agriculture Programmes, International Institute for Environment and Development, London.

of individual farmers, but on action taken by groups or communities as a whole. This makes the task more challenging. Simple extension of the message that sustainable agriculture can match conventional agriculture for profits, as well as produce extra benefits for society as a whole, will not suffice.

Sustainability is commonly seen as a property of an ecosystem. But sustainability can be seen from other perspectives, which are more relevant for extension. Environmental issues emerge from the human use of natural resources. Sustainability can therefore be defined in terms of human reasons, activities, and agreements. The definition of sustainability then becomes part of the problem because people need to agree on how they define sustainability and what priority they will give it (Pretty, 1994b).

In this approach, sustainability is not a scientific, "hard" property which can be measured according to some objective scale, or a set of practices to be fixed in time and space. Rather, sustainability is a quality that emerges when people individually or collectively apply their intelligence to maintain the long-term productivity of the natural resources on which they depend (Sriskandarajah, Bawden, & Packham, 1989). In other words, sustainability emerges out of shared human experiences, objectives, knowledge, decisions, technology, and organization. Agriculture becomes sustainable only when people have reason to make it so. They can learn and negotiate their way towards sustainability. In any discussions of sustainability, it is important to clarify what is being sustained, for how long, for whose benefit and at whose cost, over what area, and measured by what criteria. Answering these questions is difficult, because it means assessing and trading off values and beliefs. Campbell (1994) has put it this way: "[Attempts to define sustainability miss the point that, like beauty, sustain ability is in the eye of the beholder. . . . It is inevitable that assessments of relative sustainability are socially constructed, which is why there are so many definitions."

It is therefore crucial to focus on more than one system level (Fresco, Stroosnijder, Bouma, & van Keulen, 1994). At the farm level, there is the farm household. At the above-farm level, there are the collective stakeholders, who might or might not be organized for sustainable use of the whole resource unit. In an irrigation scheme, it is common for an irrigators' association collectively to manage water use at the scheme level. But when it comes to watersheds or other vulnerable resource units, it is usually impossible to identify an appropriate "platform" for decision making (Röling, 1994a, 1994b).

A key example is the Indonesian programme for integrated pest management (IPM) in irrigated rice (FAO, 1994; Van de Fliert, 1993; Röling & Van de Fliert, 1994; Kenmore, 1991). At the farm level, this programme involves farmer field schools teaching individual farmers to manage their rice plots as ecosystems, carefully maintaining the balance between pests and their natural predators and only reverting to pesticides when observation shows that the situation is running out of hand. But IPM also needs collective management of resources comprising several farms. Thus nematodes can effectively be controlled by interrupting the cultivation of wet rice by a dryland crop such as soybeans. This requires decision making at the irrigation block level. The population dynamics of rats, the most important pest in irrigated rice, cannot be controlled at the farm level. Integrated rat management requires collective action at the village level (Van de Fliert, van Elsen, & Nangsir Soenanto, 1993).

Resource-Conserving Technology Development and Transfer

Although many resource-conserving technologies and practices have been widely proven on research stations to be both productive and sustainable, the total number of farmers using them is still small. This is because these technologies involve the substitution of management skills, knowledge, and labour for external inputs. The modern approach to agricultural research and extension, however, has been to emphasize comprehensive packages of technologies. Few farmers are able to adopt the whole modern packages of production or conservation technologies without considerable adjustments. Part of the problem is that most agricultural research still occurs on the research station, where scientists experience conditions quite different from those experienced by farmers.

This is true of many sustainability-enhancing innovations. Even though resource-conserving technologies are productive and sustainable, if they are imposed on farmers, then they will not be adopted widely. Alley cropping, an agroforestry system comprising rows of nitrogen-fixing trees or bushes separated by rows of cereals, has long been the focus of research (Kang, Wilson, & Lawson, 1984; Attah-Krah & Francis, 1987; Young, 1989; Lal, 1989). Many productive and sustainable systems, needing few or no external inputs, have been developed. They stop erosion, produce food and wood, and can be cropped over long periods. But the problem is that very few, if any, farmers have adopted these alley cropping systems as designed. Despite millions of dollars of research expenditure over many years, systems that have been produced are suitable only for research stations.

Where these systems have had some success, however, farmers have taken one or two components of alley cropping and adapted them to their own farms. In Kenya, for example, farmers planted rows of leguminous trees next to field boundaries, or single rows through their fields; and in Rwanda, alleys planted by extension workers soon became dispersed through fields (Kerkhof, 1990). But the prevailing view tends to be that farmers should adapt to the technology. Of the Agroforestry Outreach Project in Haiti, it was said:

> Farmer management of hedgerows does not conform to the extension program. . . . Some farmers prune the hedgerows too early, others too late. Some hedges are not yet pruned by two years of age, when they have already reached heights of 4-5 metres. Other hedges are pruned too early, mainly because animals are let in or the tops are cut and carried to animals. . . . Finally, it is very common for farmers to allow some of the trees in the hedgerow to grow to pole size. These trees are not pruned but are harvested when needed for house construction or other activities requiring poles. (Bannister & Nair, 1990)

Farmers were clearly making their own adaptations according to their own needs.

Just occasionally, however, an environmentally beneficial technology is developed that appears to require no knowledge of farmers' conditions. The IPM programme to control cassava mealybug (CMB) (*Phenacoccus manihoti*) in west and central Africa is one example. CMB was first recorded in Africa in 1973, and an effective natural enemy, the wasp *Epidinocarsis lopezi*, was found in 1981. Since releases began, it has become established in twenty-five countries, providing good control of CMB. It is to some extent a "perfect technology" for scientists, because it is released from the air without the knowledge of farmers. It is, however, not necessarily a perfect technology for farmers. The contrast with another IPM programme in Togo is significant when it comes to issues of sustainability (Box 1).

Incorporating Farmer Experimentation

The problem with agricultural science and extension is that it has poorly understood the nature of "indigenous" and rural people's knowledge. For many, what rural people know is assumed to be "primitive," "unscientific," or overtaken by development, and so formal research and extension must "transform" what they know so as to "develop" them. An alternative view is that local knowledge is a valuable and underused resource, which can be studied, collected, and incorporated into development activities. Neither of these views, though, is entirely satisfactory because of the static view of knowledge implied (Chambers, Pacey, & Thrupp, 1989; Röling & Engel, 1989; Warren, 1991; Long & Long, 1992; Scoones & Thompson, 1994). It is more important to recognize that local people are always involved in active learning, in (re)inventing technologies, in adapting their farming systems and livelihood strategies. Understanding and supporting these processes of agricultural innovation and experimentation have become an important focus in facilitating more sustainable agriculture with its strong locality-specific nature.

The problem with modern agricultural science is that technologies are finalized before farmers get to see them. If new technologies are appropriate and fit a particular farmer's conditions or needs, then they stand a good chance of being adopted. But if they do not fit and if farmers are unable to make changes, then they have only the one choice. They have to adapt to the technology, or reject it entirely.

Box 1. Comparison of Farmers' Involvement in Two IPM Programmes.

A: Cassava mealybug (CMB) control with *Epidinocarsis lopezi*

The programme has involved close collaboration between IITA and NARSs, involving training of local technicians to participate in releases. Now mass rearing of the wasp *E. lopezi* is done in Benin; from there they are transported by air for air release. According to IITA, an important component of success has been that farmers and extension agents have not had to be involved. Farmers do not, therefore, know anything about the releases. One survey of farmers in Ghana and Côte d'Ivoire found that they recognized CMB and how it was a devastating pest. All those where *E. lopezi* had been introduced at least six months before had observed a significant decline in CMB. But because none of them knew about the programme, they attributed the decline to recent heavy rains and other climatic factors.

B: Mango mealybug control in Togo

The CMB programme is in contrast to the successful introduction of the parasitoid *Gyranusoides tebyii* to Togo in 1987 to control the mango mealybug (*Rastrococcus invadens*). The parasitoid was found in India, and following testing, rearing, and release, it rapidly spread over the whole of Togo. By 1989, no mango trees could be found on which mango mealybug was present without being parasitized. But success would be threatened without public interest, as any use of chemical control methods would kill the parasites. A great deal of publicity was given through radio, TV, and advisory leaflets. Considerable economic losses are now being prevented by the biological control system.

Source: Kiss and Meerman (1991).

The alternative is to seek and encourage the involvement of farmers in adapting technologies to their conditions. This constitutes a radical reversal of the normal modes of research and technology generation, because it requires interactive participation between professionals and farmers. Participatory technology development (PTD) is the process in which the knowledge and research capacities of farmers are joined with those of scientific institutions, whilst at the same time strengthening local capacities to experiment and innovate (Jiggins & De Zeeuw, 1992; Reijntjes, Haverkort, & Waters-Bayer, 1992; Haverkort, van der Kamp, & Waters-Bayer, 1991). Farmers are encouraged to generate and evaluate indigenous technologies and to choose and adapt external ones on the basis of their own knowledge and value systems.

But, of course, researchers and farmers participate in different ways, depending on the degree of control each actor has over the research process. The most common form of "participatory" research is researcher designed and implemented, even though it might be conducted on farmers' fields. Many on-farm trials and demonstration plots represent nothing better than passive participation (Pretty, 1994b). Less commonly, farmers may implement trials designed by researchers. But greater roles for farmers are even rarer. Fujisaka (1991) describes researcher-designed experiments on new cropping patterns in the Philippines. Even though farmers "participated" in implementing the trials, there was widespread uncertainty about what researchers were actually trying to achieve. Farmers misunderstood experiments and rejected the new technologies. The reason, as Fujisaka explains, was that "cooperation between farmers and researchers implies two groups continually listening carefully to one another. Claveria farmers are avid listeners to . . . researchers. The challenge is for all on-farm researchers to complete the circle."

Although this means that technology development must involve farmers, it does not mean that scientific research has no place. Research will have to contribute on many fronts, such as in the development of resistant cultivars, knowledge about the life cycles of pests, biological control methods, suitable crops for erosion control, and processes in nitrogen fixation. Such research also gives insight into complex processes such as the movement of nutrients in the soil and their accessibility for plants. But all these contributions must be seen as providing choices for farmers as they make farm-specific decisions and move the whole farm towards greater sustainability.

From Teaching to Learning and a Whole New Professionalism

The central principle of sustainable agriculture is that it must enshrine new ways of learning about the world. But learning should not be confused with teaching. Teaching implies the transfer of knowledge from someone who knows to someone who does not know. Teaching is the normal mode of educational curricula and is also central to many organizational structures (Ison, 1990; Argyris, Putnam, & Smith, 1985; Russell & Ison, 1991; Bawden, 1992, 1994; Pretty & Chambers, 1993). Universities and other professional institutions reinforce the teaching paradigm by giving the impression that they are custodians of knowledge which can be dispensed or given (usually by lecture) to a recipient (a student). Where these institutions do not include a focus on self-development and on enhancing the ability to learn, they do not allow students to grasp an essential skill in the sustainable management of a complex agroecosystem. In that case, "teaching threatens sustainable agriculture" (Ison, 1990).

The problem for farmers is that they cannot rely on routine, calendar-based activities if they engage in sustainable farming. Their interventions must be based on observation and anticipation. They require instruments and indicators which make more visible the ecological relationships on and among farms. Technology for sustainable farming must emphasize measurement and observation equipment or services that help individual farmers assess their situations, such as soil analysis, manure analysis, and pest identification (Röling, 1993). It also has to focus on higher system levels. Predators and parasitoids to control pests often require a larger biotope than that of a small farm. Erosion control, water harvesting, biodiversity, access to biomass, recycling waste between town and countryside and between animal and crop production, all require local cooperation and coordination.

What becomes important is the social transition, or new learning path, that farmers and communities must take to support sustainable agriculture. This is much less obvious and often remains unrecognized by extensionists. Learning for sustainable agriculture involves a transformation in the fundamental objectives, strategies, theories, risk perceptions, skills, labour organization, and professionalism of farming. This learning path has four key elements:

1. *The information system.* Sustainable agriculture must be responsive to changing circumstances, so farmers need to invest in observation, observation equipment, record keeping, and monitoring procedures.
2. *Conceptual framework.* Sustainable agriculture is knowledge intensive, and so farmers must know about life cycles of pests and disease organisms and their recognition, biological controls, ecological principles, soil life processes, nutrient cycles.
3. *Skills.* Sustainable farming requires a whole set of new skills, including observation and monitoring, compost making, mechanical weed control, spot application of pesticides, and risk assessment.
4. *Higher system-level management.* Generally, sustainable management of the farm is not enough,

and it is necessary to think at system levels higher than the farm and take part in the collective management of natural resources at those levels.

In educational systems, therefore, the fundamental requirement for sustainable agriculture is for universities to evolve into communities of participatory learners. Such changes are very rare, an exception being Hawkesbury College, which is now part of the University of Western Sydney, Australia (Bawden, 1992, 1994). However, a regional consortium of NGOs in Latin America concerned with agroecology and low-input agriculture recently signed an agreement with eleven colleges of agriculture from Argentina, Bolivia, Chile, Mexico, Peru, and Uruguay to help in the joint reorientation of curriculum and research agendas towards sustainability and poverty concerns (Altieri & Yuryevic, 1992; Yuryevic, 1994). The agreement defines collaboration to develop more systemic and integrated curricula, professional training and internship programmes, collaborative research efforts, and the development of training materials.

Box 2. The Key Principles of Farmer Field Schools.

1. What is relevant and meaningful is decided by the learner and must be discovered by the learner. Learning flourishes in a situation where teaching is seen as a facilitating process that assists people to explore and discover the personal meaning of events for themselves.
2. Learning is a consequence of experience. People become responsible when they have assumed responsibility and experienced success.
3. Cooperative approaches are enabling. As people invest in collaborative group approaches, they develop a better sense of their own worth.
4. Learning is an evolutionary process, and is characterized by free and open communication, confrontation, acceptance, respect, and the right to make mistakes.
5. Each person's experience of reality is unique. As people become more aware of how they learn and solve problems, they can refine and modify their own styles of learning and action.

Sources: Adapted from Kingsley and Musante, 1994; Van de Fliert, 1993; Kenmore, 1991; Stock, 1994.

A move from a teaching to a learning style has profound implications for agricultural development institutions. The focus is less on *what* we learn, and more on *how* we learn and *with whom* (see Box 2 for principles of farmer field schools used in the FAO IPM programme in Southeast Asia). This implies new roles for development professionals, leading to a whole new professionalism with new concepts, values, methods, and behaviour. Typically, normal professionals are single-disciplinary, work largely or only in agencies remote from people, are insensitive to diversity of context, and are concerned with themselves generating and transferring technologies. Their beliefs about people's conditions and priorities often differ from people's own views. The new professionals, by contrast, are either multidisciplinary or work in close connection with other disciplines, are not intimidated by the complexities of close dialogue with rural and urban people, and are continually aware of the context of interaction and development.

From Directive to Participatory Extension

Extension has long been grounded in the diffusion model of agricultural development, in which technologies are passed from research scientists via extensionists to farmers (Rogers, 1962, 1983). This approach is exemplified by the training and visit (T&V) system. It was first implemented in Turkey in 1967 and later widely adopted by governments (Benor, 1987; Roberts, 1989). It was designed to be a management system for energizing extension staff, turning desk-bound, poorly motivated field staff into effective extension agents. Extension agents receive regular training to enhance their technical skills, which they then hope will pass to all farmers through regular communication with small numbers of selected contact farmers.

But the contact farmers are usually selected on the basis of literacy, wealth, readiness to change, and "progressiveness," and so this sets them apart from the rest of the community. The secondary transfer of the technical messages, from contact farmers to community, has been much less successful than predicted, and adoption rates are commonly very low among noncontact farmers. Without a doubt, T&V is now widely considered as ineffective (Axinn, 1988; Howell, 1988; Moris, 1990; Antholt, 1992, 1994; Hussain, Byerlee, & Heisey, 1994).

Important lessons have been learned from the problems associated with T&V, and there is clearly a need to address the systemic issues facing extension (Zijp, 1993; Antholt, 1994). Extension will need to build on traditional communication systems and involve farmers themselves in the process of extension. Incentive systems will have to be developed to reward staff for being in the field and working closely with farmers. There must be a "well-defined link between the well-being of field officers and the extension system, based on the clients' view of the value of extension's and field workers' performance" (Antholt, 1992, P.). Participation, if it is to become part of extension, must clearly be interactive and empowering. Any pretence to participation will result in little change. Allowing farmers just to come to meetings or letting a few representatives sit on committees will be insufficient.

There have been some recent innovations in introducing elements of farmer participation and group approaches into extension. Differences in impact between individual and group approaches have been well documented in both Nepal and Kenya. In western Nepal, Sen (1993) compared the rate of adoption of new technologies when extension worked with individuals or with groups. With groups, better communication between farmers and extensionists led to more adoption. When the individual approach was resumed after the experiment, adoption rates fell rapidly in succeeding years.

In Kenya, the Ministry of Agriculture is increasingly adopting a community-oriented approach to soil and water conservation. This is steadily replacing the former individual approach of the T&V system. Where extension staff interact closely with communities in developing joint action plans, and local people freely elect members to a local catchment committee, then the impact on agricultural growth is immediate and sustained. Strong local groups mobilize the interest of the wider community and sustain action well beyond the period of direct contact with external agents. Recent studies comparing the impact of the catchment approach with the individual T&V approach have shown that, for a wide range of indicators, farmers' livelihoods were more improved where the community approach was implemented (SWCB, 1994; Pretty, Thompson, & Kiara, 1994; MALDM, 1988-1994; Eckbom, 1992).

There have been similar successes in IPM, which requires a level of analytical skill and certain basic training in crop monitoring and ecological principles. Where farmers have been trained as experts, such as in Honduras (Bentley, Rodríguez, & González, 1993) and in the rice-IPM programmes of Southeast Asia (Kenmore, 1991), then the impacts are substantial. Ordinary farmers are capable of rapidly acquiring and applying the principles and approaches. Fewer programmes are now teaching farmers new technologies and knowledge; rather, they are concerned with developing farmers' own capacity to think for themselves and develop their own solutions. These are producing substantial reductions in insecticide use, whilst maintaining yields and increasing profits (Table 1). But where extension continues to use the conventional top-down approach, then few farmers adopt, let alone learn, the principles. As Matteson (1992) put it: "[F]ew IPM programmes have made a lasting impact on farmer knowledge, attitudes or practice."

There are three major lessons for extension. First, it is important to make new things visible. An important role of extension is to make visible the state of the environment and the extent to which present farming practices are untenable. In addition, extension can demonstrate the feasibility of sustainable practices. Even more important is to give farmers the tools for observation and to train them to monitor the situation on their own farms.

The second lesson is the use of farmers' knowledge. The location-specific nature of sustainable agriculture implies that extension must make use of farmers' knowledge and work together with farmers. Often, indigenous practices which have been ignored under the impact of chemical farming can be fruitfully revived. Indigenous technology development practices and farmer experimentation can be an important "entry point" for introducing sustainable farming practices (Brouwers & Röling, in press).

The third lesson is an emphasis on facilitating learning. Instead of "transferring" technology, extension workers must help farming "walk the learning path" (Box 3). Extension workers should seek to understand the learning process, provide expert advice where required, convene and create learning groups, and help farmers overcome major hurdles in adapting their farms.

Challenges for Supportive Policy Processes

Policy making is commonly considered the prerogative of some central authority that formulates a policy, which is then decreed, imposed, and implemented regardless of conflicting knowledge and concerns. But policy is, in practice, often the net result of the actions of different interest groups pulling in complementary and opposing directions. This is particularly true with environmental problems because they are marked by uncertainty, complexity, and high stakes complexity, and high stakes (Funtowicz & Ravetz, 1993). There is therefore a growing tendency to see policy as a negotiated agreement resulting from interaction among citizens, in which central authorities play a facilitating role (Van der Poel & Van Woerkum, 1994). Policy is only effective if it is based on a widely shared consensus. From this perspective, it is easy to see why so many environmental policies which rely on coercion, control, and transfer have failed (Pretty & Shah, 1994; Pimbert & Pretty, 1994).

Box 3. The First Steps on the Learning Path in the Netherlands.

Predator mites were introduced into Dutch fruit orchards to control the red spider mite, which had become resistant to chemical controls. The use of this biological control meant that growers had to learn how to manage their orchards as biotopes for the predator mite. Soon they were carrying magnifying glasses to study the progress of their little helpers. This made them much more observant and accustomed to investing in regular observation. Furthermore, the health of the predator mites precluded use of broad-spectrum pesticides against other pests. As a result, growers also had to learn alternative controls for those pests.

Table 1. Impact of IPM Programmes Involving New Participatory Approaches to Farmer Learning on Pesticide Use and Crop Yields

Country and crop	Average changes in pesticide use (as % of conventional treatments)	Changes in yields (as % of conventional treatments)
Togo, cotton [1]	50%	90-108%
Burkina Faso, rice [1]	50%	103%
Thailand, rice [2]	50%	no data
Philippines, rice [2]	62%	110%
Indonesia, rice [2]	34-42%	105%
Nicaragua, maize [3]	25%	93%[a]
USA, nine commodities [4]	no. of applications up volume applied down	110-130%
Bangladesh, rice [5]	0-25%	113-124%
India, groundnuts [6]	0%	100%
China, rice [2]	46-80%	110%
Vietnam, rice [2]	57%	107%
India,rice [2]	33%	108%
Sri Lanka, rice [2]	26%	135%

[a] Even though yields are lower, net returns are much higher.

Sources: (1) Kiss and Meerman, 1991; (2) Kenmore, 1991; Winarto, 1993; van der Fliert, 1993; Matteson et al, 1992; FAO, 1994; (3) Hruska, 1993; (4) NRC, 1989; (5) Kamp et al, 1993; Kenmore, 1991; (6) ICRISAT, 1993

For sustainable agriculture to succeed, policy formulation must arise in a new way. Policy processes must be enabling and participatory, creating the conditions for sustainable development based more on locally available resources and on local skills and knowledge. Effective policy processes will have to bring together a range of actors and institutions for creative interaction and address multiple realities and unpredictability. What is required is the development of approaches that put participation, negotiation, and mediation at the centre of policy formulation so as to create a much wider common ownership in the practices. This is a central challenge for sustainable agriculture. The management of higher level systems, whether common grazing lands, coastal fisheries resources, communal forests, national parks, polders, or watersheds, requires social organization comprising the key stakeholders. All successful moves to more sustainable agriculture have in common coordinated action by groups or communities at the local level (Pretty, 1995). But the problem is that platforms for resource use negotiation generally do not exist, and so need to be created and facilitated (Brinkman, 1994; Röling, 1994a, 1994b).

Different methodologies are emerging to help stake-holders achieve collective resource management capacity. Well known are participatory rapid appraisal (PRA) and related methodologies (see chapter 6). In addition, the soft system methodology (SSM) developed for corporate environments is highly promising for resource use negotiation (Checkland, 1981; Checkland & Scholes, 1990). For stakeholders who have come to appreciate the fact that they share a problem, SSM takes them through a number of steps which allows them to create a "rich picture" on the basis of their multiple perspectives, reach some accommodation with respect to major causes of the problem, and hence decide on collective action. "Rapid appraisal of agricultural knowledge systems" (RAAKS) (Engel, 1995) is a related methodology for facilitating innovation as an emergent property of a knowledge network, comprising such actors as farmers, extension workers, researchers, NGO workers, and policy makers. This system provides stakeholders with different "windows" (such as mission, task differentiation, integration, articulation, coordination, performance) on their own collective practices which allow them to capture the potential synergy of their contributions to innovative performance.

A fundamental requirement if such approaches are to work is that stakeholders in a particular natural resource learn to appreciate that they have a common problem (Box 4). Extension has an important role to play here by making visible the interdependence between stakeholders and the extent to which the resource unit on which they depend has been destroyed by their uncoordinated action and the collective impact of their individual activities. It is within policy contexts thus made conducive for sustainable agriculture that technology development and extension can be especially effective.

Box 4. Resource Mapping by Farmers in Landcare Programme, Australia.

Landcare in Australia provides examples of learning to care for natural resources at higher system levels. Consider resource mapping. Farmers from a subcatchment (usually a subgroup of a Lancare group) are convened by the facilitator of the group to discuss the soils and their susceptibility to erosion. First, a soil typology is established by the farmers through field visits, digging soil pits, and so forth. After a suitable classification (which might deviate considerably from the official scientific one) has been agreed upon, farmers receive an air photo mosaic of the entire subcatchment with their property drawn in. They are also provided with a transparent overlay on which to map the soils and main features of their own properties.

These farmer maps are digitized and fed into GIS software, which allows the property resource maps to be combined into one consolidated subcatchment map. Following meetings to discuss the results, farmers agree on the resource map of the subcatchment and now have a firm grasp of the interaction between their property and the subcatchment. They also realize that vulnerable soils span several properties and that measures to prevent further soil erosion and solination require alignment of fences, roads, vegetation belts, and other features.

References

Altieri, M. A., & Yurjevic, A. (1992). Changing the agenda of the universities. *ILEIA Newsletter, 2*, 39.

Antholt, C. H. (1992). *Relevancy, responsiveness and cost-effectiveness: Issues for agricultural extension in the 21st century*. Washington, DC: Asia Region, Technical Department, World Bank.

Antholt, C. H. (1994). *Getting ready for the twenty-first century: Technical change and institutional modernisation in agriculture*. World Bank Technical Paper 217. Asia Technical Department Series. Washington, DC: World Bank.

Argyris, C., Putnam, R., & Smith, D. M. (1985). *Action science*. San Francisco and London: Jossey-Bass.

Attah-Krah, A. N., & Francis, P. A. (1987). The role of on-farm trials in the evaluation of composite technologies: The case of alley farming in southern Nigeria. *Agricultural Systems, 23*, 133-152.

Axinn, G. H. (1988). T & V (tragic and vain) extension. *INTERPAKS, INTERCHANGE, International Agriculture, 5* (3). Urbana: College of Agriculture, University of Illinois at Urbana-Champaign.

Bagadion, B. U., & Korten, F. F. (1991). Developing irrigators' organizations: A learning process approach. In M. M. Cernea (Ed.), *Putting people first* (2nd ed.). Oxford: Oxford University Press.

Bannister, M. E., & Nair, P. K. R. (1990). Alley cropping as a sustainable agricultural technology for the hillsides of Haiti: Experience of an agroforestry outreach project. *American Alternative Agriculture, 5*(2), 51-59.

Bawden, R. (1992, October). Creating learning systems: A metaphor for institutional reform for development. Paper presented at the joint conference of IIED/IDS. IDS, University of Sussex: IIED, London.

Bawden, R. (1994). *A learning approach to sustainable agriculture and rural development: Refelections from Hawkesbury (mimeo)*. Australia: Hawkesbury College.

Benor, D. (1987). Training and visit extension: Back to basics. In Rivera and Schram, *Agricultural Extension Worldwide* (p. 137-149). New York: Croom Helm.

Bentley, J. W., Rodríguez, G., & González, A. (1993). Science and the people: Honduran campesinos and natural pest control inventions. In D. Buckles (Ed.), *Gorras y sombreros: Caminos hacia la colaboración entre técnicos y campesinos*. El Zamarano, Honduras: Department of Crop Protection.

Brinkman, R. (1994). Recent developments in land use planning. Keynote address at the 75-year Anniversary Conference of the Wageningen Agricultural University. In L. O. Fresco et al. (Eds.), *Future of the land: Mobilising and integrating knowledge for land use options*. Chichester: John Wiley and Sons.

Brouwers, J., & Röling, N. (in press). Living local knowledge for sustainable development. Accepted for publication in D. M. Warren, S. Fujisaka, & G. Prain (Eds.), *Biological and cultural diversity: The role of indigenous experimentation in development*. London: Intermediate Technology Publications.

Bunch, R. (1991). *Low input soil restoration in Honduras: The Cantarranas farmer-to-farmer extension programme*. Sustainable Agriculture Programme Gatekeeper Series SA23. London: IIED.

Campbell, A. (1994). *Landcare: Communities shaping the land and the future*. Sydney: Allen and Unwin.

Chambers, R., Pacey, A., & Thrupp, L-A. (Eds.). (1989). *Farmer first: Farmer innovation and agricultural research*. London: Intermediate Technology Publications.

Checkland, P. B. (1981). *Systems thinking, systems practice*. Chichester: John Wiley.

Checkland, P., & Scholes, J. (1990). *Soft systems methodology in action*. Chichester: John Wiley and Sons.

Dobbs, T. L., Becker, D. L., & Taylor. (1991). Sustainable agriculture policy analyses: South Dakota on-farm case studies. *Journal of Farming Systems Research-Extension, 2*(2), 109-124.

Eckbom, A. (1992). *Economic impact assessment of implementation strategies for soil conservation: A comparative analysis of the on-farm and catchment approach in Trans Nzoia, Kenya*. Sweden: Unit for Environmental Economics, Department of Economics, Gothenburg University.

El Titi, A., & Landes, H. (1990). Integrated farming system of Lautenbach: A practical contribution toward sustainable agriculture. In C. A. Edwards, R. Lal, P. Madden, R. H. Miller, & G. House (Eds.), *Sustainable agricultural systems*. Ankeny: Soil and Water Conservation Society.

Engel, P. (1995). *Facilitating innovation: An action-oriented approach and participatory methodology to improve innovative social practice in agriculture*. Published doctoral dissertation, Agricultural University, Wageningen.

Faeth, P. (Ed.). (1993). *Agricultural policy and sustainability: Case studies from India, Chile, the Philippines and the United States*. Washington, DC: WRI.

Fresco, L. O., Stroosnijder, L., Bouma, J., & van Keulen, H. (Eds.). (1994). *The future of the land*. Chichester: John Wiley and Sons.

Fujisaka, S. (1991). Thirteen reasons why farmers do not adopt innovations intended to improve the sustainability of agriculture. In *Evaluation for sustainable land management in the developing world: Volume 2*. Technical Papers. IBSRAM Proceedings, *12*(2), 509-522. Thailand: IBSRAM.

Funtowicz, S. O., & Ravetz, J. R. (1993). Science for the post-normal age. *Futures, 25*(7, September), 739-755.

GTZ (1992). *The spark has jumped the gap*. Eschborn: Deutsche Gessellschaft für Technische Zusammenarbeit (GTZ).

Hanson, J. C., Johnson, D. M., Peters, S. E., & Janke, R. R. (1990). The profitability of sustainable agriculture on a representative grain farm in the mid-Atlantic region, 1981-1989. *Northeastern Journal of Agriculture and Resource Economics, 19*(2), 90-98.

Haverkort, B., van der Kamp, & Waters-Bayer, A. (Eds.). (1991). *Joining farmers' experiments: Experiences in participatory development*. London: IT Publications.

Howell, J. (Ed.). (1988). *Training and visit extension in practice*. London: Overseas Development Institute.

Hruska, A. J. (1993). Care policy promotes IPM. *International Ag-Sieve, 5*(6), 4-5.

Hussain, S. S., Byerlee, D., & Heisey, P. W. (1994). Impacts of the training and visit extension system on farmers' knowledge and adoption of technology: Evidence from Pakistan. *Agricultural Economics, 10*, 39-47.

ICRISAT. (1993). Will the pod-borer become the farmer's pall bearer? *SAT News*, p. 7-10.

Ison, R. (1990). *Teaching threatens sustainable agriculture*. Gatekeeper Series SA21. London: IIED.

Jiggins, J., & de Zeeuw, H. (1992). Participatory technology development in practice: Process and methods. In C. Reijntjes, B. Haverkort, & A. Waters-Bayer (Eds.), *Farming for the Future*. Netherlands: Macmillan and ILEIA.

Jordan, V. W. L., Hutcheon, J. A., & Glen, D. M. (1993). *Studies in technology transfer of integrated farming systems: Considerations and principles for development*. Bristol: AFRC Institute of Arable Crops Research, Long Ashton Research Station.

Kamp, K., Gregory, R., & Chowhan, G. (1993). Fish cutting pesticide use. *ILEIA Newsletter, 2*, 22-23.

Kang, B. T., Wilson, G. F., & Lawson, T. L. (1984). *Alley cropping: A stable alternative to shifting agriculture*. Ibadan: IITA.

Kenmore, P. (1991). *How rice farmers clean up the environment, conserve biodiversity, raise more food, make higher profits: Indonesia's IPM—A model for Asia*. Manila, Philippines: FAO.

Kerkhof, P. (1990). *Agroforestry in Africa: A survey of project experience*. London: Panos Institute.

Kingsley, M. A., & Musante, P. (1994, July). *Activities for developing linkages and cooperative exchange among farmers' organisations, NGOs, GOs and researchers: Case study of IPM in Indonesia*. Paper presented at IIED symposium, Brighton.

Kiss, A., & Meerman, F. (1991). *Integrated pest management in African agriculture*. World Bank Technical Paper 142. African Technical Department Series. Washington, DC: World Bank.

Krishna, A. (1993). *Watershed development in Rajasthan: The new approach*. Jaipur: Watershed Development, Government of Rajasthan.

Lal, R. (1989). Agroforestry systems and soil surface management of a tropical alfisol. I: Soil moisture and crop yields. *Agroforestry Systems, 8*, 7-29.

Liebhardt, W., Andrews, R. W., Culik, M. N., Harwood, R. R., Janke, R. R., Radke, J. K., & Rieger-Schwartz, S. L. (1989). Crop production during conversion from conventional to low-input methods. *Agronomy Journal, 81*(2), 150-159.

Lobo, C., & Kochendörfer-Lucius, G. (1992). *The rain decided to help us: An experience in participatory watershed development in Maharashtra State, India*. Ahmednagar: Social Centre.

Long, N., & Long, A. (Eds.). (1992). *Battlefields of knowledge: The interlocking of theory and practice in social research and development*. London: Routledge.

MALDM (1988-1994). Reports of catchment approach planning and rapid catchment analyses. Nairobi: Soil and Water Conservation Branch, Ministry of Agriculture, Livestock Development and Marketing.

Moris, J. (1990). *Extension alternatives in Africa*. London: ODI.

Matteson, P. C. (1992). "Farmer First" for establishing IPM. *Bulletin of Entomological Research, 82*, 293-296.

Matteson, P. C., Gallagher, K. D., & Kenmore, P. E. (1992). Extension of integrated pest management for planthoppers in Asian irrigated rice: Empowering the user. In R. F. Denno & T. J. Perfect (Eds.), *Ecology and Management of Planthoppers*. London: Chapman and Hall.

NRC (1989). *Alternative agriculture*. Washington, DC: National Research Council; National Academy Press.

Ornes, F. (1988). Community training centres for organic agriculture and appropriate technology, Dominion Republic. In C. Conroy & M. Litvinoff (Eds.), *The greening of aid.* London: Earthscan Publications.

Pimbert, M., & Pretty, J. N. (1994). *Parks, people, and professionals: Putting "participation" into protected areas management.* Geneva: UNRISD and WWF; London: IIED.

Pretty, J. N. (1994a). Policies that work for a more sustainable agriculture. London: IIED.

Pretty, J. N. (1994b). Alternative systems of inquiry for sustainable agriculture. *IDS Bulletin, 25*(2), 37-48. University of Sussex: IDS.

Pretty, J. N. (1995). *Regenerating agriculture: Policies and practice for sustainability and self-reliance.* London: Earthscan Publications.

Pretty, J. N., & Chambers, R. (1993). *Towards a learning paradigm: New professionalism and institutions for sustainable agriculture. IDS Discussion Paper* DP 334. Brighton: IDS.

Pretty, J. N., & Howes, R. (1993). *Sustainable agriculture in Britain: Recent achievements and new policy challenges.* London: IIED.

Pretty, J. N., & Shah, P. (1994). *Soil and water conservation in the 20th century: A history of coercion and control.* Rural History Centre Research Series No. 1. Reading: University of Reading.

Pretty, J. N., Thompson, J., & Kiara, J. K. (1994). Agricultural regeneration in Kenya: The catchment approach to soil and water conservation. *Ambio.*

Reijntjes, C., Haverkort, B., & Waters-Bayer, A. (1992). *Farming for the future: An introduction to low-external-input and sustainable agriculture.* The Information Centre for Low-External-Input and Sustainable Agriculture (ILEIA). London: Macmillan.

Roberts, N. (Ed.). (1989). *Agricultural extension in Africa.* World Bank Symposium. Washington, DC: World Bank.

Roche, C. (1992). *Operationality in turbulence: The need for change.* London: ACORD.

Rogers, E. M. (1962 and 1983). *Diffusion of innovations*: New York: Free Press.

Röling, N. (1992, October). *Facilitating sustainable agriculture: Turning policy models upside down.* Paper presented at IIED/IDS Conference, Brighton.

Röling, N. (1993). Agricultural knowledge and environmental regulation: The crop protection plan and the Koekoekspolder. *Sociologia Ruralis, 33*(2), 212-231.

Röling, N. (1994a). Platforms for decision making about ecosystems. In L. Fresco (Ed.), *The future of the land.* Chichester: John Wiley and Sons.

Röling, N. (1994b). Creating human platforms to manage natural resources: First results of a research programme. In *Proceedings of the International Symposium on Systems Oriented Research in Agriculture and Rural Development* (p. 391-395). Montpellier, France, November 21-25, 1994.

Röling, N., & Engel, P. (1989). IKS and knowledge management: Utilizing indigenous knowledge in institutional knowledge systems. In D. Warren et al. (Eds.), *Indigenous knowledge systems: Implications for agriculture and international development.* Studies in Technology and Social Change, 11. Ames: Technology and Social Change Program, Iowa State University.

Röling, N., & Jiggins, J. (1993). Policy paradigm for sustainable farming. *European Journal of Agricultural Education and Extension, 1*(1), 23-44.

Röling, N., & Van de Fliert, E. (1994). Transforming extension for sustainable agriculture: The case of integrated pest management in rice in Indonesia. *Agriculture and Human Values, 11*(2, 3), 96-108.

Russell, D. B., & Ison, R. L. (1991, April). *The research-development relationship in rangelands: An opportunity for contextual science.* Plenary paper presented at the 4th International Rangelands Congress, Montpellier, France.

Scoones, I., & Thompson, J. (Eds.). (1994). *Beyond Farmer First.* London: IT Publications.

Sen, C. K. (1993). Nepal: Group extension. *Rural Extension Bulletin, 3,* 17-23. University of Reading, UK.

Shah, P. (1994). Participatory watershed management in India: The experience of the Aga Khan Rural Support Programme. In I. Scoones & J. Thompson (Eds.), *Beyond Farmer First.* London: IT Publications.

Sriskandarajah, N., Bawden, R. J., & Packham, R. G. (1991). Systems agriculture: A paradigm for sustainability. *Association for Farming Systems Research-Extension Newsletter, 2*(2), 1-5.

Stock, T. (1994). *The impact of integrated pest management farmer field schools in the Cordillera region of the Philippines.* Unpublished master's thesis, University of Reading.

SWCB (1994). The impact of the catchment approach to soil and water conservation: A study of six catchments in Western, Rift Valley and Central provinces of Kenya. Nairobi: Ministry of Agriculture, Livestock Development and Marketing.

UNDP (1992). *The benefits of diversity: An incentive towards sustainable agriculture.* New York: UNDP.

Van de Fliert, E. (1993). *Integrated pest management: Farmer field schools generate sustainable practices: A case study in central Java evaluating IPM training.* WU Papers 93-3. Published doctoral dissertation. Wageningen: Agricultural University.

Van de Fliert, E., van Elsen, K., & Nangsir Soenanto, F. (1993). Integrated rat management: A community activity. Results of a pilot programme in Indonesia. *FAO Plant Protection Bulletin, 41*(3).

Van der Poel, M. M., & Van Woerkum, C. J. M. (1994). *De positionering van communicatie in Beleid.* Deelrapport 1 (literatuuronderzoek) en Deel 2 (Weergave van bijeenkomsten, conclusies van het onderzoek en aanbevelingen voor vervolgonderzoek). Wageningen: Agricultural University, Department of Communication and Innovation Studies, carried under a contract from the Government Information Office.

van der Werf, E., & de Jager, A. (1992). Ecological agriculture in South India: An agro-economic comparison and study of transition. Den Haag: LEI and ETC Foundation, Leusden.

Van Weeperen, W., Röling, N., Van Bon, K., & Mur, P. (1995). Introductie geïntegreerde akkerbouw: Het veranderingsproces. Ervaringen van akkerbouwers bij het omschakelen naar een geïntegreerde bedrijfsvoering. Lelystad: Informatie en Kenniscentrum Landbouw, afdeling AGV.

Vereijken, P. (1992). A methodic way to more sustainable farming systems. *Netherlands Journal of Agricultural Science, 40,* 209-223.

Vereijken, P., Wijnands, F., Stol, W., & Visser, R. (1994). Progress reports of research network on integrated and ecological arable farming systems for EU and associated countries. Wageningen: DLO Institute for Agrobiology and Soil Fertility.

Warren, D. (1991, May). The role of indigenous knowledge in facilitating a participatory approach to agricultural extension. Paper presented at the International Workshop on Agricultural Knowledge Systems, Bad Boll, Germany.

Winarto, Y. (1993). Farmers' agro-ecological knowledge construction: The case of integrated pest management among rice farmers on the north-west coast of West Java. *Rural People's Knowledge, Agricultural Research and Extension Practice, 1* (3), London: IIED.

Young, A. (1989). *Agroforestry for soil conservation.* Wallingford: CAB International.

Yuryevic, A. (1994, July). *Community-based sustainable development in Latin America: The experience of CLADES.* Paper presented at IIED Symposium, Brighton.

Zijp, W. (1993). Being a good communicator doesn't solve all of extension's problems. *Development Communication Report,* 80, 20.

Chapter 21
Establishing and Strengthening Farmer Organizations

Shankariah Chamala and P. M. Shingi

Historical Perspective

Historically, extension has mainly involved technology transfer, with the village extension worker (VEW) transferring knowledge from research stations to farmers by using individual, group, and mass media methods. More recently, extension has been asked to play a "technology development role" by linking research with community group needs and helping to facilitate appropriate technology development. It is in the historical context that many government agencies developed national policies for rural development and designed a policy framework to help rural people become organized so that the delivery of services could be channelled through the various types of farmer organizations or groups. Well-meaning policies also provided blueprint structures for farmer organizations (FO) in the form of cooperatives and commodity organizations in order to provide various input, marketing, and educational services to the farmers. Targets for forming groups and farmer organizations were given to VEWs without training them properly in the theory and principles of community organization. VEWs did not have many skills and not much experience in the process of establishing these organizations. Some countries such as Thailand had VEWs for establishing cooperatives in rural areas, while other countries like Malaysia and Indonesia developed "nucleus estates" for small rubber producers where smallholders bring their rubber to process it. India has introduced dairy cooperatives with some success.

However, because VEWs in many countries lack knowledge, skill and blueprint policies, and high targets, they resorted to shortcut methods to establish farmer organizations and groups. Many VEWs presented government policies in an oversimplified way to rural communities, suggesting that unless they are organized into cooperatives or associations or groups, they will not get government subsidies or access to credit and technical services. As a result, several FOs were established overnight on paper. Many FOs remained active during the period that government subsidies were distributed, but did not actively create cooperatives or partnerships and mobilize local resources to help achieve agricultural development. Mostly the elite of rural communities captured all of the services and resources, while the poor and women were left out or received little benefit. Very few attempts were made to develop the management capacities of FO leaders, their members, and VEWs. Community organization and facilitation skills were not part of staff training programmes.

The traditional approaches to organizing farmers and forming cooperatives need to be revised to meet the following development challenges of the twenty-first century:

- The increasing absolute and relative poverty in many countries
- The degradation of natural resources such as soil, water, flora, and fauna
- The low involvement of women in health, agriculture, and other development programmes
- The poor health and education facilities in rural areas
- The increasing sociopolitical unrest among the communities

New Roles for Extension

Several extension roles can be conceptualized to help rural communities get organized, but we will focus on four important roles.

Empowerment Role

The empowerment role can be a cornerstone of the new approach to extension. Extension personnel need to develop a new philosophy where their role is to help farmers and rural communities organize themselves and take charge (empowerment) of their growth and development. Telling adults what to do provokes reaction, but showing them triggers the imagination, involving them gives understanding, and empowering them leads to commitment and action (Chamala, 1990).

The term *empower* means to enable, to allow, or to permit and can be viewed as both self-initiated and initiated by others. For VEWs, empowering is an act of helping communities to build, develop, and increase their power through cooperation, sharing, and working together. The power in empowerment comes from releasing the latent energy hidden in the community and building collective actions for the common good, rather

Shankariah Chamala is an Associate Professor of agricultural extension and management in the Department of Agriculture, The University of Queensland, St. Lucia, Australia. **P. M. Shingi** is a Professor in the Indian Institute of Management, Ahmedabad, India.

than from merely redistributing power from the haves to the have-nots. FOs can help harness this synergetic power for its members' survival, growth, and development. Empowered FOs can act as convergent points or platforms for solving local problems and mobilizing human and financial resources for sustainable development (also see Manalili, 1990).

Community-Organizing Role

Village extension workers must learn the principles of community-organizing and group management skills (Chamala & Mortiss, 1990) in order to help the community, especially the poor or weaker sections, to organize itself for development. Understanding the structures, by-laws, rules, and roles will help leaders to plan, implement, and monitor their programmes and to perform this new role effectively. Skills in conflict resolution, negotiation, and persuasive communication help VEWs to develop FO leaders and members.

Human Resource Development Role

The human resource development approach empowers people and gives new meaning to all other roles. Development of technical capabilities must be combined with management capability. Training modules are now available (Chamala & Mortiss, 1990; Mortiss & Chamala, 1991) to help develop individual and group management skills. The entire philosophy of human capacity building is to encourage rural communities to understand their personal and group styles of managing themselves and to improve their planning, implementation, and monitoring skills.

Problem-Solving and Education Role

Problem solving is an important role, but the role is changing from prescribing technical solutions to empowering FOs to solve their own problems. This is achieved by helping them to identify the problems and seek the right solutions by combining their indigenous knowledge with improved knowledge and by using their resources properly. Similarly, there is a shift in the education role from lectures, seminars, and training to learning by doing and encouraging farmers and FOs to conduct experiments and undertake action-learning projects.

Farmer Organizations

Types of Organizations

Farmer organizations can be grouped into two types: one is the community-based and resource-orientated organization; the other is the commodity-based and market-orientated organization.

Community-Based, Resource-Orientated Farmer Organizations. This type could be a village-level cooperative or association dealing with inputs needed by the members, the resource owners, to enhance the productivity of their businesses based on land, water, or animals. These organizations are generally small, have well-defined geographical areas, and are predominantly concerned about inputs. However, the client group is highly diversified in terms of crops and commodities.

There are many primary-level agricultural cooperatives in the developing world, but the majority of them have been financially vulnerable and ineffective. Strategies have been developed to strengthen these organizations (see the section on how to strengthen existing farmer organizations). This group of organizations can generate income from the sale of inputs and outputs. The income can then be put back into the organization by spending it on extension, data generation, business planning, and administration. It is essential to have professional and honest management with constant monitoring and periodic rounds of evaluation (Gupta, 1989).

Commodity-Based, Market-Orientated Farmer Organizations. These organizations specialize in a single commodity and opt for value-added products which have expanded markets. They are designated as output-dominated organizations. Not specific to any single community, they can obtain members from among the regional growers of that commodity who are interested in investing some share capital to acquire the most recent processing technology and professional manpower. These FOs are generally not small and have to operate in a competitive environment. Research, input supply, extension, credit, collection of produce, processing, and marketing are all integrated to maximize the returns on the investments of the members who invested in the collective enterprise. Several successful cases are found in India, such as Anand Milk and other dairy FOs.

The rate of success of these organizations is determined by their capacity to arrange for major investments and a continuous flow of raw materials. This requires the competent and convincing management of both enterprise-related and member-related aspects. The profits generated are used to provide supplementary and supportive services at reduced cost to encourage members to use them. To do this requires a high calibre of representative and enlightened leadership from among the grower members. It is a challenging and demanding task to conceive, design, build, and nurture this type of FO.

VEWs can consult and work with other governmental agencies and nongovernmental organizations. However, each farmer organization will need to define its own BASE (basic activity sustaining the enterprise). In India, or for that matter in many developing countries, there is tremendous potential for expansion of commodity-based FOs. One rule of thumb suggests that any commodity which accounts for more than 50 per cent of the costs of the raw material can be considered for value-added processing by a farmer organization (V. R. Gaikwad, personal communication, 1994).

Issues Influencing Participation in FOs

The following issues will influence the extent of participation:

- The degree of the farmer's dependence on the outputs of the organized activity.
- The degree of certainty of the availability of the outputs.
- The extent to which the outputs will be available only as a result of collective action.
- The extent to which the rewards associated with the collective action will be distributed equitably.
- The extent of availability of rewards within a reasonable time frame.
- The extent to which the rewards are commensurate with the costs associated with continued participation (Shingi & Bluhm, 1987).

The Role of Extension in Farmer Organizations

The role of extension will vary with the role of the organization, the sectors in which the organization operates, the services offered, and the organizational form used. In community-based organizations, extension is used as a supplementary or supportive activity to realize the objectives of the BASE function of the organization. In commodity-based organizations, extension is integrated with all the other aspects of the organization to maximize the returns on the investment of the collective enterprise. Extension is taken seriously by both the organization and its members because both derive direct and measurable benefits from it.

The following issues need to be considered when developing the extension role, especially for farmer organizations which are set up to specialize in the extension function:

- Is there an identifiable need for extension in specific commodities in the area covered by the FO?
- Would the FO be able to generate enough revenue from the extension activity alone (with farmers willing to pay for these services) to meet the FO's expenses and to provide satisfactory rewards to its members for their monetary and nonmonetary contributions? It will be important to anticipate the high potential for unresolved conflict over the issues of equity and charges for extension services.
- How sustainable will the extension activity be over time, and therefore how sustainable will the organization be? It is possible for advice to be converted to freely transmittable knowledge which can be transferred to anyone without payment. The cost of the extension advice limits access to this knowledge. Therefore, the revenue earned from the extension activity could decline, especially with a client group which has a low resource base and is primarily concerned with a subsistence economy.
- Can the advice given be actually put into practice and produce tangible benefits to the FO members? The FO would need to control or arrange for the supply of necessary inputs to ensure this; otherwise, the extension organization will fail, as has happened in the past in many developing countries. This means that the extension function needs to be integrated backwards with research recommendations and forward with the supply of inputs.
- The organization will need to provide specific information in addition to the general information available from research centres. To do this and to survive, the organization will need a research linkage with government and university research institutions. It is beneficial if the FO can employ qualified and committed scientists who have active contacts in research organizations or who can act as consultants to groups of members. This would increase the cost of extension advice to members if the FO is supposed to be financially self-supporting.
- It is necessary to appreciate that "extension markets" are governed by factors such as agroclimatic variations, infrastructure development, and the strength of market forces. FOs operating in desert regions, single-crop rainfed areas, and predominantly irrigated areas will have different occupational and extension needs; therefore, variable response patterns to extension have to be anticipated (Gupta, 1981, 1985). Similarly, FOs operating in food-deficit and food-surplus stages will have different roles, expectations, and returns.

Steps in Establishing Farmer Organizations

Some of the principles and steps for social action models, community organizations, and management are used in designing the simplified step-wise approach in establishing FOs (Chamala, 1990). Care needs to be taken in selecting a person or a group of people who will take the lead in establishing community-based or commodity-based farmer organizations. This person—from another FO or government or other organization, such as a village extension worker—needs to be acquainted with and convinced of how the FO can be used to increase the income of small producers. In the following steps, the term VEW will be used to refer to the person or group of people selected to establish the FO.

Step 1. Understanding the Village Community

VEWs should enter the community with an open mind and understand the community structure. They must understand the community power structure, problems, and opportunities for development. Walking around and talking to key people can help ease their entry into the community. VEWs should also develop an under-standing of the entire community, including the poor and marginal farmers and women.

Some of the techniques that can help to understand the community are using key informant techniques, participating in the community meeting and tea stalls, and walking around. Understanding the ownership of community resources and people's attitude, knowledge, and skills in the development of agricultural production will help the VEW learn about the community. The VEW can also understand the situation in more depth by collecting both qualitative and quantitative information on the levels of income and productivity, costs of cultivation, post-harvest losses, output utilization, and the likelihood of making striking improvements to each of these factors.

Step 1 is crucial and cannot be done in a hurry. VEWs must spend some time (up to a month or so) to get the feel of the village community structure, politics, groups, and past experience in FOs.

Step 2. Identifying Potential Leaders in the Community

The usual tendency is for the VEWs to talk to formal leaders and commercial sections of the community. By using the sociometry method (see Box 1 for details on the methodology), they can identify potential leaders. However, it is also important to be sensitive to the leadership structures operating in that culture and to the knowledge and skills needed to be a successful leader of a farmer organization.

In some cultures, it is wise to search for and contact middle-aged leaders of the area who are not too young or too old. Preferably, they should come from better-off families, have social status and respect, and be from farming households. Case studies of leaders who have established farmer organizations show that these leaders initially made considerable sacrifices, experienced financial losses, and had to be supported by their families. These leaders also had to struggle and negotiate with bankers, bureaucrats, politicians, critics, and others including their own farmer members (Seetharaman & Shingi, 1992). This family background and grooming helps these leaders to deal with situations which they will have to face with confidence and without being cowed. Leadership of an FO is not a job for a sincere but ordinary farmer.

Box 1. Sociometry.

Ask a few people informally the following questions:

1. I am new to the village. Could you tell me three names of leaders of the village? (These are positional leaders who are currently holding leadership positions.) Write down their names or remember them and write them down later.

   Positional Leaders

   ....
   ....
   ....

2. Name three leaders whose opinions or ideas on agriculture have influence on their friends and community leaders. (Make sure to get these names from each section of the society: the larger, medium, and small farmers because opinion leaders exist in each socioeconomic strata of the community.)

   Opinion Leaders

   Higher ....
   Medium ....
   Lower or poorer ....

Step 3. Talking to the Identified Leaders and Seeking Cooperation from Other Agencies

VEWs might talk to these leaders on general agricultural development and get ideas and information on Fos in the village. They might also seek cooperation from government agencies and especially from NGOs (if there are any in the area) to help establish FOs and support them in achieving success.

Does the community have a farmer organization? If not, is there a need for such an organization? If the community has an FO, what is its structure and history of performance. How could the FO play a role in village or community development? VEWs can canvass for ideas from the community and add their own ideas on the need for and the role of FOs in the entire process of broad-based agricultural development. They can explain how FOs in other communities helped them in development. It is important to provide facts and figures to convince potential leaders of the possibilities and approaches for increasing the income of a sizeable number of farmers and for contributing to the economic

> **Box 2. Steps in Developing an Organizational Structure**
>
> The core group leaders' main goal is to understand the appropriate organizational structure, composition, and working rules for the efficient management of their farmer organization.
>
> **Stage 1 Leaders should secure the relevant guidelines of other FOs and study them carefully.** An FO can be initiated by the government or by farmers and the community. If it is government ini-tiated, the leaders should obtain copies of guidelines and rules. If it is community initiated, they should talk to other FOs and get their guidelines and constitution. The core group should study them carefully and discuss how they can fit into their community needs.
>
> **Stage 2. Leaders should then draw up a tentative organizational structure and working rules for their farmer organization.** They should consider various models or types of organizations that serve their special needs for agricultural development and discuss them with other experienced leaders from that village or neighbouring villages. The structure should serve the functions. Are subgroups needed to achieve each task?

development of the region. Good FO leaders might also receive political advantages through satisfied farmers and people living in the region. The leaders' support should be secured and key people encouraged to consult among themselves about the pros and cons of forming FOs in the community.

Step 4. Helping Local Leaders to Call Community Meetings.

VEWs can help enthusiastic local leaders to call for community meetings. Sometimes more than one meeting may be needed to discuss the need for and the role of FOs in agricultural development. FO leaders from neigh-bouring villages can be invited to speak at these meetings. Farmer-to-farmer information exchange helps them. Sometimes smaller meetings can be held for low-resource (small and marginal) farmers and minority groups. They may prefer having a separate FO to serve their specific needs. There is no harm in having more than one FO in a village.

It is important that producers from all sectors in the village participate in these meetings because the success of an enterprise-based FO depends on the volume of raw material procured from member producers. Every small or big producer contributes to this volume. The volume itself is more important than the socioeconomic status of the supplier. Prospective members need to be convinced that everyone benefits in proportion to his or her contributions, not just the big farmers, as is widely perceived.

Step 5. Nominating Core Group Leaders to Develop or Establish the FO

From the community meetings, core group leaders are elected or nominated to design the FO with further community consultation. In some cultures, however, this approach may lead to the appointment of leaders who are unable to cope with the complexities of a farmer organization, as mentioned in step 2. If leaders are elected or nominated, VEWs should consider the issues raised in step 2 when assessing each candidate for election or nomination.

In situations where it is more appropriate to appoint a leader, the search for a suitable person is critical and requires time and patience. It is also critical that the selected leader be involved in the search for a potential agro-based enterprise to act as a BASE (basic activity sustaining the enterprise) for economic development. This process requires time, patience, and interactive and intellectual inputs. The VEW can help the appointed FO leader in this process.

Step 6. Developing an Organizational Structure for the FO

The VEW can help the core group of leaders in developing an organizational structure for their farmer organization. In the past, the "blueprint" approach was taken without understanding the function that the FO's structure plays in its performance. Group discussions help to highlight the need for careful planning. See Box 2 for details.

The structure should serve the organization's functions and goals. Understanding various types of FOs is useful. Should they be commodity-based organizations, cooperatives, partnerships, groups, or syndicates? Should they be multipurpose? Should there be one FO for the entire village or several to cater to the needs of special-interest groups (low-resource farmers, women, craftsmen, small businesses, and the like)? Should they have subgroups and an advisory committee? It is the farmers' organization, so they must go through the process explained in Box 2 and design it carefully by describing roles, responsibilities, rewards, and punishments for the people who perform tasks in the FO. At this stage, the VEW should, as far as possible, play a passive role because the leaders are the ones who are building the FO. NGOs also may share their experience and help leaders to develop an organizational structure.

Several less exacting chores also need to be taken care of at this stage. These include locating premises and negotiating for land, money, technology, personnel, construction, and infrastructure services such as power lines and telephones (if available).

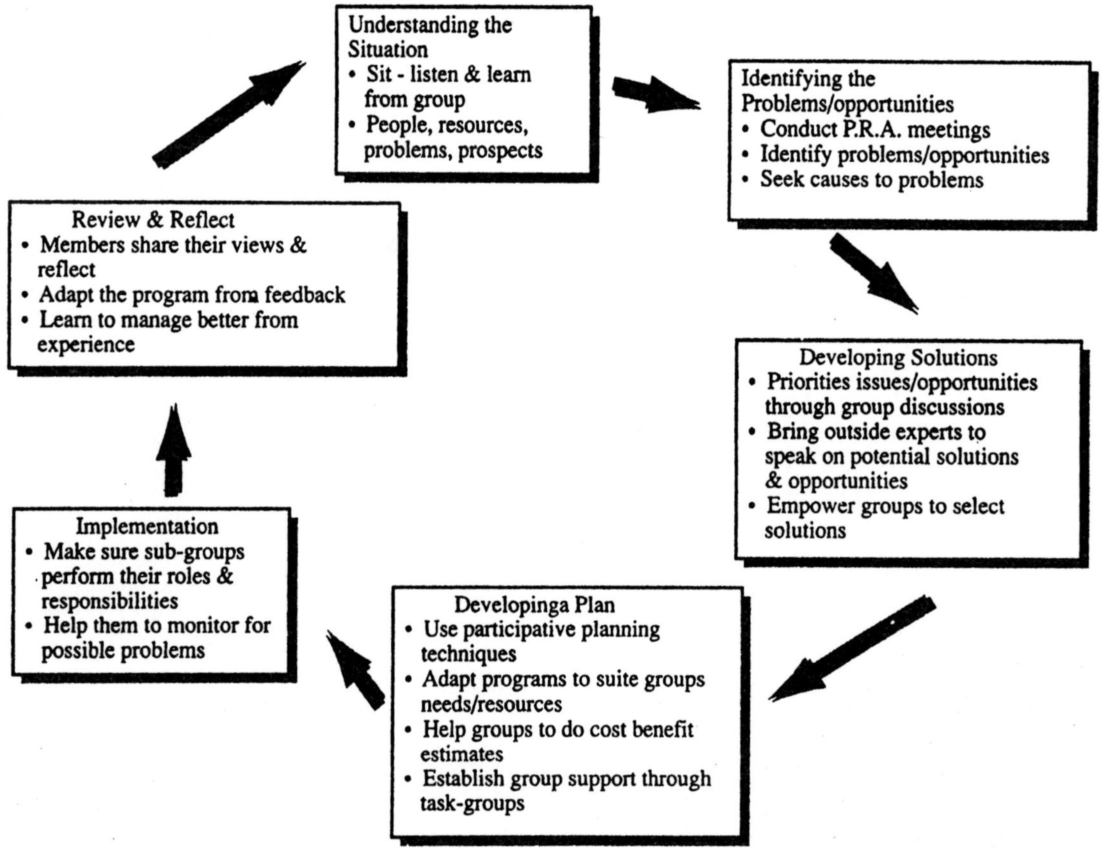

Figure 1. Detailed six-stage PAM planning cycle. Source: Chamala (1995a).

Step 7. Developing the FO's Management through Education and Action Learning

An essential part of community empowerment is to help educate the leaders and members in management principles covering planning, implementing, and monitoring their projects and programmes. The following empowerment methods may be useful:

- *Educating*. Organize formal and informal learning activities.
- *Leading*. Help the leaders to lead and to learn from their actions by reflection as a team.
- *Mentoring and supporting*. Help the members intial ly by mentoring or supporting them in their planning and implementation stages.
- *Providing*. Obtain the services of other stake holders, FOs, and VEWs in providing various services to nurture the FO in the early stages of development.
- *Structuring*. Help the FO to structure its meetings and various participative planning activities and to learn from their experience through reflection.
- *Actualizing*. Help them to reflect on the process of managing their FO. Learning by doing can help them in self-actualization.(For more details on empowerment, see Vogt and Murrell (1990).

Step 8. Gearing up for Action

In this step, FOs examine their action plans, and task groups are set up to mobilize human and financial resources. Understanding the participative action management (PAM) planning process is useful (for details see Chamala, 1995a). A detailed six-stage PAM planning cycle (Figure 1) can help FO leaders in designing an inclusive and participative planning process. This is the stage to start considering the timing, scale, and content of the extension and research input of the FO.

Step 9. Implementing Selected Projects

In this step, the village extension worker can help the FO leaders implement the projects they have chosen.

The following process may be useful:

- Start implementing the selected project(s).
- Secure resources and allocate tasks.
- Develop a calendar of activities to achieve the goal.
- Develop monitoring processes for reflecting on events and activities regularly, either formally or informally.
- Ask committee members to meet to discuss actions periodically and report to general members regularly to keep them informed and involved.

Step 10. Monitoring and Evaluating the FO's Progress

Usually evaluation is done annually to meet formal requirements. But VEWs can encourage FOs to reflect on their activities more frequently so that they learn and improve their management skills. They need to watch for people who want to take over the FOs for their self-interest. It is important to take action against anynegative influences. These monitoring or reflection processes help strengthen FOs and avoid self-defeating problems. Learning organizations are created through collective reflection and openness on financial and the other managerial matters (Senge, 1990).

The success of the farmer organization can be evaluated by measuring tne increase in the members' productivity, the increase in their net income, and the net reduction in the cost of cultivation due to bulk purchasesof inputs by the organization. It is essential to conduct monitoring and periodic evaluations.

How to Strengthen Existing FOs

Many farmer organizations that go through a high activity phase become inactive or defunct over a period of time. This is a normal process for many groups or FOs. VEWs can help the community to understand the reasons and causes for decline.

By reviewing the literature on groups and management, Chamala (1995b) identified twenty-six factors that factors, (2) service agency factors, (3) community factors, and (4) other external factors.

VEWs need to understand the factors that influence group or organization effectiveness or success (see details in Box 3). Many FOs have failed because of corruption, mismanagement, conflict, and lack of clear goals. It is important that VEWs understand these forces that influence their functioning. Several steps are suggested below to revive or revitalize existing FOs.

Box 3. Factors Influencing Community Group Effectiveness.

Group (FO) Internal Factors. At least ten factors are identified under this category:

1. Group composition
2. Group structure and size
3. Group atmosphere
4. Cohesion
5. Group standards and norms
6. Leadership styles
7. Balance between group maintenance needs, individual needs, and task needs
8. Level "group think" characteristics in the group
9. Development phase of the group
10. Group culture: empowering or controlling or a balance

Service Agency Factors. Government and nongovernment agencies can influence the effectiveness:

11. Technical capabilities of extension staff
12. Staff's "people skills" in managing groups
13. Staff attitudes and commitment to groups
14. Types of planning methods used: directive or participative, top-down or bottom-up, or balance of methods to maximize participation
15. Means or ends distinction: some groups are formed as means for development, while others are formed to harvest government subsidies. A group could get help, but it needs to mobilize its resources too.
16. Support for field extension officers

Community Factors. Groups and organizations are part of the community in which they exist. Hence the community influences a group's success.

Step 1. Understanding the Village Community and Gaining Their Trust

This step is similar to step 1 for establishing farmer organizations. The key factor here is for the VEW to gain the community's trust in order to help them develop.

Step 2. Talking to Leaders about Why FOs Are Less Active

Getting the multiple perspectives of identified leaders in the community informally is the next step. Putting pieces of the jigsaw puzzle together to understand why FOs are inactive requires diplomatic and detecting skills. The historical perspective of the FOs and various local factors that influenced their inaction must be obtained. The factors influencing community group effectiveness (Box 3) can be used as a guideline by selecting the relevant factors and ignoring the irrelevant ones.

Step 3. Seeking Ideas on Strengthening and Revitalizing Community FOs

Again, VEWs can talk to key leaders and seek their ideas on how to strengthen or revitalize the existing FOs or how to create new ones. It is the community's organization for their development, so the leaders' opinions and support should be obtained.

Step 4. Encouraging Local Leaders to Call Community Meetings

The VEW can help the leaders in setting up a community meeting to strengthen FOs or to form a new one.

The VEW can unobtrusively provide the format and make suggestions on how to conduct meetings and how to strengthen FOs.

Step 5. Establishing a Core Group of Leaders to Draft a Proposal

A core group of leaders or a steering committee will further deliberate, using the suggestions made by the members to design and develop a strategy to strengthen or revive the farmer organization.

Step 6. Getting Comments on and Community Endorsement of the Strategy

Leaders should be careful not to let the meetings and the community mood for "head hunting" or "apportioning the blame for the FO's inefficiency" to take control. It is necessary to guide the community in a positive direction to develop working strategies. Several methods are available, such as SWOT (strengths, weaknesses, opportunities, and threats) analysis, prioritizing, and action-planning methods. See Chamala and Mortiss (1990) and Carman and Keith (1994) for techniques on participative planning and community consultation.

Useful procedures are outlined in steps 7 to 10 in the section on establishing farmer organizations. It is important to help the community to understand that it is in everybody's interest to strengthen FOs with win-win strategies. Helping them to see beyond petty politics and personal jealousies and to develop inclusive principles for development is one of the main roles for extension personnel. Acquiring facilitation skills is important.

Policy Issues in Establishing and Strengthening Farmer Organizations

Several macro-level policy issues influence the establishment of farmer organizations. Some were discussed earlier, but two issues need clarification and discussion (see also Korten, 1989; Esman & Uphoff, 1984; Burkey, 1993; Garforth, 1993).

Increased Demand for Services

Will strong farmer organizations create more demand for services, and if so, how can this issue be resolved? Some policy makers believe that strong farmer organizations could create more demand on research and extension. This is partly true, but if the overall policy framework emphasizes empowerment, self-help, or user-pay systems, then more local resources could be mobilized and help increase the correct way of using current natural resources on a sustainable basis. Several low-resource farmer groups working with other stakeholders like banks, agribusiness, NGOs, private consultants, and religious and philanthropic agencies will bring more resources to the development process. Retired professional teachers, nurses, and agricultural scientists could be encouraged to work as volunteers with FOs. In many developed countries, volunteers are a major force in providing services to the community.

Competition for Services

Will organized low-resource farmers compete for extension services with the existing commercial farm sector, and if so, how can this issue be resolved? In principle, farmer organizations with full empowerment reduce the pressure on "routine extension" activities, which take up a lot of the VEW's time. For example, "Landcare" groups in Australia and some successful FOs in Malaysia, the Philippine, Thailand, India, and Africa are helping other farmer groups or farmer organizations to be effective not for egalitarian or welfare reasons, but because they see that the commercial sector is dependent on low-resource farmers or upland farmers in solving their salting, erosion, water quality, and pest and weed management problems. Ecologically, the commercial sector is dependent on resource-poor or other minority tribes and groups for their survival. Just as in farmer-to-farmer extension methods, VEWs should help link the commercial sector with the low-resource farm sector so that win-win projects can be developed.

Similarly, the commercial sector needs peace and prosperity to avoid social unrest and to stop thefts and other problems. VEWs can play a positive role in developing a common or shared vision for sustainable development. Again, several techniques like future research, problem census, and SWOT analysis will help develop a shared vision for the commercial sector as well as the low-resource farming sector or minority groups. Collective action is crucial for survival and sustainable development.

Community organization is essential for rural development. One should not take a blueprint approach, but rather a learning-process approach within an action-learning framework (Korten, 198).

References

Burkey, S. (1993). *People first: A guide to self-reliant participatory development*. London: Zed Books.

Carman, K., & Keith, K. (1994). *Community consultation techniques: Purposes, processes and pitfalls—A guide for planners and facilitators*. Brisbane: Department of Primary Industries, Queensland.

Chamala, S. (1990). Establishing a group: A participative action model. In S. Chamala & P. D. Mortiss, *Working together for Landcare: Group management skills and strategies*, p. 13-38. Brisbane: Australian Academic Press.

Chamala, S. (1991). Establishing a group: A model for participative action management. In P. D. Mortiss & S. Chamala, *Group management skills for Landcare: A trainers' guide*, p. 33-60. Brisbane: Australian Academic Press.

Chamala, S. (1995a). Overview of participative action approaches in Australian land and water management. In S. Chamala & K. Keith (Eds.), *Participative approaches to Landcare: Perspectives, policies and programs*. Brisbane: Australian Academic Press.

Chamala, S. (1995b). Group effectiveness: From group extension methods to participative community Landcare groups. In S. Chamala & K. Keith (Eds.), *Participative approaches to Landcare: Perspectives, policies and programs*. Brisbane: Australian Academic Press.

Chamala, S., & Mortiss, P. D. (1990). *Working together for Landcare: Group management skills and strategies*, p. 368. Brisbane: Australian Academic Press.

Esman, M. J., & Uphoff, N. T. (1984). *Local organizations: Intermediaries in rural development*. Ithaca, NY: Cornell University Press.

Garforth, C. (1993). Rural people's organizations and extension communication in northern Thailand. *Journal of Extension System, 9*(2), 33-64.

Garkovich, L. E. (1989). Local organizations and leadership in community development. In J. A. Christenson & J. W. J. R. Robinson (Eds.), *Community development in perspective*, p. 196-218. Ames: Iowa State University Press.

Gupta, A. K. (1981). Small farmers' credit constraints: A paradigm. *Agricultural System, 7*(4), 157-161.

Gupta, A. K. (1985). On organizing equity: Are solutions really the problem? *Journal of Social and Economic Studies, 4*, 295-312.

Gupta, V. K. (1989). *Guide to monitoring and evaluation of small farmers' cooperatives in Asia*, p. 121. Rome: FAO.

Gupta, V. K., & Gaikwad, V. R. (1986). *Guide to management of small farmers' cooperatives*, p. 172. New Delhi: Concept Publishing Company.

Korten, D. C. (1980). Community organization and rural development: A learning process approach. *Public Administration Review, 40*, 480-510.

Korten, D. C. (1989). *Getting to the 21st century: Voluntary action and the global agenda*. Manila: Bookmark.

Manalili, A. G. (1990). *Community organizing for people's empowerment*. Manila: Kapativan-Kaunlaran Foundation.

Mortiss, P. D., & Chamala, S. (1991). *Group management skills for Landcare: A trainers' guide*, p. 407. Brisbane: Australian Academic Press.

Noppakun, M., & Chamala, S. (1992). Adoption of livestock practices by successful and less successful extension groups of small farmers. In P. W. Daniels, S. Holden, E. Lewin, & S. Dadi (Eds.), *Livestock services for smallholders: A critical evaluation*. Proceedings of a seminar held in Yogyakarta, Indonesia, November 10-15, 1992. Jakarta: Directorate General of Livestock Services, Indonesia.

Seetharaman, S. P., & Shingi, P. M. (1992). *Agribusiness cooperatives*, p. 210. ICA Domust Trust. New Delhi: Oxford & IBH Publishing Compan.

Senge, P. M. (1990). The leader's new work: Building learning organizations. *Sloan Management Review*, p. 7-22.

Shingi, P. M., & Bluhm, L. H. (1987). Participation in irrigation projects: Changing patterns in northwestern India. In H. K. Schwarzweller (Ed.), *Research in Rural Sociology and Development. Volume 3*, p. 65-84. Greenwich: Jai Press.

Chapter 22
Privatizing Agricultural Extension

William M. Rivera and John W. Cary

Diverse agricultural extension funding and delivery arrangements have been undertaken since the mid-1980s by governments worldwide in the name of "privatization." This chapter reviews these actions and their implications.

When agricultural extension is discussed, *privatization* is used in the broadest sense—of introducing or increasing private sector participation, which does not necessarily imply a transfer of designated state-owned assets to the private sector. In fact, various cost-recovery, commercialization, and other so-called privatization alternatives have been adopted to improve agricultural extension.

This chapter is organized into five main sections. The first introduces the major forces catalysing governments to structure new extension funding and delivery arrangements, or what is generally referred to as "privatization." The second section discusses the resultant strategies for change. The third discusses and categorizes the major privatization types of institutional changes in funding and delivering agricultural extension. The fourth section considers the rationales underlying extension's institutional changes, and the fifth examines the implications of these changes. The chapter concludes with the recommendation that countries examine the institutional changes made by others, then undertake an analysis of their own situation and, on that basis, develop guidelines for the future development of agricultural extension services.

Forces for Change

The evolution of public agricultural extension arrived at a worldwide turning point in the 1980s, one that represented the end of a major phase in the growth of publicly funded extension in both the developed and developing world. Agricultural extension increasingly has become defined as one or other of (apparently) differentiated activities of technology transfer or rural development. In many situations, the transfer of technology, heretofore considered the purview of public sector systems, has been reconceived. Such changes suggest a refocussing of paradigms for the delivery of public sector extension.

In developed industrialized countries, which often provide models for extension service delivery elsewhere, the declining relative importance of agriculture for economic growth, the increasing education and affluence of smaller populations of rural producers, and the increasing use of externally purchased inputs have changed the nature of publicly funded extension services and led to a questioning of the means of delivery of extension services by governments (Cary, 1993a). In developing countries, where publicly funded extension is often more important, there has been considerable questioning of the structure and forms of extension delivery.

Global Competition

The consequence of the ratification of the General Agreement on Tariffs and Trade (GATT) is that countries will have to more actively develop comparative agricultural advantages in the production and marketing of food and fibre. Coincidental with a shift toward more conservative political ideologies and free-market economics, global developments suggest increased competition in agriculture. While countries will focus more on their comparative advantages, they also, in many cases, still face national food security concerns and abject rural poverty.

Reassessment of Public Extension

While "modern" extension has existed since the nineteenth century, agricultural extension is quite young worldwide as a formal institution, with the majority of countries initiating such services since the 1950s and 1960s. Even in high-income countries where extension began at earlier dates, fiscal commitment took significant upswings following World War II when a backlog of science and technology had accumulated. In an FAO survey of 207 agricultural extension organizations in 115 countries (Swanson, Farner, & Bahal, 1990), 50 per cent of these organizations had been established or were reorganized in the previous two decades.

Against this background, governments in recent times have found that they are less able to continue providing all the services previously provided. With costs rising, limited resources available, and changes in the prevailing philosophy of the appropriate extent of government intervention, governments have been slow to

---

William M. Rivera is an Associate Professor in the College of Agriculture at the University of Maryland at College Park, Maryland. **John W. Cary** is an Associate Professor in the Department of Agriculture at the University of Melbourne.

increase appropriations for many publicly funded activities. Some functions of government have been curtailed, and others have been privatized. Such changes have been particularly significant in the formerly centrally managed economies.

Because extension worldwide has large numbers of staff, the recurrent costs of extension are of significant magnitude. In the FAO *Report of the Global Consultation on Agricultural Extension*, Swanson et al. (1990) reported that there were approximately 600,000 agricultural extension personnel worldwide, with 95 per cent of these working in public agricultural extension systems. In the United States, there are about 9,000 extension agents, 4,000 subject-matter specialists, and 1,000 directors and administrative support personnel (USDA, 1993 data).

While the unit cost of extension staff in many countries is low, large staff sizes translate into large government outlays. As a result of financial concerns, many countries have examined alternative structural arrangements, including the feasibility of reducing public sector extension expenditures (with associated staff reductions), changes in tax raising, charges for government extension services, and commercialization and privatization (Howell, 1985). A number of countries have moved towards reducing, recovering, or shifting the burden of the costs associated with provision of public sector agricultural extension, particularly transferring "private good" functions to private industry.

Concerns about the costs of extension need to be judged against the economic and social returns associated with successful extension. While more research is needed on measuring the economic payoff from investment in public sector extension services, available research tends to indicate, in contrast to some current criticisms, that extension in many instances provides high rates of return and is, therefore, a profitable public investment (see chapter 3, plus Evenson, 1987; Birkhauser, Evenson, & Feder, 1988). In addition, not all extension expenditure can be measured by benefits from technology transfer; the benefits of extension concerned with human development are difficult to quantify in the short term.

Strategies for Change

Public sector extension, facing criticism for its cost and its lack of efficiency and for not pursuing programmes that foster equity, is confronted with a number of possibilities for change.

There has been a trend, perceptible throughout various extension systems undergoing adjustment, of greater flexibility and multiple partners in funding agricultural advisory services (OECD, 1989; Le Gouis, 1991). Le Gouis observed three major policies adopted by government and farm organizations regarding privatization of extension:

1. Public financing by the taxpayer only for the kinds of services that are of direct concern to the general public
2. Direct charging for some individual services with direct return (in the form of improved income)
3. Mixed funding shared between public and private professional association contributions for some services where the benefits are shared (Le Gouis, 1991, p. 32)

A pervading development in new forms of financial support for extension is the trend to mixed sources of funding, reflecting strategies to gain access to additional sources of funding. In several developing countries, public-private extension coordination is already established. Alternative patterns indicate a fostering of private corporate initiative, encouraging cooperative ventures by farmers, coordinating public-private extension services, and privatizing the public system (Wilson, 1991).

The need for improved and expanded extension activities, together with a strengthening philosophical view of less government involvement in national economies, has led to a number of strategies for changing the way extension services are delivered.

Revitalization

The United States Cooperative Extension Service, when criticized for lack of relevance and vision (Dillman, 1986), regrouped and reviewed the criticisms. Its Extension Committee on Organization and Policy (ECOP) organized a Futures Task Force to review issues and put forward recommendations with a view to revitalizing the system (ECOP, 1987), which has led to various alterations structurally and programmatically.[1] Meanwhile, the advancement of electronic information systems is resulting in increased privatization, with important implications for the future structure of U.S. agriculture (Goe & Kenney, 1988).

Commercialization

New Zealand's Ministry of Agriculture and Fisheries' (MAF) agricultural advisory service now operates under user-pay, commercial criteria (Hercus, 1991). The MAF advisory service, renamed MAF Consulting and, subsequently, Agriculture New Zealand, has remained (temporarily) a public agency, although its employees have given up a number of public employment benefits and now receive commissions for consulting work undertaken. The agency depends for its annual budget on consulting fees received from farmers and contractual arrangements with government for the supply of policy information and rural intelligence to government.[2]

Cost Recovery

Other public extension systems have moved toward cost-recovery approaches. Mexico has developed a fee-based system among large-scale farmers in the north-west region and plans the development of a similar arrangement among small-scale farmers in the south central region (Wilson, 1991). The Agricultural Development and Advisory Service (ADAS) in England and Wales, notionally "commercialized," operates on a partial cost-recovery basis. Clients of ADAS pay a fee for advice which formerly was free of charge. This process of cost recovery, introduced in 1987, was directed towards the agency receiving 50 per cent of its income from commercial fees by 1993-94 (Bunney & Bawcutt, 1991; Harter, 1992).

Voucher Systems

Some countries have replaced public extension delivery systems with vouchers, distributed by government services, for farmers to use in hiring private extension consultants (as in Chile). Coupons attached to agricultural bank loans, committing a certain percentage of the loan for extension services, have been used in Colombia.

Gradual "Privatization"

In 1990 The Netherlands "privatized" approximately one-half of its public extension service by transferring field extension personnel, with initial government financial support, to the farmer associations. The elements of the extension service responsible for linking research and the privatized extension services, policy preparation, implementation, and promotion and regulatory tasks remained under the aegis of the Ministry of Agriculture (Le Gouis, 1991). The "privatized" extension service is governed by a board on which farmers' organizations and the government are equally represented (Proost & Röling, 1991).

Dutch farmers make a partial contribution to the cost of the new organization through membership subscriptions to farmer associations, as well as through direct payment for individual analyses. Farmers will eventually contribute 50 per cent of the cost of the service; special services such as individual analyses will be fully paid for by the farmer clients. The Dutch government has established new government-funded structures for integrating subject-matter specialists into extension teams to facilitate the transfer of information and knowledge and for the provision of information on government policy (Bos, Proost, & Kuiper, 1991; Proost & Röling, 1991).

A gentler form of "privatization" has been proposed for the delivery of government extension services in the Australian state of Victoria. A review of extension services determined that, for government-provided services conferring essentially private benefits to individuals, rather than cost recovery by government fee charging, it is more desirable and more efficient that private advisers deliver such services. However, because of the complexities of extension service delivery and the varying nature and levels of development of different agricultural sectors, a number of constraints were identified which precluded universal application of such a principle (Watson et al., 1992; Cary, 1993a).

In order for rural industry organizations to take a greater responsibility for technology transfer, the Victorian government has proposed "outsourcing" for delivery of future extension programmes. Outsourcing means that the government extension agency will retain a core pool of extension project staff and "buy in" private sector professional services with skills that the agency considers unnecessary to maintain. Agricultural consultants and contract staff will be employed to help deliver services in specific projects funded by rural industry and the federal government. Such projects are likely to be broad and industrywide and not tailored to individual farm circumstances.

In most cases, governments have not actually "privatized" their agricultural extension services. In its pure sense, privatization implies a full transfer of ownership (usually by way of sale) from government to a private entity, with that entity meeting all costs and receiving any profits. In the case of extension, governments have followed a number of distinct pathways such as commercializing the service while retaining it as a public agency, shifting public sector delivery services to private sector delivery of the service while maintaining oversight and basic funding of delivery, or pursuing cost-recovery measures to pay for the service. Thus the phrase "privatization of agricultural extension" generally is misleading.

Other Arrangements

Some countries have never developed public sector agricultural extension services, leaving the function of agricultural extension to private sector commodity enterprises or industry agencies, albeit often with some government financial subsidy. In France, while chambers of agriculture and private sector companies provide extension services, the former are substantially supported financially by public funds. In New Zealand, extension services to the dairy industry for many years have been delivered by the Dairy Board consulting service, financed by the dairy industry.

In other cases, nongovernmental organizations have been used to supplement public sector extension services, especially in the area of rural development (Amanor & Farrington, 1991). This arrangement has certain advantages for increasing extension coverage and encouraging farmer participation in technology systems, but it also has certain inherent limitations.

In most countries, private sector companies are already important contributors to technology transfer and the advancement of agricultural development through, mainly, contract arrangements with farmers. Rightfully, the private sector has come to be acknowledged as a major information provider to both large and small farmers involved in monocropping (Cary & Wilkinson, 1992). The characteristic of "privatized" extension systems is a focus on commercial farms. It is salutary to state the obvious in relation to decisions regarding private and public provision of extension: when extension is delivered privately, it represents a commercial decision; when extension is delivered publicly, it is a political or bureaucratic decision. In determining whether to privatize, it is important, in the first instance, to establish whether an extension programme is designed to help commercial enterprises or small-scale farming and rural development.

Alternative Funding and Delivery

Diverse directions have been taken and multiple means of payment (public and private) have emerged as governments have opted for alternative financial and delivery arrangements to pay for and deliver public sector agricultural extension services. Extension provision is often multi-institutional and organized in ways that are not necessarily independent.

Where the public sector provides extension, the alternative funding arrangements include:

1. *General tax-based public funding* for agriculture, including funding of agricultural extension services, that is, the traditional public sector mode of funding extension
2. *Commodity tax-based public funding* (through cess or parafiscal tax), for example on an agricultural commodity such as coffee, as in El Salvador
3. *Fee-based public funding*, in which fees are charged, usually to large farmers for extension service, for instance in Mexico's grainrich northern region
4. *Contract-based commercialization of public services*, whereby contract-based arrangements are made between farmer and public sector extension services, as in New Zealand

Where the *private sector* provides extension, the alternative funding arrangements include:

1. *Government revenue-based vouchers*, provided to farmers who then contract with private sector agents for extension information provision, as in Chile
2. *Public credit revenue-based coupon schemes* attached to agriculture loans, obligating the farmer-borrower to use a percentage of the loan for extension advising purposes
3. *Membership and fee-based, including commodity tax-based funding*, whereby farmers pay member ship and service fees, and the private organization (e.g., a chamber of agriculture) also receives funds through a public cess or parafiscal tax charged on agricultural commodities, which funds are then transferred to the private sector organization;

Who Pays

Who Delivers	**Public**	**Private**
Public	Total public funding and delivery (traditional Min. of Ag. extension systems)	Fee-based or contract-based services paid by farmers (e.g., Mexico, New Zealand)
Private	Public funding (through vouchers or coupons to small farmers) for private consultant delivery (Chile, Costa Rica); also parastatal arrangements	Farmer associations, private entities (extension delivery by agricultural development agents, production consultants-input suppliers), NGOs

Figure 1. Government institutional changes in funding and delivering agricultural extension. Modified from Neilson (1993).[4]

the private sector then provides the extension services—although public sector officials generally sit on the chamber's governing board

4. *Membership fee plus commercial sponsorship by groups of input suppliers*, where farmer groups are provided nonadvisory, educational extension services by a consortium of privately employed agricultural consultants with partial financial support from rural sector commercial sponsors—such groups can operate on a large scale, with coordinated extension objectives[3]
5. *Privatization*, whereby provision and, eventually, agent salary payments are shifted to a farmers' association or other private entity

In most countries, extension involves a complex of public and private activities. Thus multiple arrangements exist for the funding and the delivery of agricultural extension. Figure 1 illustrates combinations of public and private delivery and payment for agricultural extension services.[4] Actual arrangements are often more complex than implied by this framework. In France, for example, the chambers of agriculture are considered private institutions operating on farmer fees for membership and services, but the government of France contributes sizeable funding (upwards of 49 per cent) for operational and programme costs.

The Context for Extension "Privatization"

The debate on the role of the public sector is not limited to the context of agricultural extension, but encompasses the larger concerns of public policy and institutional and organizational development. Indeed, the degree of government versus private involvement in an economy is an enduring philosophically and politically vexing question. The move toward privatization and efforts to decentralize government functions relate to this theme.

The Privatization Debate

There are two themes in the broader privatization debate: first, a "political economy" consideration of the role and size of government in an economy, which focusses on whether or not there is a failure of private markets; and, secondly, an expressed need to reduce government outlays. While many reassessments of publicly funded extension have reflected the second theme, it is worth considering the rationale for public versus private activity in an economy.

In mixed economies, the prevailing economic justification for government involvement in an activity such as agricultural extension is market failure, whereby the market mechanism alone cannot perform all economic functions for appropriate resource allocation. Market failure may arise because some goods or services are public goods (such as publicly funded agricultural research knowledge) which can be consumed in a nonrival fashion by all members of society without any individual's consumption reducing the amount available for other individuals. Because the benefit of providing such goods cannot be appropriated by individuals, individuals generally will not provide such goods in a society even though there may be significant gains for producers and consumers. Some extension activities are clearly concerned with public goods subject to market failure. Other activities (such as individually tailored advice) confer appropriable private benefits which could be adequately supplied by private markets.

Private goods sometimes are subject to market failure, whereby the operation of private markets does not provide certain services at a socially optimal level or where external costs or benefits are accrued by others rather than the provider of the goods. Market failure also may arise when current generations place insufficient value on preservation of resources for future generations. These latter circumstances are particularly characteristic of land and water degradation (Cary, 1983). Publicly funded conservation extension is often directed to overcoming such market failures (Barr & Cary, 1992).

Government support for the provision of extension services may reflect that such services would be inadequately provided without intervention or, for reasons of equity, because services would not be available to the extent thought socially desirable. Some situations for agricultural extension clearly reflect private goods; other situations clearly are characterized as public goods. There is a lot of fuzzy ground in the middle where it is not particularly clear that an extension activity is conferring a public or private good. In such situations, the extent of publicly funded extension is likely to be determined by the political influence brought to bear by relevant interest groups (Cary, 1993b).

The philosophical thrust of the general privatization debate has centred, on the one hand, on whether certain government activities could be performed more efficiently by private agencies operating in private markets and, on the other hand, on whether inequities may arise because not all individuals have access to resources to purchase privately supplied services.

The Debate with Respect to Extension

While much of the public policy debate related to extension has focussed on either so-called privatization or commercialization as means of reducing government out-lays, other aspects need consideration. The commercialization experience of Agriculture New Zealand (Walker, 1993), while not without its problems, provides examples of some of the arguments for commercialization. Commercialization is perceived to have had a positive effect on moving "beyond the farm gate" into an involvement of the extension staff in the entire production-processing-transporting-marketing chain. There also has been the shift in focus to a client orientation and a concern to identify and produce results rather than simply to engage in activities (Hercus, 1991).

In economically developed countries with a predominance of larger-scale commercial farming, increasingly the technologies of modern, industrialized farming are being developed by nongovernment industrial institutions; such technologies are appropriable for private marketing and generally have little need for government extension. In developed economies, it is more difficult to argue for publicly funded extension for rural industries containing fewer producers who are closely linked and integrated with research systems.

The weaknesses of privatization are more apparent in the context of developing countries, where the situation may be quite different. For instance in African agriculture, funding by user fees may not be viable. An erroneous assumption may be that recipients of government services are generally being subsidized by the government. According to Leonard (1985), this is far from the case with African agricultural producers, who instead are usually subsidizing the rest of society. The most obvious shortcoming may be the difficulty of collecting user fees and establishing cost-accounting procedures to set charges at appropriate levels. The subsistence nature of most African farming leads to a much stronger case for state intervention in support of food production than in developed countries.

Institutional Considerations

The search for appropriate institutional arrangements for different situations echoes the larger debate currently under way on creative use of the private sector for supplanting or supplementing public services. Privatization represents one position in the debate over how public functions should be organized. Wise (1990, p. 152) has observed that "privatization . . . is not necessarily a simplifying strategy . . . the responsibilities of public organizations do not disappear, they merely change." The primary issue may not be whether a certain function should be entrusted to public or private organizations, but, rather, what configuration of organizations, both public and private, is needed and what arrangements between them provide the most effective outcomes.

In some instances, central government bureaucracies are seen as unresponsive and inefficient, and the diffusion of responsibility arises out of a concern that the public sector should be reduced in size. In other cases, however, the emphasis is less on reducing the size of the public sector and more on sharing authority among different units (White, 1989). The question of what role the government should play within an increasingly complex institutional arena is itself complex and not one to which, necessarily, there are simple answers.

Summary of Rationales

The rationale for private sector provision of agricultural extension services is generally based on an expectation of increased efficiency with the operation of private markets and with the resulting efficiencies contributing to the growth of a country's GNP. In contrast, the rationale for public provision of agricultural extension services is based on the following points: (1) much agricultural information is a public good; (2) only government extension services are likely to promote concern for natural resources management; (3) public sector extension may enhance the education of farmers who often lack adequate access to educational institutions; (4) the public service often provides information that reduces risk to farmers; (5) the service may provide information that reduces transaction costs; and (6) an extension service may be concerned with community health issues related to possible human hazards such as accidents and poisonings linked to agricultural chemicals.

The argument for privatization is based upon:

- More efficient delivery of services
- Lowered government expenditures
- Higher quality of services

Privatization may have some attendant disadvantages because of unequal access to resources and because of a diversity of "agencies" and the associated difficulty of coordinating external groups and other government departments. Private delivery agents will be less responsive to government policy direction, and there may be linkage problems with public applied research organizations.

While the process of information transfer amongst farmers traditionally has been characterized by a cooperative, free exchange of information, industrial information traditionally has been a private good characterized by patent rights, process licensing, the use of paid consultants, and differentiated production and marketing processes. In developed economies with commercialized agriculture sectors, many of these features of industrial information transfer are becoming more common in agriculture. The trend to privatization will be stronger the more such circumstances exist. The range of different circumstances prevailing in agricultural extension worldwide suggests that a wide variety of approaches should prevail.

Implications of Extension "Privatization"

In general, a more commercialized approach broadens the focus of extension personnel and makes an extension service more responsive to client needs and changing economic and social conditions. But other immediate implications of privatization appear to include (1) the tendency toward a reduction of linkages both among organizations and among farmers in the exchange of agricultural and other relevant information; (2) the tendency to enhance large-scale farm enterprise to the detriment of small-scale farming; (3) the diminishing emphasis on public-good information and the advancement of knowledge as a saleable commodity; and (4) the trend toward agricultural development services that cater primarily to large-scale farming.

The Netherlands' experience in moving to a partially privatized system highlights some of the implications for agricultural extension, particularly in developed countries. The Netherlands' approach reduced government outlays as well as the government agency role conflict between concern for farmers' interests and the implementation of increasingly stringent environmental policies (Bos et al., 1991; Proost & Röling, 1991). With farmers paying for an increasing share of the extension services, their representatives have more influence on the direction of the extension service. New organizational structures and linkages have had to be established to link the "privatized" and private extension services with the research institutes, experiment stations, and regional experiment farms.

Consequent upon, or in parallel with, the changed Dutch arrangements, other changes have taken place in the Netherlands' extension system. There is some evidence, at least for the vegetable greenhouse sector, that the high level of cooperation among extension information organizations in both the public and private sectors no longer exists (Huang, 1992). The more commercial orientation of the system appears to be creating tensions between extension workers and their clients in a less "open" knowledge and information system, with farmers who used to share information during study-group meetings now being more reluctant to do so.

The New Zealand Ministry of Agriculture and Fisheries advisory service, now fully commercialized and receiving no direct government funding, if sold will be the first extension service fully privatized from government ownership. In 1994 the number of consultants employed in this agency was about half of the peak number of advisers employed in 1987. Some of these advisers will have retired or departed voluntarily; others have established private consulting businesses. The consequence of the changes in New Zealand has been an increase in fee-for-service consulting (the number of farm and horticultural consultants has approximately doubled), with the traditional "advisory" extension no longer existing on a large scale. While, in most cases, the changes seem to have been readily acceped, there remains concern over the effective transfer of scientific findings to agriculture (Walker, 1993). Wider structural changes have sharpened the focus and efficiency of research agencies and advisory consulting work. Traditional technology transfer extension is now largely confined to agricultural commodity boards. Agriculture New Zealand engages in some specific "public good" technology transfer projects on a contract basis to commodity research agencies and the national Foundation for Research, Science and Technology.

There has been no formal assessment of the impact of the New Zealand changes. However, there does appear to be less interaction among organizations, reduced feedback from farmers to science providers, and more limited information distribution, particularly to less well-off and poorer performing farmers (Walker, 1993).

Those extension services that have adopted a commercialization or privatization strategy most vigorously have traditionally employed an advisory approach to extension delivery. The advice given is more likely to be a private good. As well, the extension advisers are more likely to be able to adapt to providing services commercially. However, some staff will not make such a transition easily, new commercial skills will be required by newly commercialized advisers, and the dynamics of any change will have to be planned carefully. Le Gouis (1991) has noted that government "commercial" fees should be set at the market rate so as not to compete unfairly with existing private consultants.

Institutional Implications

The new developments highlight greater institutional pluralism. Extension, interpreted broadly, now is often a mixed system or a "complex" where services are provided by private and public sector entities. The larger context in which a mix of public and private services operates presents a new challenge with new potential roles and responsibilities for the public sector. A major premise of this chapter is that policy makers must consider the entire agricultural extension complex when planning to allocate funds or seeking alternative funding arrangements for the public sector.

By taking into account the complex of extension services, their purposes, and audiences, governments can better reconsider the public sector role. In particular, to what extent should the public sector:

1. Attend targeted audiences unserved by the private sector?
2. Coordinate multiple extension providers?
3. Serve as the final reference or arbitrator of conflicting information?
4. Assure accountability of both public and private extension services to the public?
5. Facilitate the operation of the complex through regulation and information provision?

Considering public sector extension as only one or one set among many providers of extension services facilitates analysis and informed decisions on how best to provide extension within the complex of diffused responsibility.

Conclusion

This chapter concludes with a recommendation for consideration of the various "privatization" arrangements mentioned herein, but also stresses the importance of individual country situational analysis and independent political and technical determinations, not the use of implantation or formulas, in developing funding and delivery arrangements to provide for agricultural extension.

Tunisia, for instance, has recognized the need to move slowly and study carefully the existing developments regarding privatization of extension. In 1990, Tunisia created an Agency for Agricultural Extension and Training, first to improve extension services and then gradually move them toward privatization. A national workshop was held in 1994 to examine alternative funding and delivery arrangements of other extension services worldwide and to clarify the direction in which Tunisia might move in privatizing extension.

The diverse financial arrangements adopted in the last two decades by governments worldwide to fund agricultural extension services provide a valuable menu of options for consideration by other countries confronting the "privatizing" of public sector services.

Still, several countries have resisted the trend toward privatization of agricultural extension, concerned perhaps by the implications reviewed in this chapter. In both developed and developing countries, renewed debate and experimentation around extension is certainly needed, but not only around allocation decisions and how best to develop cooperative arrangements with the private sector.

In most countries, government-funded extension is likely to focus its activities more selectively on public-good activities which exist and on areas where the marketplace is unlikely to provide services at a socially optimal level. Such areas will include "broad" rather than "specific" technology transfer, dissemination of environmental and resource technology, and human resource development.

The move in the public sector toward privatization and efforts to decentralize government functions can serve to highlight the continuing and key role of the public sector and focus the operative question on its responsibility as a coordinating agent. Its roles of regulation and providing service for priority audiences unserved by the private sector will be undiminished.

Notes

1. The Future Task Force report called for (1) the institutionalization of anticipatory issues-based planning mechanisms, (2) the formation of issues-oriented interdisciplinary teams, (3) the development of innovative linkages needed with additional university departments and new funding sources, and (4) clarification of the function of extension education as "a vehicle for problem solution versus only technology transfer."
2. At the end of 1994, the MAF consulting service (Agriculture New Zealand) was self-funding. The New Zealand government is currently seeking to fully privatize the agency by offering it for sale.
3. In southeastern Australia, the Farm Management 500 programme comprises 450 farmer members, 15 privately employed consultants, and five commercial input or service supplier sponsors. Farmer membership fees contribute about half of the operating expenses (Cary & Wilkinson, 1994).
4. David Neilson, World Bank, Latin American Bureau, presented this schema at a World Bank seminar held at Harper's Ferry, West Virginia, 10 September 1993.

References

Amanor, K., & Farrington, J. (1991). NGOs and agricultural technology development. In W. M. Rivera & D. J. Gustafson (Eds.), *Agricultural extension: Worldwide institutional evolution and forces for change*. Amsterdam: Elsevier.

Barr, N. F., & Cary, J. W. (1992). *Greening a brown land: The Australian search for sustainable land use.* Melbourne: Macmillan.

Birkhauser, D., Evenson, R. E., & Feder, G. (1988). *The economic impact of agricultural extension: A review.* Washington, DC: The World Bank.

Bos, J. T. M., Proost, M. D. C., & Kuiper, D. (1991). Reorganizing the Dutch agricultural extension service: The IKC in focus. In D. Kuiper & N. G. Röling (Eds.), *Proceedings of the European seminar on knowledge management and information technology.* Wageningen, The Netherlands: Agricultural University, Department of Extension Service.

Bunney, F. M. G., & Bawcutt, D. E. (1991). Making a business of an extension service. *Agricultural Progress, 66,* 36-43.

Buttel, F. (1991). The restructuring of the American public agricultural research and technology transfer system: Implications for agricultural extension. In W. M. Rivera & D. J. Gustafson (Eds.), *Agricultural extension: Worldwide institutional evolution and forces for change.* Amsterdam: Elsevier.

Cary, J. W. (1983). Social dilemmas in catchment management for pollution control. In *Non-point sources of pollution in Australia.* Australian Water Resources Council, Canberra: Australian Government Publishing Service.

Cary, J. W. (1993a). Changing foundations for government support of agricultural extension in economically developed countries. *Sociologia Ruralis, 33* (3/4), 334-345.

Cary, J. W. (1993b). Three eras of extension in Australia: From public folly to private good. *Australia-Pacific extension conference proceedings: Vol. 1*. Brisbane: Department of Primary Industries.

Cary, J. W., & Wilkinson, R. L. (1992). *The provision of government extension services to the Victorian farming community*. Parkville: University of Melbourne, School of Agriculture and Forestry.

Cary, J. W., & Wilkinson, R. L. (1994). *An evaluation of farm management 500: A programme promoting better business management*. Parkville: University of Melbourne, School of Agriculture and Forestry.

Dillman, D. A. (1986). Cooperative extension at the beginning of the 21st century. *The Rural Sociologist, 6* (2), 102-119.

Evenson, R. E. (1987). *Economic issues in agricultural extension policy*. New Haven, CT: Yale University, Economic Growth Center.

Extension Committee on Organization and Policy (ECOP), Futures Task Force (1987). *Extension in transition: Bridging the gap between vision and reality*. Washington, DC: NASULGC.

Goe, W. R., & Kenney, M. (1988). The political economy of the privatization of agricultural information: The case of the United States. *Agricultural Administration and Extension, 28* (2), 81-99.

Gustafson, D. J. (1990). Developing extension within a complex institutional arena. *Journal of Extension Systems, 8* (1), 87-99.

Harter, D. (1992). Commercialization in Britain. *Interpaks Interchange, 9* (1), 5-6.

Hercus, J. M. (1991). The commercialization of government agricultural extension services in New Zealand. In W. M. Rivera & D. J. Gustafson (Eds.), *Agricultural extension: Worldwide institutional evolution and forces for change*. Amsterdam: Elsevier.

Howell, J. (1985). *Recurrent costs and agricultural development*. London: Overseas Development Institute.

Huang, R. Q. (1992, August). *The level of cooperation among agricultural extension organizations in the greenhouse vegetable sector in the westland area in the Netherlands*. Paper presented at the eighth World Congress for Rural Sociology, University Park, Pennsylvania State University.

Le Gouis, M. (1991). Alternative financing of agricultural extension: Recent trends and implications for the future. In W. M. Rivera & D. J. Gustafson (Eds.), *Agricultural extension: Worldwide institutional evolution and forces for change*. Amsterdam: Elsevier.

Leonard, D. K. (1985). African practice and the theory of user fees. In J. Howell (Ed.), *Recurrent costs and agricultural development* (p. 130-145). London: Overseas Development Institute.

Neilson, D. (1993, September). *Institutional reform and privatization in research and extension*. Paper presented at the Latin American Agricultural Retreat, Harpers Ferry, WV. Washington, DC, The World Bank.

OECD (Organization of Economic Cooperation and Development) (1989). *Survey on effects and consequences of different forms of funding agricultural services*. Paris: OECD doc. AGR/REE 89, 7.

Proost, J., & Röling, N. (1991). "Going Dutch" in extension. *Interpaks Interchange, 9* (1), 3-4.

Rivera, W. M. (1989). An overview of agricultural extension systems. In A. M. Kesseba (Ed.), *Technology systems for small farmers: Issues and options*. Boulder, CO: Westview Press, with the International Fund for Agricultural Development.

Rivera, W. M. (1993). Impacts of extension privatization. *Journal of Extension*, Fall, 28-29.

Rivera, W. M. (1994). *Rapport de la deuxième mission du consultant coordinateur*. Projet PNUD/FAO-T. UN/90/005, Assistance technique au développement de la vulgarisation agricole en Tunisie. Rome: FAO.

Rivera, W. M., & Gustafson, D. J. (1991). New roles and responsibilities for public sector agricultural extension: The impact of multi-institutional activities. In W. M. Rivera & D. J. Gustafson (Eds.), *Agricultural extension: Worldwide institutional evolution and forces for change*. Amsterdam: Elsevier.

Rivera, W. M., & Gustafson, D. J. (1991). *Agricultural extension: Worldwide institutional evolution and forces for change*. Amsterdam: Elsevier.

Rivera, W. M., & Schram, S. G. (1987). *Agricultural extension worldwide: Issues, practices and emerging priorities*. London: Croom Helm.

Swanson, B. E., Farner, B. J., & Bahal, R. (1990). The current status of agricultural extension worldwide. In B. E. Swanson (Ed.), *Report of The Global Consultation on Agricultural Extension*. Rome: FAO.

U.S. Department of Agriculture (USDA) (1993). Agriculture Data. Washington, DC.

U.S. Office of Technology Assessment (OTA) (1986). *Technology, public policy and the changing structure of American agriculture*. Washington, DC.

Walker, A. B. (1993). Recent New Zealand experience in agricultural extension. *Australia-Pacific extension conference proceedings: Vol. 1*. Brisbane: Department of Primary Industries.

Watson, A. S., Hely, R., O'Keeffe, M., Cary, J. W., & Clark, N. (1992). *Review of field-based services in the Victorian Department of Food and Agriculture*. Melbourne: Agmedia.

White, L. G. (1989). Public management in a pluralistic arena. *Public Administration Review, 49*(6), 522-531.

Wilson, M. (1991). Reducing the costs of public extension services: Initiatives in Latin America. In W. M. Rivera & D. J. Gustafson (Eds.), *Agricultural extension: Worldwide institutional evolution and forces for change*. Amsterdam: Elsevier.

Wise, C. R. (1990). Public service configurations and public organizations: Public organization design in the post-privatization era. *Public Administrations Review, 50*(2), 141-155.

Chapter 23
The Role of Nongovernmental Organizations in Extension

John Farrington

In recent years, many observers have suggested that agricultural and rural development strategies would benefit from increased collaboration between government research and extension organizations and nongovernmental development organizations, hereafter called GOs and NGOs, respectively (Carroll, 1992; de Janvry et al., 1989; Jordan, 1989; Korten, 1987). Donors in particular have begun to call for more NGO involvement in programmes that have traditionally been implemented through the public sector, and there has been a recent upsurge of donor interest in direct-funding south-based NGOs (World Bank, 1991a, 1991b; Farnworth, 1991; Bebbington & Riddell, 1994).

These advocates of closer NGO-GO collaboration have tended to underemphasize:

- The wide range of interaction that currently exists, not all of it collaborative; much involves pressure by one side or the other.
- The limitations facing efforts to work together.
- The preconditions for successful collaboration; in particular, the prior informal contacts necessary to build up mutual trust.
- Limitations as well as successes of NGO action.
- The extent to which certain functions relating to, for example, "public goods" will remain more cost-effectively performed by the public sector than by NGOs. Analysis of how GOs might work with NGOs must be accompanied by continuing attention to ways of improving public sector management, an area in which structural adjustment reforms have not had the success expected.

This chapter draws on a recent major study of the role of NGOs in sustainable agricultural development and the potential for collaborative links with GOs (Farrington & Bebbington, 1993; Bebbington & Thiele, 1993; Farrington & Lewis, 1993; Wellard & Copestake, 1993). The following will be reviewed: the characteristics of NGOs, their strengths and weaknesses in relation to agricultural technology, and the practical ways in which they and public sector extension services might collaborate more fully in the future.

NGO Characteristics

NGOs are defined here as nonmembership development-oriented organizations. Our concern here is with the stronger of the south-based NGOs that provide services either directly to the rural poor or to grass-roots membership organizations, and with the local branches of international NGOs that enjoy varying degrees of autonomy. They are therefore distinct from (but, as discussed below, often linked with) formal and informal membership organizations such as farmers' associations. But even within this definition, there exists wide diversity of origins and philosophy. Some NGOs were set up by left-leaning professionals or academics in opposition to the politics of government or its support for or indifference to the prevailing patterns of corruption, patronage, or authoritarianism. Some are based on religious principles, others on a broadly humanitarian ethos, and yet others were set up as quasi-consultancy concerns in response to recent donor-funding initiatives. Some NGOs reject existing social and political structures and see themselves as engines for radical change; others focus on more gradual change through development of human resources (usually through group formation) to meet their own needs or to make claims on government services; yet others focus more simply on the provision of services (e.g., advice, input supply) largely within existing structures. Their ideological orientations also differ widely in relation to agricultural technology: many are concerned with low external input agriculture, 1 others pursue fundamentally organic approaches, 2 and, especially in the case of Andean societies, some are concerned to strengthen or reinstate traditional agricultural practices which formed the basis of social organization (CAAP, 1991).

Of crucial importance when considering NGO-GO links is that NGOs are *independent*: they are not mandated to collaborate with research and extension services in the way that government departments might be. They will therefore collaborate only if GOs have something useful to offer.

NGO Strengths

- The majority of NGOs are small and horizontally structured with short lines of communication and are therefore capable of responding flexibly and rapidly to clients' needs and to changing circumstances. They are also characterized by a work ethic conducive to generating sustainable processes and impacts.
- NGOs' concern with the rural poor means that they often maintain a field presence in remote locations, where it is difficult to keep government staff in post.

John Farrington is an agricultural economist and Director of the Rural Resources and Poverty Research Programme at the Overseas Development Institute in London.

Box 1. Chile: INDAP contracts NGOs to carry out extension.

In its Programme for Technology Transfer, the part of the Chilean NARS responsible for technology transfer (INDAP) provides extension to farmers through a number of private consulting companies. The activities of the companies are financed by a government subsidy of about $US330 for each subsistence farmer who participates. INDAP plans the allocation of resources and monitors the extension programme.

Following the restoration of electoral democracy and civilian rule in 1990, INDAP began to allow NGOs to participate as consulting companies. NGOs have criticized various aspects of the programme on the grounds that it is designed for individual farmers, covers only technical assistance and is not related to credit programmes, is top-down, and is short term. Even so, many of them have chosen to participate in the programme, in part to influence it, and in part to gain access to funds. AGRARIA presented itself as a consulting company and, by 1991, had seven separate INDAP programmes (this has since increased). Extension methods were based on the prior experience of AGRARIA in the region. Although the programme has become more flexible, its rigidity still affects the way extension is carried out. Nonetheless, since the state is providing resources for transfer activities, AGRARIA feels it can no longer countenance funding transfer work with its own funds.

GIA also decided to enter the INDAP programme, and several of its technical teams have formed themselves into a consulting company. GIA's motivations for doing so are, however, slightly different. By working within the programme in an action research mode, GIA's objective is to examine its impact on peasant farmers and to make recommendations to INDAP on how the programme can be strengthened.

To date, GIA has done this by contracting for three modules within the programme and by experimenting with different ways of implementing them. For instance, it has used group instead of individual approaches in extension and, on the basis of broadly positive experiences, has made recommendations to INDAP on how these might be introduced more widely.

Source: Aguirre and Namdar-Irani (1992); Sotomayor (1991).

AGRARIA is a Chilean NGO, best rendered in English as "Food and Agrarian Development." GIA is also a Chilean NGO: "Grupo de Investigaciones Agrarias," or Agrarian Research Group.

- One of NGOs' main concerns has been to identify the needs of the rural poor in sustainable agricultural development. They have therefore pioneered a wide range of participatory methods for diagnosis[3] and, in some contexts, have developed and introduced systems approaches for testing new technology, for example in Chile (Sotomayor, 1991). In some cases, these approaches have extended beyond farming systems into processing and marketing, as with soya in Bangladesh (Buckland & Graham, 1990), sesame in the Gambia (Gilbert, 1990), and cocoa in Bolivia (Trujillo, 1991).
- NGOs' rapport with farmers has allowed them to draw on local knowledge systems in the design of technology options and to strengthen such systems by ensuring that the technologies developed are reintegrated into them (Chaguma & Gumbo, 1993).
- NGOs have also developed innovative dissemination methods, relying on farmer-to-farmer contact, whether on a group or individual basis (e.g., Sollows, Thongpan, & Leelapatra, 1993).
- In some cases, NGOs have *developed* new technologies such as soya production in Bangladesh (Buckland & Graham, 1990) or management practices such as the sloping agricultural land technology in the Philippines (Watson & Laquihon, 1993), but more often they have sought to *adapt* existing technologies, such as PRADAN's efforts in India to scale down technologies developed by government for mushroom and raw silk production and so make them accessible to small-scale farmers (Vasimalai, 1993).
- Undoubtedly, one of the main strengths of NGOs has been their work in group formation. This has been in response to perceived needs at several levels: (1) To meet the technical requirements of certain types of innovation. Thus, Action for World Solidarity in India worked with grass-roots organizations to achieve simultaneous action in an integrated pest management programme (Satish & Vardhan, 1993). In the Gambia and Ethiopia, NGOs helped farmers to organize local informal seed production in ways to avoid undesirable cross-pollination (Henderson & Singh, 1990). (2) To manage "lumpy" assets. In Bangladesh, NGOs have helped to organize landless labourers to acquire and operate water-pump ing technology (Mustafa, Rahman, & Sattar, 1993). (3) To manage common property resources. Many examples exist of formal and informal associations, often supported by NGOs, which manage irrigation water. In other cases, NGOs have supported group efforts in soil and water conservation, whether on private land or on a micro-watershed basis involving both private

and common land (Fernandez, 1993a). They have also helped in managing common grazing and forest land in a sustainable fashion in relation both to technology and the creation of a capacity to make demands on government over, for example, access issues (Fernandez, 1993b).

NGO Weaknesses

- NGOs' small size means that their projects rarely address the structural factors that underlie rural poverty. Small size, independence, and differences in philosophy also militate against learning from each other's experience and against the creation of effective forums, whether at national or provincial levels.
- Some "fashionable" locations have become so densely populated by a diversity of NGOs that problems have arisen not merely of competition for the same clientele, but of some undermining the activities of others (Ayers, 1992).
- NGOs have limited capacities for agricultural technology development and dissemination and limited awareness of how to create effective demand-pull on government services.
- Some NGOs are more accountable to external funding agencies than to the clientele they claim to serve. Donor pressure to achieve short-term impacts, combined with a lack of cross-learning, has led in some cases to the promotion of inappropriate technology, such as protected horticultural systems in the Bolivian Andes (Kohl, 1991).
- Many NGOs place great emphasis on voluntarism. Whilst such concepts as "volunteer extension workers" have great intuitive appeal and reflect widely commended values, they are some times promoted at the expense of financially sustainable alternatives.. This was evident in SIDA's farm-level forestry project in North Vietnam, for instance, where the scope for supporting an emerging private nursery sector in the provision of technical advice was ignored, and complex and largely voluntary advisory services at the village level were promoted instead (author's observation, April 1994).

Examples of Potentially Replicable NGO-GO Interaction

The examples that follow illustrate the types of NGO-GO configurations that offer potential for replication and adaptation in three areas: providing technical advice and feedback, training, and working with groups.

NGO-GO Configurations for Providing Technical Advice and Feedback

This chapter argues that the extreme institutional position in which all extension services are provided by the public sector is likely to be inefficient. At the other extreme, only in very specific circumstances can government hand over large parts of the extension function entirely to NGOs. This has been done in Chile, where government has contracted private technology companies to cater to the larger commercial farmers, and NGOs for small subsistence-oriented farmers (Box 1).

However, similar attempts in India have been largely unsuccessful. A proposal in the Eighth Plan to hand over the entire range of technology transfer and training activities to NGOs in parts of Gujarat, Rajasthan, Orrisa, Kerala, and West Bengal, with some technical support from the state agricultural universities and departments of agriculture, has generated only a lukewarm response from NGOs. One reason for this is the NGO concern that many of the technical recommendations from GOs that they would be expected to disseminate are not relevant to small-scale farmers. Another reason is that mechanisms for bottom-up feedback in existing technologies and for the articulation of demands for new technologies remain weak.

An attempt by the secretary of state for agriculture in Rajasthan to hand over responsibility for extension to NGOs in Udaipur District—renowned for its high density of NGOs—provoked a reaction that is likely to be typical of NGOs in many countries: namely, that by doing so the state is abrogating what is properly its responsibility to ensure a regular supply of technologies relevant to small-scale farmers.

Models developed elsewhere to provide a division of tasks more closely corresponding to the respective comparative advantages of the two sides have been more successful. One of the best-known models is in eastern Bolivia, where public sector extension services have long been characterized by chronic weakness (Thiele, Davies, & Farrington, 1988). Under a new strategy devised in 1989, Centre for Tropical Agricultural Research (CIAT) established a coordination unit—the Technology Transfer Department (DTT)—whose role was not to work directly with farmers, but with various intermediate users of technologies who had their own local extension services. NGOs are one of the most important types of intermediate users.

The DTT has subject-matter specialists and zonal specialists whose work is supported by a communications section. The subject-matter specialists (SMSs) are in regular contact with their corresponding CIAT researcher and collaborate on some research work. They package research information for delivery to intermediate users and are mandated to transmit feedback on farmer needs to the researcher.

Frequently, technologies developed in the experimental centres are still not ready for transfer. SMSs therefore carry out on-farm adaptive trials, in addition to ensuring that extensionists pass on the appropriate messages to farmers. Other duties of the SMSs include preparing technical bulletins for extensionists, enhancing feedback and advice to extensionists (for instance, on the establishment of demonstration plots), and explaining how to give talks to farmers.

Informal collaborative arrangements rely heavily on the initiative of GO staff to feed lessons back into the next round of the research and extension agenda. In Bolivia, feedback was encouraged through a range of instruments, including NGO representation on the research planning committee of the local research station, and consultation with a number of zonal substations, part of whose function was to assemble users' views on the technologies being made available.

A different type of formal arrangement being developed in Udaipur District of India is a quarterly forum hosted by a "hybrid" NGO-GO institution—a government Farm Science Centre located in an NGO—in which interested NGOs and GOs participate. In essence, it is intended to promote familiarization by allowing cross-visits to be set up to each other's programmes, to allow training courses to be designed to meet NGOs' requirements, and to allow NGOs to "feed back" on currently available technologies and to voice further technology needs that are not currently being met. The forum is also intended to develop lessons for future interaction from efforts to monitor the process and impacts of current NGO-GO interaction within the district. The forum is beginning to make progress in some of these areas, but a major difficulty remains in the form of pre-programmed "targets" that the Department of Agricultural Extension has to meet each season in terms of, for example, the number of demonstrations of a given type. The forum has had the added benefit of facilitating substantive interaction among GOs (departments of agricultural extension, of animal production and health, and of horticulture) which otherwise meet on a monthly basis, but only for administrative purposes, and among the large number of NGOs in the district.

What remains to be addressed is the scope for similar formal links between NGOs, the membership organizations they are working with, and the GOs (such as the revenue and forest departments) which control large areas of land suitable for grazing or fodder collection. Although the central government has approved wider access by villagers (often with the assistance of NGOs) to such land under "joint management" arrangements, parts of the land have been encroached upon by wealthier farmers, and local-level officials find it difficult to rectify this so as to make the land accessible to village groups.

NGO-GO Configurations in Training

Some of the farmer training conducted by GOs is linked more strongly with GOs' programmes and targets than with farmers' needs. Much training is given in a classroom environment, without the practical content necessary to engage farmers' interests. NGOs have sought to work with GOs to address these shortcomings in several contexts:

- In Gujarat, India, the Aga Khan Rural Support Project (AKRSP) identified village training needs through discussions with farmer groups (Shah & Mane, 1993). Initially, AKRSP organized government provision of this training, but the courses were formal in style (lectures in a classroom), and farmers' evaluations showed that they had learned little of practical value from them. In response, AKRSP developed an alternative needs-based training and dissemination methodology which it tested over several areas. Government staff were then brought in to observe, participate in, and finally adopt the methodology. Successful adoption was reinforced by informal networks and exchange of experience at workshops and consul tations. AKRSP, along with Myrada (Fernandez, 1993a), has also been instrumental in training GO staff in participatory methods.
- In a different context, the International Institute for Rural Reconstruction in the Philippines brought together resource people from NGOs and GOs at a one-week workshop, the objective of which was to produce a completed *Agroforestry Resources Training Manual.* The manual is now widely used (Gonsalves & Miclat-teves, 1993).
- Clearly, there are also many instances in which NGO staff benefit from the skills which GO staff are able to impart; training in grafting techniques, for instance, has been found useful by a number of NGOs (Giordano, Satish, & Farrington, 1993).

NGO-GO Configurations in Group Formation

Substantial scope exists for GOs to benefit from NGOs' group-organizing skills. In India, for instance, recent modifications to the training and visit extension sytem now require village-level extension workers to interact with groups of approximately twenty farmers instead of with individual "contact farmers." However, extension workers are not trained in group formation skills, and groups that they form are unlikely—if they survive at all—to become interested in anything other than the testing of immediately available technology. The examples cited above illustrate how NGOs can effectively organize groups around integrated pest management, soil and water management, and the management of common property resources and capital assets.

Box 2. Technologies for women and the landless: improved poultry production promoted by the Bangladesh Rural Advancement Committee (BRAC).

In Bangladesh, almost 50 per cent of rural households are landless or near-landless, and women face cultural restrictions on work outside the household compound. Livestock production is one activity which can be conducted within the compound. Poultry production alone is estimated to account for 23 per cent of per capita animal protein consumption in the form of both meat and eggs, but mortality is high and productivity low.

After unsuccessful efforts to upgrade poultry production between 1979 and 1983, by the mid-1980s, BRAC had devised a complementary set of technical and local institutional innovations. By 1990, these efforts had been replicated by government and other NGOs in 7,400 villages, affecting some 10 per cent of poultry production. The innovations comprised:

- One poultry worker (female) per 1,000 birds, trained in rearing techniques, health care, and vaccination
- Vaccines for the poultry worker provided by the Department of Livestock (DoL) and training provided jointly by DoL and BRAC; worker's remuneration covered largely from vaccination fees
- The establishment of systems to allow poultry keepers access either to day-old chicks from government breeding farms or, if they felt confident enough to handle such young birds, to two-month-old chicks reared at special units set up by BRAC and DoL, but managed by local key rearers, and capable of rearing batches of 250 chicks from day-old to two months
- A feed production centre serving several villages to provide a balanced feed supplement for cross-bred stock

Scaling up the scheme means that more than 33,000 key rearers are now operating commercially, and almost 5,500 poultry workers have been trained. Demand for day-old chicks from government hatcheries has risen from 0.5 million per year in the mid-1980s to almost two million currently. However, the system remains crucially dependent on the capacity of government to deliver inputs, especially vaccines, down to the local level.

Source: Mustafa et al. (1993).

What Extension Services Can Do to Further Collaborate with NGOs

The foregoing has several implications for extension services which aim to develop closer links with NGOs:

- Explicit recognition of the wide diversity of NGO types will be necessary. Not all Many NGOs seek to support the establishment and growth of membership organizations capable of meeting their technology requirements over the longer term either from their own resources or by creating demands on government services or by a combination of both. Thus in seed supply, Cromwell and Wiggins (1993), for instance, quote numerous examples of ways that NGOs have supported local groups to produce seed, including vegetable and soya bean seed production in Bangladesh, and the multiplication of planting material for potatoes in the Ecuadorian Andes. In other countries (e.g., Nepal, the Gambia) local seed production initiatives have arisen plant breeding focused more fully on the needs of the rural poor, and the facilitation of linkages among the various agencies concerned with seed production and distribution.
Some of these efforts see viable commercial arrangements as an essential feature of longterm sustainability. Thus PRADAN, in India, in an effort spanning several years to support the introduction of chrome-leather tanning by a local group, encouraged links with commercial lending organizations and private leather traders, not least because the latter could give accurate feedback on product quality. In a more complex example of experimenting over several years with several types of women's groups for poultry production, the Bangladesh Rural Advancement Committee finally devised a multitier structure embracing rearing, local feed production, and health care by women paravets. These women drew on government for the necessary vaccines, earned a living by charging for injections, and provided elements of the extension function by giving advice on management and nutrition to those who paid for vaccinations (Box 2) will be willing to enter into a collaborative "service delivery" relationship with government, and those that do will do so only if GOs have something to offer appropriate to their clients' needs. Extension services therefore need to search for links with NGOs from a position of confidence that the research-extension system already has, or at least has the capacity to generate, something useful to NGOs and their clients.
- Close interaction will be impossible if extension departments expect NGOs merely to assist in fulfilling preset targets such as the achievement of a given number of demonstrations of a given kind each season. GOs will have to bring an open agenda into the relationship, where possible keeping some resources "unallocated" in order to be able to respond to needs as they are articulated by NGOs.
- Very specific efforts will have to be made to con-

vey both feedback on existing technologies and NGOs' requirements for new technologies to researchers. In many GOs, reward systems pro vide no incentive among either researchers or extensionists to respond to feedback.

- GO and NGO staff can jointly participate in training courses (ideally led by a joint team) in the action-oriented methods such as participatory rural appraisal favoured by NGOs. The relevance of these to individual NGO staff will vary, but their capacity to enhance awareness of farmers' perspectives is important.
- Depending on their philosophy, NGOs are concerned to develop local capacities for experimentation which build solely on farmers' indigenous knowledge or on this and relevant "outside" ideas. This strategy may contribute to rural advancement in its own right, and the capacity it creates may prove a useful independent source of innovations in the absence of usable technologies from government. Alternatively, where GOs are willing and able to work with the poor, it will be a useful complement to what extension service can offer.

Conclusions

The examples now becoming available offer wider lessons on ways that NGOs and extension services can work in a mutually reinforcing fashion. Predictably enough, it is the group-organizing and human resource development skills of NGOs which have tended to complement the technical skills and facilities available to government. Less predictable are the types of interaction that might evolve in specific settings: much appears likely to develop on an ad hoc basis in response to the individual characteristics of NGOs and the settings in which they work. However, it is becoming increasingly clear that a formal forum is required for certain types of interaction, including training, the joint planning of research and extension agenda, and the securing of joint management agreements for soil and water, fodder, and grazing resources. The task for the coming decade will be to develop these in ways which are nonthreatening both to the organizations involved and to the informal interaction they already undertake and, as a prior requirement, to develop the mutual trust and awareness of each other's activities on which formal interaction depends.

Notes

1. For examples, see the Newsletter of the Information Centre for Low External and Sustainable Agriculture (ILEIA), based at PO Box 64, 3830 AB Leusden, Netherlands.
2. Such as organizations affiliated to the International Federation of Organic Agriculture Movements (IFOAM).
3. For reviews of these methods, see, for example, Farrington and Martin, 1988; Chambers, Pacey, & Thrupp, 1989; Scoones and Thompson, 1994. For a recent review of the concepts and practices of farmer participatory research, see Okali *et al.*, 1994.

References

Aguirre, F., & Namdar-Irani, M. (1992). Complementarities and tensions in NGO-state relations in agricultural development: The trajectory of AGRARIA (Chile). *Agricultural Research and Extension Network Paper No. 32.* London: Overseas Development Institute.

Ayers, A. J. (1992). Conflicts or complementarities? The state and NGOs in the colonization zones of San Julian and Berlin, eastern Bolivia. *Agricultural Research and Extension Network Paper No. 37.* London: Overseas Development Institute.

Bebbington, A. J., & Riddell, R. (1994). *New agendas and old problems: Issues, options and challenges in direct funding of southern NGOs.* Draft, mimeo. London: Overseas Development Institute.

Bebbington, A. J., & Thiele, G. (Eds.). (1993). *NGOs and the state in Latin America: Rethinking roles in sustainable agricultural development.* London: Routledge.

Buckland, J., & Graham, P. (1990). The Mennonite Central Committee's experience in agricultural research and extension in Bangladesh. *Agricultural Research and Extension Network Paper No. 17.* London: Overseas Development Institute.

CAAP (1991, December). Generación y transferencia de tecnología agropecuaria: Sistematización de experencias en el CAAP. Paper presented at *Taller regional para América del Sur: Generación y transferencia de tecnología agropecuaria; el papel de las ONGs y el sector público*, Santa Cruz, Bolivia. (Summarized in English as "El centro Andino de acción popular" in A. J. Bebbington & G. Thiele [Eds.], *NGOs and the state in Latin America: Rethinking roles in sustainable agricultural development.* London: Routledge, 1993.)

Carroll, T. (1992). *Intermediary NGOs: The supporting link in grassroots development.* West Hartford, CT: Kumarian Press.

Chaguma, A., & Gumbo, D. (1993). ENDA-Zimbabwe and community research. In K. Wellard & J. G. Copestake (Eds.), *NGOs and the state in Africa: Rethinking roles in sustainable agricultural development.* London: Routledge.

Chambers, R., Pacey, A., & Thrupp, L. A. (Eds.). (1989). *Farmer first: Farmer innovation and agricultural research.* London: Intermediate Technology Publications.

Cromwell, E., & Wiggins, S. (1993). *Sowing beyond the state: NGOs and seed supply in developing countries.* London: Overseas Development Institute.

de Janvry, A., Marsh, R., Runsten, D., Sadoulet, E., & Zabin, C. (1989). Impacto de la crisis en la economía campesina de América Latina y el Caribe. In F. Jordan (Ed.), *La economía campesina: Crisis, reactivación, políticas* (p. 91-205). San José, Costa Rica: Instituto Interamericano de Cooperación para la Agricultura.

Farnworth, E. G. (1991, June). *The Inter-American Development Bank's interactions with non-governmental environmental organizations.* Paper presented at the Third Consultative Meeting on the Environment,Caracas, Venezuela.

Farrington, J., & Bebbington, A. J. (1993). *Reluctant partners? Non-governmental organisations, the state, and sustainable agricultural development.* London: Routledge.

Farrington, J., & Lewis, D. (Eds.). (1993). *NGOs and the state in Asia: Rethinking roles in sustainable agricultural development.* London: Routledge.

Farrington, J., & Martin, A. (1988). Farmer participation in agricultural research: A review of concepts and practices. *Occasional Paper No. 11.* London: Overseas Development Institute.

Fernandez, A. (1993a). NGOs and government: A love-hate relationship. In J. Farrington & D. Lewis (Eds.), *NGOs and the state in Asia: Rethinking roles in sustainable agricultural development.* London:Routledge.

Fernandez, A. P. (1993b). *The Myrada experience: The interventions of a voluntary agency in the emergence and growth of people's institutions for sustained and equitable management of micro-watersheds.* Bangalore, India: Myrada.

Fernandez, A. P., & Mascarenhas, J. (1993). MYRADA-participatory rural appraisal and participatory learning methods. In J. Farrington & D. Lewis (Eds.), *NGOs and the state in Asia: Rethinking roles in sustainable agricultural development.* London: Routledge.

Gilbert, E. (1990). Non-governmental organisations and agricultural research: The experience of the Gambia. *Agricultural Research and Extension Network Paper No. 12.* London: Overseas Development Institute.

Giordano, E., Satish, S., & Farrington, J. (1993). Greenwork at Auroville: From survival to inter-institutional collaboration. In J. Farrington & D. Lewis (Eds.), *NGOs and the state in Asia: Rethinking roles in sustainable agricultural development.* London: Routledge.

Gonsalves, J., & Miclat-Teves, A. (1993). The Interna-tional Institute for Rural Reconstruction (IIRR): Developing an agroforestry kit. In J. Farrington & D. Lewis (Eds.), *NGOs and the state in Asia: Rethinking roles in sustainable agricultural development.* London: Routledge.

Henderson, P. A., & Singh, R. (1990). NGO-government links in seed production: Case studies from the Gambia and Ethiopia. *Agricultural Research and Extension Network Paper No. 14.* London: Overseas Development Institute.

Jordan, F. (Ed.). (1989). *La economía campesina: Crisis, reactivación, políticas.* San José, Costa Rica: Instituto Interamericano de Cooperación para la Agricultura.

Kohl, B. (1991). Protected horticultural systems in the Bolivian Andes: A case study of NGOs and inappropriate technology. *Agricultural Research and Extension Network Paper No. 29.* London: Overseas Development Institute.

Korten, D. (1987). Third generation NGO strategies: A key to people-centred development. *World Development, 15* (Supplement), 145-159.

Mustafa, S., Rahman, S., & Sattar, G. (1993). Bangladesh Rural Advancement Committee (BRAC): Backyard poultry and landless irrigators programmes. In J. Farrington & D. Lewis (Eds.), *NGOs and the state in Asia: Rethinking roles in sustainable agricultural development.* London: Routledge.

Okali, C., Sumberg, J., & Farrington, J. (1994). *Farmer participatory Research: Rhetoric and reality.* London: IT Publications.

Satish, S., & Vardhan, T. J. P. S. (1993). Action for world solidarity and the red-headed hairy caterpillar. In J. Farrington & D. Lewis (Eds.), *NGOs and the state in Asia: Rethinking roles in sustainable agricultural development.* London: Routledge.

Scoones, I., & Thompson, J. (1994, October). *Beyond farmer first: Rural people's knowledge, agricultural research and extension practice: Towards a theoretical framework.* Paper presented at the Beyond Farmer First Workshop, Institute of Development Studies, University of Sussex, Brighton.

Shah, P., & Mane, P. M. (1993). Networking in participatory training and extension: Experiences of the Aga Khan Rural Support Programme in Gujarat State. In J. Farrington & D. Lewis (Eds.), *NGOs and the state in Asia: Rethinking roles in sustainable agricultural development.* London: Routledge.

Sollows, J., Thongpan, N., & Leelapatra, W. (1993). NGO-governmental interaction in ricefish culture in NE Thailand. In J. Farrington & D. Lewis (Eds.), *NGOs and the state in Asia: Rethinking roles in sustainable agricultural development.* London: Routledge.

Sotomayor, O. (1991). GIA and the new Chilean public sector: The dilemmas of successful NGO influence over the state. *Agricultural Research and Extension Network Paper No. 30.* London: Overseas Develop-ment Institute.

Thiele, G., Davies, P., & Farrington, J. (1988). Strength in diversity: Innovation in agricultural technology development in eastern Bolivia. *Agricultural Research and Extension Network Paper No. 1.* London: Overseas Development Institute.

Trujillo, E. (1991, December). *Investigación y extensión por El Ceibo en el Alto Beni.* Paper presented at the Regional Workshop for South America, "The generation and transfer of agricultural technology: The role of NGOs and of the public sector." Santa Cruz, Bolivia.

Vasimalai, M. (1993). Professional Assistance for Development Action (PRADAN): An NGO de-mystifies and scales down technology. In J. Farrington & D. Lewis (Eds.), *NGOs and the state in Asia: Rethinking roles in sustainable agricultural development.* London: Routledge.

Watson, H. R., & Laquihon, W. A. (1993). The Mindanao Baptist Rural Life Center's sloping agricultural land technology (SALT) research and extension in the Philippines. In J. Farrington & D. Lewis (Eds.), *NGOs and the state in Asia: Rethinking roles in sustainable agricultural development.* London: Routledge.

Wellard, K., & Copestake, J. G. (Eds.). (1993). *NGOs and the state in Africa: Rethinking roles in sustainable agricultural development.* London: Routledge.

World Bank (1991a). *World Development Report 1991: The challenge of development.* Oxford: Oxford University Press.

World Bank (1991b). *Trends in developing countries 1991.* Washington, DC: World Bank.

DTP by P.B. / IKONOS